METHODS IN MOLECULAR BIOLOGY

Series Editor
John M. Walker
School of Life and Medical Sciences
University of Hertfordshire
Hatfield, Hertfordshire, AL10 9AB, UK

For further volumes:
http://www.springer.com/series/7651

Unconventional Protein Secretion

Methods and Protocols

Edited by

Andrea Pompa and Francesca De Marchis

Instituto di Bioscienze e Biorisorse, Perugia, Italy

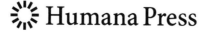 Humana Press

Editors
Andrea Pompa
Instituto di Bioscienze e Biorisorse
Perugia, Italy

Francesca De Marchis
Instituto di Bioscienze e Biorisorse
Perugia, Italy

ISSN 1064-3745   ISSN 1940-6029   (electronic)
Methods in Molecular Biology
ISBN 978-1-4939-8144-1   ISBN 978-1-4939-3804-9   (eBook)
DOI 10.1007/978-1-4939-3804-9

This Humana Press imprint is published by Springer Nature
The registered company is Springer Science+Business Media LLC New York

Preface

> The envious billows sidelong swell to whelm my track; let them; but
> first I pass (*Captain Ahab; Moby Dick*)

The essential function of the secretory proteins is to sense the environment and to drive the correct development of monocellular and, especially, pluricellular organisms. These polypeptides mediate fundamental processes in all the living organisms, such as cell-to-cell communication, defense response, and many others. Protein secretion is essentially achieved by two mechanisms. In the classic secretory pathway, proteins travel from the endoplasmic reticulum (ER), where they are inserted thanks to an amino acidic N-terminal signal peptide (SP) sequence, through the Golgi apparatus and finally reach the extracellular space or the lysosome/vacuole. Alongside this well-characterized route, an unconventional protein secretion (UPS) has been described which incorporates all the mechanisms that do not follow the "classical way." UPS has grown in importance in cell biology studies due to the increasing number of SP lacking proteins (leaderless proteins, LSPs) recovered in the extracellular space of many organisms. Even though some mechanisms that underlie this type of protein traffic have already been described, there is still much to be discovered and probably many new routes will be described in the next future. This book has the purpose to present the relevant background and methodologies nowadays available for UPS study. It has been thought and written with the aim to explore the latest techniques and protocols that have been successfully applied for UPS analysis in different laboratories around the world. Detailed chapters include an overview of conventional and unconventional secretory pathways along with multidisciplinary approaches and methods used for UPS analysis in different organisms. This book will be useful for all the researchers interested in the secretory pathway field as well as for studies in cell biology, cell development, biomedical research, and healthcare.

Perugia, Italy

Andrea Pompa
Francesca De Marchis

Acknowledgments

The editors gratefully acknowledge all contributing authors for their collaboration which made this project possible; the series editor Prof. John Walker whose help and guidance have been instrumental; Mr. Patrick Marton, Ms. Monica Suchy, Mr. David Casey, and Ms. Anna Rakovsky for making the production of this book possible. We would like to thank, also, Ms. Serena Deodato for her help in the editorial process management.

Contents

Preface.. *v*

Contributors... *xi*

PART I   BACKGROUND

1   ER to Golgi-Dependent Protein Secretion: The Conventional Pathway......   3
    Corrado Viotti

2   Unconventional Protein Secretion in Animal Cells.....................   31
    Fanny Ng and Bor Luen Tang

3   Unconventional Protein Secretion in Plants.........................   47
    *Destiny J. Davis, Byung-Ho Kang, Angelo S. Heringer, Thomas E. Wilkop,
    and Georgia Drakakaki*

PART II   UPS CONTEMPLATES MULTIDISCIPLINARY APPROACHES
          FOR PLANTS

4   Chemical Secretory Pathway Modulation in Plant Protoplasts   67
    Francesca De Marchis, Andrea Pompa, and Michele Bellucci

5   From Cytosol to the Apoplast: The Hygromycin Phosphotransferase (HYG[R])
    Model in Arabidopsis   81
    Haiyan Zhang and Jinjin Li

6   Following the Time-Course of Post-pollination Events
    by Transmission Electron Microscopy (TEM): Buildup of Exosome-Like
    Structures with Compatible Pollinations   91
    Darya Safavian, Jennifer Doucet, and Daphne R. Goring

PART III   UPS CONTEMPLATES MULTIDISCIPLINARY APPROACHES
           FOR ANIMALS AND FUNGI

7   Investigating Alternative Transport of Integral Plasma Membrane
    Proteins from the ER to the Golgi: Lessons from the Cystic Fibrosis
    Transmembrane Conductance Regulator (CFTR)   105
    *Margarida D. Amaral, Carlos M. Farinha, Paulo Matos,
    and Hugo M. Botelho*

8   Quantification of a Non-conventional Protein Secretion:
    The Low-Molecular-Weight FGF-2 Example...........................   127
    Tania Arcondéguy, Christian Touriol, and Eric Lacazette

9   Human Primary Keratinocytes as a Tool for the Analysis
    of Caspase-1-Dependent Unconventional Protein Secretion   135
    *Gerhard E. Strittmatter, Martha Garstkiewicz, Jennifer Sand,
    Serena Grossi, and Hans-Dietmar Beer*

10 A Reporter System to Study Unconventional Secretion of Proteins
    Avoiding N-Glycosylation in *Ustilago maydis* 149
    Janpeter Stock, Marius Terfrüchte, and Kerstin Schipper

11 Stress-Inducible Protein 1 (STI1): Extracellular Vesicle Analysis
    and Quantification 161
    *Marcos Vinicios Salles Dias, Vilma Regina Martins,
    and Glaucia Noeli Maroso Hajj*

12 Analysis of Yeast Extracellular Vesicles 175
    *Marcio L. Rodrigues, Debora L. Oliveira, Gabriele Vargas,
    Wendell Girard-Dias, Anderson J. Franzen, Susana Frasés,
    Kildare Miranda, and Leonardo Nimrichter*

13 Exploring the *Leishmania* Hydrophilic Acylated Surface
    Protein B (HASPB) Export Pathway by Live Cell Imaging Methods 191
    Lorna MacLean, Helen Price, and Peter O'Toole

14 Characterization of the Unconventional Secretion of the Ebola Matrix
    Protein VP40 205
    Olivier Reynard and Mathieu Mateo

15 Role and Characterization of Synuclein-γ Unconventional Protein
    Secretion in Cancer Cells 215
    Caiyun Liu, Like Qu, and Chengchao Shou

PART IV   SECRETOME ISOLATION FROM PLANT AND ANIMAL SAMPLES
          TO IDENTIFY LEADERLESS SECRETORY PROTEINS (LSP)

16 Characterization of the Tumor Secretome from Tumor Interstitial
    Fluid (TIF) 231
    Pavel Gromov and Irina Gromova

17 Vacuum Infiltration-Centrifugation Method for Apoplastic Protein
    Extraction in Grapevine 249
    *Bertrand Delaunois, Fabienne Baillieul, Christophe Clément,
    Philippe Jeandet, and Sylvain Cordelier*

18 Isolation of Exosome-Like Vesicles from Plants by Ultracentrifugation
    on Sucrose/Deuterium Oxide (D$_2$O) Density Cushions 259
    *Christopher Stanly, Immacolata Fiume, Giovambattista Capasso,
    and Gabriella Pocsfalvi*

Index 271

Contributors

MARGARIDA D. AMARAL • *Faculty of Sciences, BioISI – Biosystems & Integrative Sciences Institute, University of Lisboa, Lisbon, Portugal*

TANIA ARCONDÉGUY • *Centre de Recherches en Cancérologie de Toulouse – CRCT UMR1037 Inserm/Université Toulouse III Paul Sabatier ERL5294 CNRS, Oncopole de Toulouse, Toulouse, France*

FABIENNE BAILLIEUL • *Laboratoire Stress, Défenses et Reproduction des Plantes, Unité de Recherche Vignes et Vins de Champagne-EA 4707, SFR Condorcet – FR CNRS 3417, Université de Reims Champagne-Ardenne, Reims, France*

HANS-DIETMAR BEER • *Department of Dermatology, University Hospital Zurich, University Zurich, Zurich, Switzerland*

MICHELE BELLUCCI • *Institute of Biosciences and Bioresources, Research Division of Perugia, National Research Council, Perugia, Italy*

HUGO M. BOTELHO • *Faculty of Sciences, BioISI – Biosystems & Integrative Sciences Institute, University of Lisboa, Lisbon, Portugal*

GIOVAMBATTISTA CAPASSO • *Division of Nephrology, Department of Cardio-Vascular Sciences, Second University of Naples, Naples, Italy*

CHRISTOPHE CLÉMENT • *Laboratoire Stress, Défenses et Reproduction des Plantes, Unité de Recherche Vignes et Vins de Champagne-EA 4707, SFR Condorcet – FR CNRS 3417, Université de Reims Champagne-Ardenne, Reims, France*

SYLVAIN CORDELIER • *Laboratoire Stress, Défenses et Reproduction des Plantes, Unité de Recherche Vignes et Vins de Champagne-EA 4707, SFR Condorcet – FR CNRS 3417, Université de Reims Champagne-Ardenne, Reims, France*

DESTINY J. DAVIS • *Department of Plant Sciences, University of California, Davis, CA, USA*

BERTRAND DELAUNOIS • *Laboratoire Stress, Défenses et Reproduction des Plantes, Unité de Recherche Vignes et Vins de Champagne-EA 4707, SFR Condorcet – FR CNRS 3417, Université de Reims Champagne-Ardenne, Reims, France*

MARCOS VINICIOS SALLES DIAS • *International Research Center, AC Camargo Cancer Center Rua Taguá, São Paulo, Brazil*

JENNIFER DOUCET • *Department of Cell & Systems Biology, University of Toronto, Toronto, ON, Canada*

GEORGIA DRAKAKAKI • *Department of Plant Sciences, University of California, Davis, CA, USA*

CARLOS M. FARINHA • *Faculty of Sciences, BioISI – Biosystems & Integrative Sciences Institute, University of Lisboa, Lisbon, Portugal*

IMMACOLATA FIUME • *Mass Spectrometry and Proteomics, Institute of Biosciences and BioResources, National Research Council of Italy, Naples, Italy*

ANDERSON J. FRANZEN • *Laboratório de Tecnologia em Cultura de Células, Centro Universitário Estadual da Zona Oeste, Rio de Janeiro, RJ, Brazil*

SUSANA FRASÉS • *Laboratório de Ultraestrutura Celular Hertha Meyer, Instituto de Biofísica Carlos Chagas Filho and Instituto Nacional de Ciência e Tecnologia em Biologia Estrutural e Bioimagem, Universidade Federal do Rio de Janeiro, Rio de Janeiro, RJ, Brazil; Laboratório de Biologia Estrutural, Diretoria de Programas, Instituto Nacional de Metrologia, Normalização e Qualidade Industrial (INMETRO), Xerém, Rio de Janeiro, RJ, Brazil*

MARTHA GARSTKIEWICZ • *Department of Dermatology, University Hospital Zurich, University Zurich, Zurich, Switzerland*

WENDELL GIRARD-DIAS • *Instituto Oswaldo Cruz, Fiocruz, Rio de Janeiro, Brazil; Laboratório de Ultraestrutura Celular Hertha Meyer, Instituto de Biofísica Carlos Chagas Filho and Instituto Nacional de Ciência e Tecnologia em Biologia Estrutural e Bioimagem, Universidade Federal do Rio de Janeiro, Rio de Janeiro, RJ, Brazil*

DAPHNE R. GORING • *Department of Cell & Systems Biology, University of Toronto, Toronto, ON, Canada*

PAVEL GROMOV • *Danish Cancer Society Research Center, Genome Integrity Unit, Cancer Proteomics Group, Copenhagen, Denmark*

IRINA GROMOVA • *Danish Cancer Society Research Center, Genome Integrity Unit, Cancer Proteomics Group, Copenhagen, Denmark*

SERENA GROSSI • *Department of Dermatology, University Hospital Zurich, University Zurich, Zurich, Switzerland*

GLAUCIA NOELI MAROSO HAJJ • *International Research Center, AC Camargo Cancer Center Rua Taguá, São Paulo, Brazil*

ANGELO S. HERINGER • *Department of Plant Sciences, University of California, Davis, CA, USA*

PHILIPPE JEANDET • *Laboratoire Stress, Défenses et Reproduction des Plantes, Unité de Recherche Vignes et Vins de Champagne-EA 4707, SFR Condorcet – FR CNRS 3417, Université de Reims Champagne-Ardenne, Reims, France*

BYUNG-HO KANG • *Center for Organelle Biogenesis and Function, School of Life Sciences, Chinese University of Hong Kong, Hong Kong, China*

ERIC LACAZETTE • *Centre de Recherches en Cancérologie de Toulouse – CRCT UMR1037 Inserm/Université Toulouse III Paul Sabatier ERL5294 CNRS Oncopole de Toulouse, Toulouse, France*

JINJIN LI • *Key Laboratory of Plant Resources, Institute of Botany, Chinese Academy of Sciences, Beijing, China*

CAIYUN LIU • *Key Laboratory of Carcinogenesis and Translational Research (Ministry of Education), Beijing, China; Department of Biochemistry & Molecular Biology, Peking University Cancer Hospital & Institute, Beijing, China*

LORNA MACLEAN • *Drug Discovery Unit, Division of Biological Chemistry and Drug Discovery, College of Life Sciences, University of Dundee, Dundee, UK*

FRANCESCA DE MARCHIS • *Institute of Biosciences and Bioresources, Research Division of Perugia, National Research Council, Perugia, Italy*

VILMA REGINA MARTINS • *International Research Center, AC Camargo Cancer Center Rua Taguá, São Paulo, Brazil*

MATHIEU MATEO • *UBIVE, Institut Pasteur, International Center for Infectiology Research (CIRI), Paris, France*

PAULO MATOS • *Faculty of Sciences, BioISI – Biosystems & Integrative Sciences Institute, University of Lisboa, Lisbon, Portugal; Department of Human Genetics, National Health Institute 'Dr. Ricardo Jorge', Lisbon, Portugal*

KILDARE MIRANDA • *Laboratório de Ultraestrutura Celular Hertha Meyer, Instituto de Biofísica Carlos Chagas Filho and Instituto Nacional de Ciência e Tecnologia em Biologia Estrutural e Bioimagem, Universidade Federal do Rio de Janeiro, Rio de Janeiro, RJ, Brazil; Laboratório de Biologia Estrutural, Diretoria de Programas, Instituto Nacional de Metrologia, Normalização e Qualidade Industrial (INMETRO), Xerém, Rio de Janeiro, RJ, Brazil*

FANNY NG • *Department of Biochemistry, Yong Loo Lin School of Medicine, National University of Singapore, Singapore, Singapore*

LEONARDO NIMRICHTER • *Instituto de Microbiologia Professor Paulo de Góes, Universidade Federal do Rio de Janeiro, Rio de Janeiro, RJ, Brazil*

PETER O'TOOLE • *Technology Facility, Department of Biology, University of York, York, UK*

DEBORA L. OLIVEIRA • *Fundação Oswaldo Cruz (Fiocruz), Centro de Desenvolvimento Tecnológico em Saúde (CDTS), Rio de Janeiro, Brazil*

GABRIELLA POCSFALVI • *Mass Spectrometry and Proteomics, Institute of Biosciences and BioResources, National Research Council of Italy, Naples, Italy*

ANDREA POMPA • *Institute of Biosciences and Bioresources, Research Division of Perugia, National Research Council, Perugia, Italy*

HELEN PRICE • *Centre for Applied Entomology and Parasitology, School of Life Sciences, Keele University, Keele, Staffordshire, UK*

LIKE QU • *Key Laboratory of Carcinogenesis and Translational Research (Ministry of Education), Beijing, China; Department of Biochemistry & Molecular Biology, Peking University Cancer Hospital & Institute, Beijing, China*

OLIVIER REYNARD • *Inserm U1111, Université Claude Bernard Lyon 1, International Center for Infectiology Research (CIRI), Lyon Cedex 7, France; Université de Lyon, Ecole Normale Supérieure de Lyon, Lyon, France*

MARCIO L RODRIGUES • *Fundação Oswaldo Cruz (Fiocruz), Centro de Desenvolvimento Tecnológico em Saúde (CDTS), Rio de Janeiro, Brazil; Instituto de Microbiologia Professor Paulo de Góes, Universidade Federal do Rio de Janeiro, Rio de Janeiro, Brazil*

DARYA SAFAVIAN • *Department of Cell & Systems Biology, University of Toronto, Toronto, ON, Canada*

JENNIFER SAND • *Department of Dermatology, University Hospital Zurich, University Zurich, Zurich, Switzerland*

KERSTIN SCHIPPER • *Heinrich Heine University Düsseldorf, Institute for Microbiology, Düsseldorf, Germany; Bioeconomy Science Center (BioSC), Forschungszentrum Jülich, Jülich, Germany*

CHENGCHAO SHOU, PH.D., M.D. • *Key Laboratory of Carcinogenesis and Translational Research (Ministry of Education), Beijing, China; Department of Biochemistry & Molecular Biology, Peking University Cancer Hospital & Institute, Beijing, China*

CHRISTOPHER STANLY • *Mass Spectrometry and Proteomics, Institute of Biosciences and BioResources, National Research Council of Italy, Naples, Italy*

JANPETER STOCK • *Institute for Microbiology, Heinrich Heine University Düsseldorf, Düsseldorf, Germany; Bioeconomy Science Center (BioSC), Forschungszentrum Jülich, Jülich, Germany*

GERHARD E. STRITTMATTER • *Department of Dermatology, University Hospital Zurich, University Zurich, Zurich, Switzerland*

BOR LUEN TANG • *Department of Biochemistry, Yong Loo Lin School of Medicine, National University of Singapore, Singapore, Singapore; NUS Graduate School for Integrative Sciences and Engineering, National University of Singapore, Singapore, Singapore*

MARIUS TERFRÜCHTE • *Institute for Microbiology, Heinrich Heine University Düsseldorf, Düsseldorf, Germany; Bioeconomy Science Center (BioSC), Forschungszentrum Jülich, Jülich, Germany*

CHRISTIAN TOURIOL • *Centre de Recherches en Cancérologie de Toulouse – CRCT UMR1037 Inserm/Université Toulouse III Paul Sabatier ERL5294 CNRS Oncopole de Toulouse, Toulouse, France*

GABRIELE VARGAS • *Instituto de Microbiologia Professor Paulo de Góes, Universidade Federal do Rio de Janeiro, Rio de Janeiro, Brazil*

CORRADO VIOTTI • *Institute of Biochemistry and Biology, Plant Physiology, University of Potsdam, Potsdam, Germany*

THOMAS E. WILKOP • *Department of Plant Sciences, University of California, Davis, CA, USA*

HAIYAN ZHANG • *Key Laboratory of Plant Resources, Institute of Botany, Chinese Academy of Sciences, Beijing, China*

Part I

Background

Chapter 1

ER to Golgi-Dependent Protein Secretion: The Conventional Pathway

Corrado Viotti

Abstract

Secretion is the cellular process present in every organism that delivers soluble proteins and cargoes to the extracellular space. In eukaryotes, conventional protein secretion (CPS) is the trafficking route that secretory proteins undertake when are transported from the endoplasmic reticulum (ER) to the Golgi apparatus (GA), and subsequently to the plasma membrane (PM) via secretory vesicles or secretory granules. This book chapter recalls the fundamental steps in cell biology research contributing to the elucidation of CPS; it describes the most prominent examples of conventionally secreted proteins in eukaryotic cells and the molecular mechanisms necessary to regulate each step of this process.

Key words ER, Ribosome, SRP, Translocon, COPII, COPI, SNARE, Golgi, TGN, Secretory vesicles, Secretory granules, Plasma membrane, Regulated secretion

1 Introduction

Cell secretion is a fundamental physiological process present both in prokaryotes and eukaryotes that delivers soluble proteins and cargoes to the outside. The need to expel substances to the extracellular space is instructive for a multitude of purposes: growth, cell homeostasis, cytokinesis, defense, structural maintenance, hormone release, and neurotransmission among others. While prokaryotic cells excrete cellular waste and other substances through translocons localized to the limiting cell membranes and secrete effector molecules to other cells through dedicated organs [1], eukaryotes rely on different cellular mechanisms. Eukaryotic cells not only have the characteristic of enclosing the genetic information into a specialized compartment (the nucleus), but they also have the peculiarity of carrying several different organelles across the cytoplasm which are functionally interconnected via a multitude of transport routes that constitute the secretory pathway. Selective cargo transport among compartments is mediated by different vesicular carriers that bud from a donor

Andrea Pompa and Francesca De Marchis (eds.), *Unconventional Protein Secretion: Methods and Protocols*, Methods in Molecular Biology, vol. 1459, DOI 10.1007/978-1-4939-3804-9_1, © Springer Science+Business Media New York 2016

membrane and fuse with another [2]. Both soluble cargoes and membrane proteins are firstly translocated in the endoplasmic reticulum (ER) from where they are transported either to other organelles or secreted to the extracellular space [3, 4]. When we focus on the latter case, the best characterized mechanism of transport in eukaryotes is the conventional protein secretion (CPS): the transport route that delivers proteins from the ER to the Golgi apparatus (GA), then to the trans-Golgi network (TGN), and subsequently to the plasma membrane (PM). The TGN is the organelle where proteins destined to be secreted are segregated from lysosomal/vacuolar enzymes and sorted in budding secretory vesicles or secretory granules [5]. When secretory vesicles and granules are released from the tubular elements of the TGN, they are transported at different rates along the cytoskeletal filaments and across the cytoplasm toward the plasma membrane with which they fuse, discharging their content to the outside. Importantly, integral PM proteins are delivered and integrated to the plasma membrane through membrane fusion by the same trafficking route. Secretory vesicles and secretory granules are distinct vesicular carriers employed in constitutive and regulated secretion, respectively. While constitutive secretion is constantly undergoing in every eukaryotic cell, regulated secretion is additionally present in special types of animal cells only (e.g., endocrine and exocrine cells, neurons), and it is exclusively triggered by extracellular stimuli [5, 6]. Both constitutive and regulated secretion are included in the CPS, and for both these types of secretion the ER-Golgi-TGN segment of the transport route is identical (Fig. 1). Although individual steps of CPS show a certain degree of variability among different organisms, the basic mechanisms hold true in every eukaryotic cell. The discovery of major principles of cell secretion started in the 1950s.

2   Conventional Protein Secretion: A Historic Perspective

The elucidation of cell secretion has been paved between the 1940s and 1950s, when major advances in electron microscopy were accomplished by Keith Porter, Albert Claude, and George Palade at the Rockefeller University. The discovery of the endoplasmic reticulum (initially called "lace-like reticulum") in culture cells from chicken embryos [7], and the evidence that in cells synthesizing secretory proteins the majority of the ribosomes is attached to the ER membrane [8, 9], led George Palade to set crucial experiments to investigate the meaning of the ER-ribosome interaction. In an elegant combination of biochemistry, cell fractionation, and electron microscopy Palade and Philip Siekevitz showed that the microsomal fraction isolated from liver or pancreatic cells is almost homogeneously composed of ribosome-bound

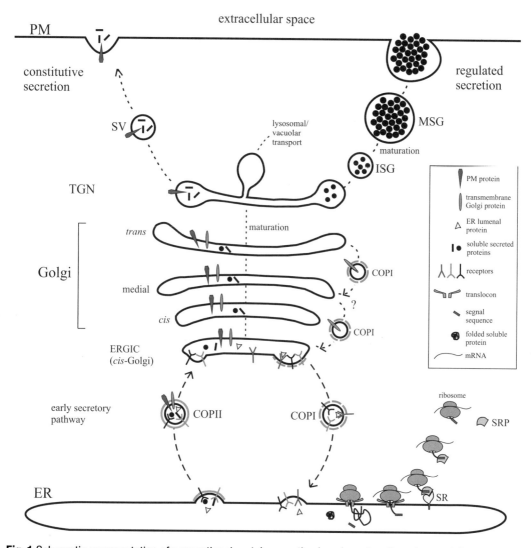

Fig. 1 Schematic representation of conventional protein secretion in eukaryotes. Secretory proteins are translocated in the ER upon (1) signal sequence recognition by the signal recognition particle (SRP); (2) SRP interaction with its receptor SR; and (3) transport through the translocon and into the ER lumen. In the ER, the signal sequence is cleaved off, and proteins are folded by molecular chaperons (not shown), and packed in COPII vesicles upon receptor-ligand interaction. COPII vesicles are delivered to the ERGIC (in animals) or to the *cis*-Golgi (in yeasts and plants). Escaped ER luminal proteins are retrotransported from the ERGIC or from the *cis*-Golgi to the ER via COPI vesicles. PM proteins and secreted proteins are transported via cisternal maturation to the TGN, whereas integral Golgi proteins are retrieved via intra-Golgi COPI-mediated transport, although another model has been proposed. At the TGN, proteins destined to be secreted are sorted in secretory vesicles (SVs) or immature secretory granules (ISGs). SVs are constitutively delivered toward the PM, whereas ISGs accumulate in the cytoplasm. Upon the arrival of specific stimuli, ISGs form mature secretory granules (MSGs) that are transported to the PM

ER-membrane vesicles [10, 11]. Because Palade recognized that exocrine pancreatic cells of guinea pig contain an exceptionally developed network of ER membranes and produce massive amounts of digestive enzymes at the same time, this system was used for following key experiments in which Palade and Siekevitz demonstrated that the ribosomes are the exclusive site of protein synthesis [12, 13]. Soon after, performing in vivo labeling with radioactive C^{14}-leucine to track the subcellular localization of newly synthesized digestive enzymes, Palade and Siekevitz showed that the pancreatic enzyme chymotrypsin is primarily detected in the microsomal fraction and synthesized by ER-bound ribosomes [14, 15]. These results led Palade to hypothesize that nascent polypeptide chains are driven across the ER-limiting membrane and into its lumen, which was demonstrated few years later by Palade, Siekevitz, and Colvin Redman using a microsomal fraction prepared from pigeon pancreas. Using radioactive amino acids, they analyzed the subcellular localization of the secreted enzyme amylase, which was initially associated with the ER-bound ribosomes. After longer incubation, microsomes were treated with sodium deoxycholate (a compound capable to solubilize membranes) and labeled amylase was detected in the soluble fraction, demonstrating that the newly synthesized enzyme was transported from the ribosomes into the microsomal lumen [16]. Similar results were obtained by Redman and David Sabatini using hepatic microsomes where secretory proteins were released upon puromycin treatment [17], and thus the rough ER (RER) was ascertained to be the site of secretory protein synthesis.

The functional link between the RER and the Golgi apparatus within the secretory pathway was demonstrated during the same years by Palade, Lucien Caro, and James Jamieson by using electron microscopic autoradiography and innovative pulse-chase experiments. These methods allowed the scientists to track in time and follow within cells the whole transport route of secretory proteins. The autoradiographic images obtained by intravenous injections of H^3-leucine showed that after ~5 minutes the labeling was localized mostly to the endoplasmic reticulum, at ~20 minutes in the elements of the Golgi complex, and after one hour in the zymogen granules [18]. Moreover, the data highlighted that the zymogen granules were formed in the Golgi region by a progressive concentration of secretory products [18]. In order to better define the role of the Golgi and its surrounding vesicular elements, Palade and Jamieson used pancreatic tissue slices incubated in vitro that allowed shorter pulse labeling and a better resolution with respect to the in vivo situation. By using isopycnic centrifugation in a linear sucrose density gradient smooth-surfaced microsomes (representing mostly the peripheral, vesicular elements of the Golgi complex) and zymogen granules were separated from the rough microsomes (consisting of RER membranes). Labeled proteins appeared initially in the rough microsomes, but shortly after they were more abundantly detected to the smooth ones, reaching the peak of concentration in this fraction after 7 minutes chase incubation. Moreover, after 17 and 37 minutes the zymogen granules were half maximally and maximally labeled, respectively [19–21]. These results not only provided the first indication that vesicles could have been the shuttling elements responsible for intracellular trafficking among compartments, they additionally proved that the Golgi apparatus (discovered in 1898 by Camillo Golgi) was an authentic cell organelle, and not just an artifact produced by cell fixation (an issue discussed at length at the time [22]), having a specific role in cell secretion. Thus, the major cellular structures involved in this process had been finally related to specific cellular functions, although the biochemical and molecular mechanisms underlying the individual steps where still unknown.

In 1971 Günter Blobel and David Sabatini postulated that protein translocation in the ER lumen was dependent on the presence of a specific amino acid sequence at the amino-terminal portion of the nascent polypeptide chain. They also speculated that the putative "signal sequence" would have been capable to recruit a "binding factor" able to guide the ribosome to the ER membrane [23]. Intriguing results were obtained in 1972 by the laboratories of Philip Leder and Cesar Milstein using cell-free translation systems producing immunoglobulin light chains that were 6–8 amino acids longer than the normal secreted version [24, 25], leading to hypothesizing the cleavage of the putative signal sequence after translation. The final proof of the existence of the signal sequence (or "signal peptide") was provided few years later by Günter Blobel and Bernhard Dobberstein. Rough microsomes isolated from canine pancreatic cells were added to a cell-free protein-synthesizing mixture supplemented by exogenous mRNA of the immunoglobulin light chain. Subsequently, ribosomes were detached from the ER membranes with a detergent and collected. The isolated ribosomes, carrying unfinished proteins, were transferred in a suitable media where they resumed synthesis of interrupted polypeptide chains without starting new rounds of translation due to the presence of aurintricarboxylic acid (an inhibitor of initiation but not elongation of polypeptide synthesis). Initially the shorter, processed chains appeared, resulting from the completion of peptides in advanced stages of translation. However, few minutes later the in vitro-synthesizing system completed longer chains too, demonstrating that the enzyme responsible for the cleavage of the signal sequence resides in the ER [26]. When rough microsomes, producing only the short version of the protein, were treated with the proteolytic enzymes trypsin and chymotrypsin (which rarely enter the microsomes) the polypeptide chains were not digested, confirming that the newly synthesized secretory proteins are immediately sequestered and driven into the microsomal lumen when translation starts. Instead, when the in vitro system was set to produce the non-secreted protein globin the digestion with trypsin and chymotrypsin occurred, indicating that this protein did not slip into the microsomes [27]. Moreover, when unprocessed light chains were added after the microsomes, they did not lose the signal sequence, demonstrating that its removal occurs during translation and not afterwards [27]. These studies showed that secretory protein precursors enclose the information for their own translocation across the ER membrane.

Since translocation across lipid bilayers was abolished by extracting the microsomal membranes with high-ionic-strength buffers, and it was rescued by adding back the salt extract [28], it became clear that there was a cytosolic component playing a crucial role in the process of protein translocation. In 1980 the signal recognition particle (SRP) was discovered by Günter Blobel and Peter Walter from canine pancreatic cell microsomes. SRP, initially named "signal recognition protein," was purified from the salt extract using hydrophobic chromatography SDS-gel electrophoresis revealed that SRP is a multimeric complex formed by six subunits of 9, 14, 19, 54, 68, and 72 kDa, respectively [29]. Moreover, SRP was shown to selectively associate with ribosomes engaged in the synthesis of secretory proteins [30, 31]. The association occurs through the binding of the 54 kDa subunit to the signal peptide (typically 7–12 hydrophobic amino acids) of nascent polypeptide chains emerging from the ribosome, which causes temporary arrest of translation [32–37]. In addition to the six different polypeptide components, SRP contains a 7S RNA molecule required for both structural and functional properties, that also represents the backbone to which the six subunits associate [38]. Thus, SRP was recognized to be a ribonucleoprotein (RNP) and was therefore renamed "signal recognition particle" [38].

The ribosome attachment to the ER membrane is mediated by the interaction between SRP and an integral ER-membrane protein, the SRP-receptor (SR), first found by Bernhard Dobberstein and David Meyer the same year of SRP discovery (i.e., 1980). Initially the cytosolic portion of SR was identified [39, 40]; afterwards the protein was intracellularly localized in vivo with a specific antibody via immunofluorescence [41], and the apparent full size determined to be 72 kDa [41–44]. Few years later, it was shown that SR actually consists of two subunits, the previously identified SRα of 72 kDa and SRβ of 30 kDa [45]. The interaction between SRP and SR is GTP dependent, and both SRP and SR are displaced from the ribosome upon GTP hydrolysis. GTP hydrolysis is additionally required by the ribosome for chain elongation, but not for the polypeptide movement across the ER membrane. Remarkably, the SRP-dependent mechanism of protein targeting is present in all three kingdoms of life. Homologues of SRP and SR have been found also in prokaryotes, where they mediate protein secretion to the periplasmic space through the translocons localized to the inner membrane [46–49].

The vectorial transfer of secretory proteins into the ER lumen can proceed as a consequence of the positional shift of ribosomes on dedicated ER membrane sites [46, 50–52]. The existence of specific locations ("aqueous channels") on the ER membrane through which secretory proteins enter the ER was already postulated in 1975 by Blobel and Dobberstein [26]. In a review of 1986, about the mechanism of protein translocation across the ER membrane, Walter and Lingappa coined the term "translocon" to identify the sites where polypeptide chains would have crossed the ER membrane to gain access to the lumen [53]. The existence of protein-conducting channels in the ER membrane was demonstrated by electrophysiological techniques. Rough microsomal vesicles were fused on one side (*cis*) of a planar lipid bilayer separating two aqueous chambers. At low puromycin concentration, single channels with a conductance of 220 picosiemens (pS) were observed. Increasing amounts of puromycin added to the *cis* side caused a large increase of membrane conductance, until it was abolished when salt concentration reached levels at which ribosomes detach from the vesicles, demonstrating that the ribosome attachment is required for the channel opening [54, 55]. The proteins that form the translocon were identified by photocross-linking using photoreactive probes that were incorporated into nascent polypeptide chains of various lengths. The chains were synthesized by an in vitro translation system supplemented with truncated mRNAs. Upon photolysis, the nascent chain was photocross-linked to specific ER membrane proteins adjacent to the nascent chain throughout translocation [56–59]. Afterwards, the translocon components that formed photoadducts with nascent chains were purified, reconstituted into proteoliposomes, and shown to execute the transfer [60–63].

3   Protein Translocation in the Endoplasmic Reticulum

The channel of the translocon is formed by the Sec61 complex, consisting of the heterotrimer Sec61α, Sec61β, and Sec61γ in mammals [62, 63]. The prefix "Sec" was chosen because the first isolated component Sec61α is homologous to the budding yeast *Saccharomyces cerevisiae* Sec61p protein, which was identified in a previous screening for *sec*retory mutants that led to the isolation of 23 fundamental genes of the secretory pathway [64, 65]. The α- and γ-subunits are highly conserved, and both are essential for the function of the channel and for cell viability, whereas the β-subunit is dispensable. The Sec61 complex is the essential element for protein translocation, and the α-subunit alone forms the pore [63]. The same holds true in yeast, where the homologous components of the Sec61 complex are Sec61p, Sbh1p, and Sss1p [66], and in prokaryotes, where the bacterial heterotrimeric translocation pore complex (subunits SecY, SecE, and SecG) of plasma membrane translocons mediates secretion of different substances to the periplasmic space [67, 68]. Several integral ER membrane proteins can associate to the Sec61 complex to perform translocation, although the function of some of them is not fully clarified. In mammals, the associated proteins that mediate translocation are: (a) the translocation-associated membrane protein (TRAM) [61]; (b) the translocon-associated protein complex TRAP, a heterotetramer consisting of subunits α, β, γ, δ [62, 69]; (c) the oligosaccharyl transferase complex (OST), responsible for N-glycosylation in the ER, whose core complex is a heterotetramer formed by ribophorin I (66 kDa), ribophorin II (63/64 kDa), OST48 (48 kDa), and DAD1 (10 kDa) [70–72]; (d) the signal peptidase complex (SPC), responsible for the cleavage of the signal sequence in the ER lumen, consisting of five subunits, whose names SPC12, SPC18, SPC21, SPC22/23, and SPC25 indicate the respective molecular size [73]; and (e) the Sec62/Sec63 complex [74, 75]. As well as in mammals, the function of Sec61, OST, and SP complexes has been well characterized in yeast [66, 76, 77]. Depending on which associated components work in concert with the Sec61 complex, two different mechanisms of protein translocation in eukaryotes occur: co- or post-translationally. The co-translational mechanism is present in all cell types and occurs both for soluble and membrane proteins. The targeting phase requires the interaction of SRP with the signal sequence of a nascent polypeptide chain. Subsequently, the interaction between SRP and SR mediates the ribosome-channel alignment. During translocation of membrane proteins, specific polypeptide sequences do not enter the channel, but protrude from the ribosome-channel junction into the cytosol, generating a cytosolic domain [78]. In several, if not all organisms, some proteins are translocated after completion of their synthesis, therefore "post-translationally," and they are not completely folded after their release from the ribosome [79]. Post-translational translocation is more frequently occurring in simpler organisms like bacteria and yeast. In *S. cerevisiae* the heterotetrameric Sec62/Sec63 complex specifically mediates post-translational translocation in concert with the cytosolic chaperon Hsp70, the Sec61 complex, and the luminal chaperone Kar2p/BiP in an ATP-dependent manner [79–84]. Instead, the co-translational mechanism requires the function of the Sec61 complex only and it is instead GTP dependent [85]. Although in mammals translocation seems to occur preferentially co-translationally [85, 86], posttrans-lational mechanisms have been shown for specific kinds of proteins. In fact, the SRP-dependent pathway, although ubiquitous, is inaccessible for those proteins carrying a single transmembrane domain (TMD) on their C-terminal portion, because they are released from the ribosome before the TMD emerges from the ribosomal tunnel. These peptides, called tail-anchored proteins (TA), are involved in a wide range of cellular processes and include the SNAREs (involved in vesicular traffic), several translocon components, structural Golgi proteins, and enzymes located in almost every membrane. Thus, TAs are inserted in the ER membrane post-translationally both in higher eukaryotes and yeast. Cross-linking experiments revealed that the cytosolic TMD recognition complex TRC40 (previously known as Asna-1) interacts post-translationally with TAs in a TMD-dependent manner and mediates their targeting to the ER membrane [87, 88]. A conserved three-protein complex composed of Bat3, TRC35, and Ubl4A facilitates the TA protein capture by TRC40 [90]. Homologues of TRC40 are conserved in many species, including *S. cerevisiae* where it is termed Get3 [90]. TRC40 delivers TAs to an ER receptor composed of the tryptophan-rich basic protein (WRB) [91] and the calcium-modulating cyclophilin ligand (CAML) [92], mammalian equivalents of the yeast components Get1 and Get2, respectively [86, 93].

4   The COPII-Mediated ER Exit

Nascent secretory and membrane proteins are translocated or inserted at the ER, eventually glycosylated, and then folded through the action of a multitude of molecular chaperons and cofactors that ensure conformation quality and fidelity. When the protein-folding capacity of the ER is unable to sustain a sufficient rate of folding, the accumulation of misfolded proteins triggers a multitude of signaling pathways collectively termed unfolded protein response (UPR) that increases the folding capacity. However, when problems persist, misfolded polypeptides are degraded through the action of the ER-associated degradation (ERAD) pathway, and the mutated and/or misfolded proteins are retro-translocated to the cytosol to be degraded by the 26S proteasome machinery [94, 95].

When membrane and soluble proteins reach the correct conformation and are not ER-resident proteins, they exit the ER. In all eukaryotic cells, the best characterized mechanism of ER exit is the COPII-mediated transport, whose components were all identified after a screening for yeast secretory mutants [64]. The coat protein complex II (COPII) assembles on specific locations of the ER membrane, called ER-exit sites (ERES), from which COPII-coated vesicles bud off [96]. ERES are also known as transitional elements (TEs) or transitional ER (tER). The number, size, and dynamics of ERES vary among cell types and organisms; however, these organized export sites are present in most eukaryotic cells [97]. The assembly of COPII starts with the recruitment of the cytosolic small GTPase Sar1 (secretion-associated RAS-related 1) to the ER membrane [98, 99], where it is activated through the action of the guanine nucleotide exchange factor (GEF) Sec12, an integral ER membrane protein that catalyzes GDP/GTP exchange [100, 101]. An activated, GTP-bound Sar1 inserts its N-terminal helix into the ER membrane, inducing initial membrane curvature [102–104] alongside with the recruitment of the cytosolic Sec23/Sec24 heterodimer [105]. The Sar1-Sec23-Sec24 complex is recognized and bound by the Sec13/Sec31 heterotetramer, which forms the outer layer of the COPII cage [106–109]. Transmembrane cargo proteins are recognized and bound by Sec24, whereas soluble cargoes bind specific receptors that span the ER membrane. Multiple adjacent Sec13/Sec31 subcomplexes drive membrane bending and vesicle fission using the energy of GTP hydrolysis [110, 111]. Sec23 serves as a bridge between Sar1 and Sec24 and is a GTPase-activating protein (GAP) that stimulates Sar1 GTP hydrolysis [99], which is additionally needed for vesicle uncoating after release [111]. There is evidence that Sec31 interacts directly with Sar1 to promote Sec23 GAP activity [112]. In addition to the six core COPII components, Sec16 is involved in ERES maintenance and COPII-mediated ER export. Sec16 localizes to the ERES independent of Sec23/24 and Sec13/31, and its localization depends on Sar1 activity [113]. Sec16 has been shown to bind several COPII components and seems to serve as scaffold protein that concentrates, organizes, and stabilizes COPII proteins [114–116]. However, the precise Sec16 function is still not fully understood.

Since most COPII subunits have one or more paralogues [117], and since COPII transport is assisted by several different accessory proteins (e.g., 14-3-3, PX-RICS, Deshavelled) depending on the cell type [118–121], the result is a high number of molecularly different COPII-coated vesicles with tissue specificities and selectivity for different cargo molecules. The number and size of ERES, together with the expression levels of COPII components, may play a major role in the secretion rate in different tissues. One of the biggest open questions regarding COPII-mediated transport is how large-sized cargoes can be lodged inside vesicles which are typically of 60–100 nm in diameter. Procollagen fibrils (PC), composed of rigid triple helices of up to 400 nm in length, represent one of the most abundant secreted cargoes in animal cells, since collagen composes approximately 25 % of the whole-body protein content, and is fundamental for almost all cell-cell interactions [122]. There are several lines of evidence indicating that collagen secretion is COPII dependent. Depletion of Sec13 [123], disruption of Sec24D [124], mutation of Sec23A [125], loss-of-function of the Sedlin gene (a TRAPPI complex component interacting with Sar1 at the ER-Golgi interface) [126], and depletion of Sar1A and Sar1B [127] all block collagen secretion, leading to severe diseases. Cryomicroscopical data suggest a significant level of flexibility of the COPII cage, which in vitro can assemble on flatter membranes, forming larger cages that could accommodate procollagen fibrils [128, 129]. Recently, a potential mechanism for giant COPII-carriers biogenesis has been proposed, which involves TANGO1-mediated packing. TANGO1/Mia3 is a transmembrane protein identified from a screening for secretory mutants in *Drosophila* S2 cells, and shown to localize to early Golgi cisternae and to the ERES [130, 131]. Knockdown of TANGO1 with siRNA severely inhibits ER export of PC VII. TANGO1 interacts with Sec23A and Sec24C through its cytoplasmic proline-rich domain (PRD), and binds PC VII via its luminal SH3 domain [132]. cTAGE5 is the partner of TANGO1 in PC VII secretion; it is anchored to the ERES and interacts via its PRD with Sec23A, Sec24C, and Sec12 [133, 134]. Cullin3 (an E3 ligase), and its specific adaptor protein KLHL12, ubiquitinates SEC31. In mouse embryonic fibroblasts, Cul3 knockdown inhibits collagen IV secretion, and overexpression of KLHL12 increases secretion of PC I in the human fibroblast cell line IMR-90. The model proposes that TANGO1-cTAGE5 pack collagens in ERES enriched with Sec23/24 to the inner coat shell, and Cul3-KLHL12 mediate the assembly of a large outer layer composed of Sec13/31-ubiquitin. The final result would be the formation of a giant COPII-carrier carrier for procollagen export from the ER [122]. However, the evidence that TANGO1 interacts with the conserved syntaxin 5-binding protein Sly1, which in turn interacts with the ER-specific t-SNAREs syntaxin-17 and syntaxin-18 (involved in membrane fusion), leads to formulate a second hypothesis: a membrane domain of the ERGIC (ER-Golgi-intermediate compartment) could be recruited to the ERES, and the resulting fusion would promote the elongation of the PC VII-enriched domain into a tubular uncoated bud, while the TANGO1-cTAGE5-Sec12-Sec23/24 complex would remain at the neck [122].

5   The ER-Golgi Interface and COPI Vesicles

Passive incorporation of soluble cargoes into COPII vesicles can occur [135–138], whereas membrane proteins and receptors require diacidic or dihydrophobic motifs in their cytosolic domains for efficient transport through the interaction with multiple binding sites of Sec24 [139–142]. It is still unclear in mammals whether COPII vesicles are transported to the ERGIC along microtubules (from the plus- to minus-end), since contrasting results have been so far collected [117]. The directionality and fidelity of COPII vesicle transport and fusion with either the ERGIC or the *cis*-Golgi (depending on the organism) are mediated by the concerted action of RAB GTPases, tethering factors, and integral membrane SNARE proteins. In mammalian cells, RAB1 and the tethering factors p115, GM130, GRASP65, and the TRAPPI complex orchestrate the tethering [143–150]. TRAPPI-mediated RAB1 activation recruits p115, generating a localized signal to tether COPII vesicles, and TRAPPI binds directly Sec23 [151, 152]. Fusion of COPII-tethered vesicles depends on a set of four SNAREs: syntaxin-5, membrin/GS27, BET1, and Sec22B [153–155]. Additionally, the syntaxin 5-binding protein Sly1 is required for this vesicle fusion step [156] and may serve to coordinate the vesicle tethering and fusion. All fusion events between membranes require the correct pairing of specific cognate SNAREs on the vesicle surface and on the acceptor membrane. SNAREs (soluble *N*-ethylmaleimide-sensitive factor adaptor protein receptors) are tail-anchored proteins that contain a conserved membrane-proximal heptad repeat sequence known as the SNARE motif. The *trans*-assembly of motifs into a four-helix bundle drives the fusion between lipid bilayers [157–161]. In mammals, COPII vesicles reach first the ER-Golgi-intermediate compartment (ERGIC), alternatively termed vesicular tubular cluster (VTC), which is a distinct organelle respect to the Golgi and is absent in yeasts and plants [97]. While in animal cells the Golgi apparatus is a relatively stationary organelle, in plant cells the Golgi is instead highly mobile and moves with a speed of up to 4 μm/sec. [162]. Golgi stacks in plant cells move extensively along both the ER tubules and actin filaments (which are aligned to each other) throughout the cytoplasm. The movement relies on actomyosin motors, and displays a distinctive stop-and-go pace [162–167]. The plant ER-Golgi interface is spatially reduced (around 500 nm), and the two compartments are tightly coupled, as demonstrated by using optical tweezers [168]. The plant Golgi receives budding COPII vesicles from the ERES in a cytoskeleton-independent manner [169] within the so called secretory unit model, in which the two compartments are embedded in a ribosome-free surrounding matrix [170–174]. While plant COPI vesicles (the retrograde Golgi-to-ER carrier) have been biochemically isolated and localized in situ [175], visualization of COPII in plant tissues is rare (although observed) even when ultra-rapid cryofixing techniques are employed [170, 176–178]. Thus, it is a matter of debate whether COPII-mediated transport in higher plants can additionally occur via coated-tubular connections [179].

COPI mediates retrograde transport of receptors and soluble proteins from the *cis*-Golgi (from the ERGIC in mammals) back to the ER along microtubules. The coat protein complex I (COPI), or "coatomer," is a heptameric (α, β, β', g, δ, ε, ζ) complex, where the γ-COP, δ-COP, ζ -COP, and β-COP subunits constitute the inner coat layer, and α-COP, β-COP, and ε-COP form the outer shell [180–182]. Upon activation by ADP-ribosylation factor guanine nucleotide exchange factors (ARF-GEFs), the myristoylated membrane-anchored ARF1 GTPase recruits the COPI subunits to the Golgi membranes [183, 184]. Subunits α-COP, β′-COP, γ-COP, and δ-COP recognize sorting motifs on the cytosolic domain of membrane cargoes and mediate the load of soluble proteins into nascent COPI vesicles. ARF GTPase-activating proteins (GAPs) bind cytoplasmic signals on cargo proteins, γ-COP, β′-COP, and ARF1. Stimulation of the GTPase activity of ARF1 by GAPs leads to the release of ARF1 from the complex and to the dissociation of GAPs and the coat subunits [185]. COPI vesicles deliver ER receptors (recycled for new rounds of transport) and luminal ER proteins that escape through bulk flow via COPII vesicles. Luminal ER proteins classically carry a KDEL motif (in animals and yeast) or an HDEL motif (in plants) within their C-terminal domain, which represent the retrograde sorting signals recognized by dedicated Golgi receptors (Erd2 in yeast and plants; KDELRs in mammals). Targeting of COPI vesicles to the ER requires the multisubunit DSL1 tethering complex, and the SNARE proteins syntaxin-18, Sec20, Slt1, and Sec22B [186, 187].

6 The Golgi Apparatus, the TGN, and the Rab GTPase-Mediated Secretory Vesicle Formation

In most eukaryotes the Golgi apparatus (or Golgi complex) consists of a series of stacked cisternae, with a *cis* to *trans* polar orientation. The cisternae are kept adjacent by structural proteins present in the surrounding ribosome-free matrix [188], and by heterotypic tubular connections [189, 190]. In mammals the Golgi includes 4-8 cisternae, each of them 0.7–1.1 μm wide and 10–20 nm thick. Multiple Golgi stacks can be laterally interconnected by tubules, forming the so-called Golgi ribbon. In several lower eukaryotes, like the budding yeasts *S. cerevisiae* and *Pichia pastoris*, or in the fruit fly *Drosophila melanogaster*, the Golgi is formed by individual cisternae scattered throughout the cytoplasm, which can occasionally associate but do not form stacks, although polar features are maintained [188]. Single stacks are present both in higher plants (e.g., *Arabidopsis thaliana*, tobacco), and algae (e.g., *Chlamydomonas reinhardtii*). Depending on the enrichment of specific enzymes, three major regions can be recognized within one Golgi complex: *cis*, medial, and *trans* [188]. Juxtaposed to the Golgi *trans*-most cisternae, a pleiomorphic, tubular-vesicular compartment is present: the trans-Golgi-network (TGN) [191, 192]. In plant cells the TGN has been shown to additionally hold the role of early endosome (EE, the first compartment reached by endocytosed molecules) [193–195], whereas in animals the TGN and EE are distinct compartments. Two models have been proposed for secretory protein transport through the Golgi complex: (1) anterograde COPI-vesicular transport between stable cisternae; and (2) cisternal progression/maturation [196]. Detection of cargoes and bidirectional transport by distinct populations for COPI vesicles support the first scenario [196]; however, exclusive retrograde transport for COPI is supported by the detection of the KDEL receptor, resident Golgi proteins, and glycosylation enzymes. The cisternal maturation model is currently preferred because, among other reasons, it explains how transport of large cargoes is achieved [196]. In this view, the cisternae continuously mature from *cis*-to-*trans*, and secretory proteins are transported along the anterograde flow, and up to the TGN. The anterograde maturation is the net result from COPII vesicle entry and secretory vesicles exit on the respective *cis* and *trans* sides. Homotypic fusion of COPII vesicles gives rise to newly formed *cis*-cisternae, while the *trans*-most cisternae mature into a TGN. Intra-Golgi retrieval of integral Golgi proteins from older to younger cisternae occurs via COPI vesicles and through the heterotypic tubular connections. The Golgi is the organelle where glycosylation of soluble cargoes, membrane proteins, and lipids is completed, and where polysaccharide synthesis occurs. The *cis*-to-*trans* polarity in the distribution of Golgi glycosylation enzymes was discovered by cytochemical staining based on different enzymatic activity among cisternae, and it reflects the sequence of oligosaccharide processing reactions [188, 196, 197].

At the TGN, proteins are sorted toward three different destinations: PM, endosomes, and lytic compartments. These trafficking routes differ in terms of adaptors, effector molecules, and sorting signals involved. Formation of secretory vesicles delivered to the PM is GTP dependent, requires either ARF GTPases or Rab GTPases, and may be mediated by clustering of specific lipids on TGN subdomains. However, the molecular mechanisms and the sorting signals for TGN-to-PM delivery are far less understood in comparison to COPII-, COPI- and clathrin-mediated vesicle transports.

The heterotetrameric adaptor protein complexes (APs) are the most well-characterized cargo adaptors at the TGN. Five APs have been identified in higher eukaryotes, and three of them (AP-1, AP-3, and AP-4) sort proteins at the TGN. APs bind membrane cargoes and receptors via their μ subunit, and contribute to form coated carriers. AP-1 and AP-3 interact with clathrin, whereas AP-4 does not [198]. While AP-3 is involved in lysosomal/vacuolar sorting and traffic, AP-1 and AP-4 mediate polar transport of basolateral-located proteins in epithelial cells [199, 200], and both AP-1 and AP-4 require the function of ARF1. The PM of epithelial cells is polarized into apical and basolateral domains, and each of them contain distinct set of proteins carrying specific functions. Protein sorting at the TGN contributes to polar delivery of apical/basolateral proteins, and to the asymmetric localization of

signaling receptors that determine planar cell polarity (PCP) of
epithelia [201]. Tyrosine-based motifs and dileucine motifs at the
C-terminal domain are canonical sorting signals for basolateral-
targeted proteins, whereas apical sorting determinants are diversi-
fied and vaguely defined [198, 201]. However, apical determinants
promote partitioning into glycosphingolipid- and cholesterol-rich
membrane microdomains (i.e., lipid rafts) at the TGN, from where
carriers arise [201–203].

In yeast, a unique adaptor complex, termed "exomer," mediates protein transport directly from the TGN to the PM. Exomer
is a heterotetramer consisting of two copies of Chs5p and two copies of the ChAPs family proteins (Chs6, Bud7p, Bch1p, and
Bch2p). Chs5p binds to the small GTPase Arf1, whereas the
ChAPs are responsible for cargo binding and sorting [204–208].
Exomer regulates trafficking of chitin synthase III (Chs3p) and
Fus1p from the TGN to the PM [204, 205, 209, 210]. No known
homologs of exomer have been found in metazoans as yet.

Secretory vesicles in yeast are transported to the cell surface
through the function of the Sec4 GTPase [211], whose homolog
in plants is RabE1 [212]. In plants, secretory vesicles deliver hemi-
celluloses and pectins to the plant apoplast from the TGN/EE
[193], a transport route mediated by the protein ECHIDNA
(ECH), which interacts with the Rab GTPases YIP4a and YIP4b
[213, 214]. On the contrary, cellulose is synthesized by plasma
membrane-localized cellulose synthase complexes [215]. ECH
also specifically mediates the targeting of the auxin influx carrier
AUX1 from the TGN to the PM, but not the transport of the
auxin influx carriers LAX1-3 and of the efflux carrier PIN3 [216].
In contrast to animals, secretion in plants is fundamental for cytokinesis, since plants have evolved a unique mechanism of cell division. Instead of forming a contractile ring that constricts the plasma
membrane, dividing plant cells target secretory vesicles to the center of the division plane, where they fuse with one another to form
the cell plate. Afterwards, the cell plate fuses with the parental PM
on both sides [217, 218]. This mechanism requires the targeting
and function of the PM-located plant-specific syntaxin KNOLLE,
the Sec1-like protein KEULE, and the t-SNARE AtSNAP-33
[219–222].

After budding, vesicles are delivered to the PM by motor-
mediated transport along a cytoskeletal track (microtubules or
actin), in which kinesins have been shown to be implicated [203,
223]. The tethering factor that mediates fusion of secretory vesicles and secretory granules with the PM is the exocyst complex,
formed by eight components: Sec3, Sec5, Sec6, Sec8, Sec10,
Sec15, Exo70, and Exo84, whose functions are conserved among
eukaryotes [224–226].

7   Secretory Granules and Regulated Secretion

Animal cells where regulated secretion is present include endocrine and exocrine cells, epithelial cells, mast cells, platelets, large granular lymphocytes, neutrophils, and neurons. Secretion of insulin from endocrine pancreatic β-cells, secretion of zymogen from exocrine pancreatic cells to digest food, secretion of growth hormone from GH cells of the pituitary gland, and the release of neurotransmitters at the synapses are only few examples of regulated secretion. Secretory granules contain massive amounts of cargoes, which accumulate first in subdomains of the TGN, and are later released as immature secretory granules (ISGs) that accumulate in the cytoplasm. In endocrine cells the concentration factor from the ER to secretory granules may be as high as 200-fold, whereas in constitutive secretory vesicles there is at most a 2-fold concentration of secretory products then in the ER [5]. Biogenesis of mature secretory granules (MSGs) involves specific mechanisms of protein sorting, pro-hormone processing, and vesicle fusion. Specific sorting signals and domains in regulated secretory proteins (RSPs) are needed to direct them into the regulated secretory pathway, and for their segregation from constitutive secreted proteins at the TGN. Cell-type-specific composition of RSPs in the TGN has an important role to determine how the RSPs are sorted into ISGs. Lipid rafts are implicated in RSP sorting at the TGN and specific SNAREs are required for either MSG formation and for their fusion with the PM [6, 227].

References

1. Wooldridge K (2009) Bacterial secreted proteins: secretory mechanisms and role in pathogenesis. Caister Academic Press, Norfolk, VA

2. Bonifacino JS, Glick BS (2004) The mechanisms of vesicle budding and fusion. Cell 116:153–166. doi:10.1016/S0092-8674(03)01079-1

3. Palade GE (1975) Intracellular aspects of the process of protein synthesis. Science 189:347–358. doi:10.1126/science.1096303

4. Kelly RB (1985) Pathways of protein secretion in eukaryotes. Science 230:25–32. doi:10.1126/science.2994224

5. Burgess TL, Kelly RB (1987) Constitutive and regulated secretion of proteins. Annu Rev Cell Biol 3:243–293. doi:10.1146/annurev.cb.03.110187.001331

6. Tooze SA, Martens GJ, Huttner WB (2001) Secretory granule biogenesis: rafting to the SNARE. Trends Cell Biol 11:116–122. doi:10.1016/S0962-8924(00)01907-3

7. Porter KR, Claude A, Fullam EF (1945) A study of tissue culture cells by electron microscopy: methods and preliminary observations. J Exp Med 81:233–246. doi:10.1084/jem.81.3.233

8. Palade GE (1955) A small particulate component of the cytoplasm. J Biophys Biochem Cytol 1:59–68. doi:10.1083/jcb.1.1.59

9. Palade GE (1955) Studies on the endoplasmic reticulum: II. Simple dispositions in cells in situ. J Biophys Biochem Cytol 1:567–582. doi:10.1083/jcb.1.6.567

10. Palade GE, Siekevitz P (1956) Liver microsomes: an integrated morphological and biochemical study. J Biophys Biochem Cytol 2:171–200. doi:10.1083/jcb.2.2.171

11. Palade GE, Siekevitz P (1956) Pancreatic microsomes: an integrated morphological and biochemical study. J Biophys Biochem Cytol 2:671–690. doi:10.1083/jcb.2.6.671

12. Siekevitz P, Palade GE (1958) A cytochemical study on the pancreas of the guinea pig.

I. Isolation and enzymatic activities of cell fractions. J Biophys Biochem Cytol 4:203–218. doi:10.1083/jcb.4.2.203

13. Siekevitz P, Palade GE (1958) A cytochemical study on the pancreas of the guinea pig. II. Functional variations in the enzymatic activity of microsomes. J Biophys Biochem Cytol 4:309–318. doi:10.1083/jcb.4.3.309

14. Siekevitz P, Palade GE (1958) A cytochemical study on the pancreas of the guinea pig. III. In vivo incorporation of leucine-1-C14 into the proteins of cell fractions. J Biophys Biochem Cytol 4:557–566. doi:10.1083/jcb.4.5.557

15. Siekevitz P, Palade GE (1960) A cytochemical study on the pancreas of the guinea pig. 5. In vivo incorporation of leucine-l-C 14 into the chymotrypsinogen of various cell fractions. J Biophys Biochem Cytol 7:619–630. doi:10.1083/jcb.7.4.619

16. Redman CM, Siekevitz P, Palade GE (1966) Synthesis and transfer of amylase in pigeon pancreatic micromosomes. J Biol Chem 241:1150–1158

17. Redman CM, Sabatini DD (1966) Vectorial discharge of peptides released by puromycin from attached ribosomes. Proc Natl Acad Sci U S A 56:608–615. doi:10.1073/pnas.56.2.608

18. Caro LG, Palade GE (1964) Protein synthesis, storage, and discharge in the pancreatic exocrine cell – an autoradiographic study. J Cell Biol 20:473–495. doi:10.1083/jcb.20.3.473

19. Jamieson JD, Palade GE (1966) Role of the Golgi complex in the intracellular transport of secretory proteins. Proc Natl Acad Sci U S A 55:424–431. doi:10.1073/pnas.55.2.424

20. Jamieson JD, Palade GE (1967) Intracellular transport of secretory proteins in pancreatic exocrine cell. I Role of peripheral elements of Golgi complex. J Cell Biol 34:577–596. doi:10.1083/jcb.34.2.577

21. Jamieson JD, Palade GE (1967) Intracellular transport of secretory proteins in the pancreatic exocrine cell. II Transport to condensing vacuoles and zymogen granules. J Cell Biol 34:597–615. doi:10.1083/jcb.34.2.597

22. Farquhar MG, Palade GE (1981) The Golgi apparatus (complex)-(1954–1981)-from artifact to center stage. J Cell Biol 91:77–103. doi:10.1083/jcb.91.3.77s

23. Blobel G, Sabatini D (1971) Dissociation of mammalian polyribosomes into subunits by puromycin. In: Manson LA (ed) Biomembranes. Springer, Berlin, pp 193–195

24. Swan D, Aviv H, Leder P (1972) Purification and properties of biologically active messenger RNA for a myeloma light chain. Proc Natl Acad Sci U S A 69:1967–1971. doi:10.1073/pnas.69.7.1967

25. Milstein C, Brownlee GG, Harrison TM, Mathews MB (1972) A possible precursor of immunoglobulin light chains. Nat New Biol 239:117–120. doi:10.1038/newbio239117a0

26. Blobel G, Dobberstein B (1975) Transfer of proteins across membranes. I Presence of proteolytically processed and unprocessed nascent immunoglobulin light chains on membrane-bound ribosomes of murine myeloma. J Cell Biol 67:835–851. doi:10.1083/jcb.67.3.835

27. Blobel G, Dobberstein B (1975) Transfer of proteins across membranes. II Reconstitution of functional rough microsomes from heterologous components. J Cell Biol 67:852–862. doi:10.1083/jcb.67.3.852

28. Warren G, Dobberstein B (1978) Protein transfer across microsomal membranes reassembled from separated membrane components. Nature 273:569–571. doi:10.1038/273569a0

29. Walter P, Blobel G (1980) Purification of a membrane-associated protein complex required for protein translocation across the endoplasmic reticulum. Proc Natl Acad Sci U S A 77:7112–7116. doi:10.1073/pnas.77.12.7112

30. Walter P, Ibrahimi I, Blobel G (1981) Translocation of proteins across the endoplasmic reticulum. I. Signal recognition protein (SRP) binds to in-vitro-assembled polysomes synthesizing secretory protein. J Cell Biol 91:545–550. doi:10.1083/jcb.91.2.545

31. Walter P, Blobel G (1981) Translocation of proteins across the endoplasmic reticulum. II. Signal recognition protein (SRP) mediates the selective binding to microsomal membranes of in-vitro-assembled polysomes synthesizing secretory protein. J Cell Biol 91:551–556. doi:10.1083/jcb.91.2.551

32. Walter P, Blobel G (1981) Translocation of proteins across the endoplasmic reticulum III. Signal recognition protein (SRP) causes signal sequence-dependent and site-specific arrest of chain elongation that is released by microsomal membranes. J Cell Biol 91:557–561. doi:10.1083/jcb.91.2.557

33. Gilmore R, Blobel G (1983) Transient involvement of signal recognition particle and its receptor in the microsomal membrane prior to protein translocation. Cell 35:677–685. doi:10.1016/0092-8674(83)90100-9

34. Kurzchalia TV, Wiedmann M, Girshovich AS, Bochkareva ES, Bielka H, Rapoport TA (1986) The signal sequence of nascent preprolactin interacts with the 54K polypep-

tide of the signal recognition particle. Nature 320:634–636. doi:10.1038/320634a0

35. Krieg UC, Walter P, Johnson AE (1986) Photocrosslinking of the signal sequence of nascent preprolactin to the 54-kilodalton polypeptide of the signal recognition particle. Proc Natl Acad Sci U S A 83:8604–8608. doi:10.1073/pnas.83.22.8604

36. Siegel V, Walter P (1988) Each of the activities of signal recognition particle (SRP) is contained within a distinct domain: analysis of biochemical mutants of SRP. Cell 52:39–49. doi:10.1016/0092-8674(88)90529-6

37. Bernstein HD, Poritz MA, Strub K, Hoben PJ, Brenner S, Walter P (1989) Model for signal sequence recognition from amino-acid sequence of 54K subunit of signal recognition particle. Nature 340:482–486. doi:10.1038/340482a0

38. Walter P, Blobel G (1982) Signal recognition particle contains a 7S RNA essential for protein translocation across the endoplasmic reticulum. Nature 299:691–698. doi:10.1038/299691a0

39. Meyer DI, Dobberstein B (1980) A membrane component essential for vectorial translocation of nascent proteins across the endoplasmic reticulum: requirements for its extraction and reassociation with the membrane. J Cell Biol 87:498–502. doi:10.1083/jcb.87.2.498

40. Meyer DI, Dobberstein B (1980) Identification and characterization of a membrane component essential for the translocation of nascent proteins across the membrane of the endoplasmic reticulum. J Cell Biol 87:503–508. doi:10.1083/jcb.87.2.503

41. Meyer DI, Louvard D, Dobberstein B (1982) Characterization of molecules involved in protein translocation using a specific antibody. J Cell Biol 92:579–583. doi:10.1083/jcb.92.2.579

42. Meyer DI, Krause E, Dobberstein B (1982) Secretory protein translocation across membranes-the role of the "docking protein". Nature 297:647–650. doi:10.1038/297647a0

43. Gilmore R, Blobel G, Walter P (1982) Protein translocation across the endoplasmic reticulum. I. Detection in the microsomal membrane of a receptor for the signal recognition particle. J Cell Biol 95:463–469. doi:10.1083/jcb.95.2.463

44. Gilmore R, Walter P, Blobel G (1982) Protein translocation across the endoplasmic reticulum. II. Isolation and characterization of the signal recognition particle receptor. J Cell Biol 95:470–477. doi:10.1083/jcb.95.2.470

45. Tajima S, Lauffer L, Rath VL, Walter P (1986) The signal recognition particle receptor is a complex that contains two distinct polypeptide chains. J Cell Biol 103:1167c1178. doi:10.1083/jcb.103.4.1167

46. Keenan RJ, Freymann DM, Stroud RM, Walter P (2001) The signal recognition particle. Annu Rev Biochem 70:755–775. doi:10.1146/annurev.biochem.70.1.755

47. Römisch K, Webb J, Herz J, Prehn S, Frank R, Vingron M, Dobberstein B (1989) Homology of 54K protein of signal-recognition particle, docking protein and two E. coli proteins with putative GTP-binding domains. Nature 340:478–482. doi:10.1038/340478a0

48. Poritz MA, Bernstein HD, Strub K, Zopf D, Wilhelm H, Walter P (1990) An E. coli ribonucleoprotein containing 4.5S RNA resembles mammalian signal recognition particle. Science 250:1111–1117. doi:10.1126/science.1701272

49. Wolin SL (1994) From the elephant to E. coli: SRP-dependent protein targeting. Cell 77:787–790

50. Connolly T, Gilmore R (1986) Formation of a functional ribosome-membrane junction during translocation requires the participation of a GTP-binding protein. J Cell Biol 103:2253–2261. doi:10.1083/jcb.103.6.2253

51. Connolly T, Gilmore R (1989) The signal recognition particle receptor mediates the GTP-dependent displacement of SRP from the signal sequence of the nascent polypeptide. Cell 57:599–610. doi:10.1016/0092-8674(89)90129-3

52. Connolly T, Rapiejko PJ, Gilmore R (1991) Requirement of GTP hydrolysis for dissociation of the signal recognition particle from its receptor. Science 252:1171–1173. doi:10.1126/science.252.5009.1171

53. Walter P, Lingappa VR (1986) Mechanism of protein translocation across the endoplasmic reticulum membrane. Annu Rev Cell Biol 2:499–516. doi:10.1146/annurev.cb.02.110186.002435

54. Simon SM, Blobel G, Zimmerberg J (1989) Large aqueous channels in membrane vesicles derived from the rough endoplasmic reticulum of canine pancreas or the plasma membrane of Escherichia coli. Proc Natl Acad Sci U S A 86:6176–6180. doi:10.1073/pnas.86.16.6176

55. Simon SM, Blobel G (1991) A protein-conducting channel in the endoplasmic retic-

ulum.   Cell   65:371–380.
doi:10.1016/0092-8674(91)90455-8

56. Krieg UC, Johnson AE, Walter P (1989) Protein translocation across the endoplasmic reticulum membrane: identification by photocross-linking of a 39-kDa integral membrane glycoprotein as part of a putative translocation tunnel. J Cell Biol 109:2033–2043. doi:10.1083/jcb.109.5.2033

57. Wiedmann M, Görlich D, Hartmann E, Kurzchalia TV, Rapoport TA (1989) Photocrosslinking demonstrates proximity of a 34 kDa membrane protein to different portions of preprolactin during translocation through the endoplasmic reticulum. FEBS Lett   257:263–268. doi:10.1016/0014-5793(89)81549-2

58. High S, Görlich D, Wiedmann M, Rapoport TA, Dobberstein B (1991) The identification of proteins in the proximity of signal-anchor sequences during their targeting to and insertion into the membrane of the ER. J Cell Biol 113:35–44. doi:10.1083/jcb.113.1.35

59. Thrift RN, Andrews DW, Walter P, Johnson AE (1991) A nascent membrane protein is located adjacent to ER membrane proteins throughout its integration and translation. J Cell Biol 112:809–821. doi:10.1083/jcb.112.5.809

60. Nicchitta CV, Blobel G (1990) Assembly of translocation-competent proteoliposomes from detergent-solubilized rough microsomes.   Cell   60:259–269. doi:10.1016/0092-8674(90)90741-V

61. Görlich D, Hartmann E, Prehn S, Rapoport TA (1992) A protein of the endoplasmic reticulum involved early in polypeptide translocation.   Nature   357:47–52. doi:10.1038/357047a0

62. Görlich D, Prehn S, Hartmann E, Kalies KU, Rapoport TA (1992) A mammalian homolog of SEC61p and SECYp is associated with ribosomes and nascent polypeptides during translocation.   Cell   71:489–503. doi:10.1016/0092-8674(92)90517-G

63. Görlich D, Rapoport TA (1993) Protein translocation into proteoliposomes reconstituted from purified components of the endoplasmic reticulum membrane. Cell 75:615–630. doi:10.1016/0092-8674(93)90483-7

64. Novick P, Field C, Schekman R (1980) Identification of 23 complementation groups required for post-translational events in the yeast secretory pathway. Cell 21:205–215. doi:10.1016/0092-8674(80)90128-2

65. Deshaies RJ, Schekman R (1987) A yeast mutant defective at an early stage in import of secretory protein precursors into the endoplasmic reticulum. J Cell Biol 105:633–645. doi:10.1083/jcb.105.2.633

66. Hartmann E, Sommer T, Prehn S, Görlich D, Jentsch S, Rapoport TA (1994) Evolutionary conservation of components of the protein translocation complex. Nature 367:654–657. doi:10.1038/367654a0

67. Brundage L, Hendrick JP, Schiebel E, Driessen AJ, Wickner W (1990) The purified E. coli integral membrane protein SecY/E is sufficient for reconstitution of SecA-dependent precursor protein translocation. Cell   62:649–657.   doi:10.1016/0092-8674(90)90111-Q

68. Akimaru J, Matsuyama S, Tokuda H, Mizushima S (1991) Reconstitution of a protein translocation system containing purified SecY, SecE, and SecA from Escherichia coli. Proc Natl Acad Sci U S A 88:6545–6549. doi:10.1073/pnas.88.15.6545

69. Hartmann E, Görlich D, Kostka S, Otto A, Kraft R, Knespel S, Bürger E, Rapoport TA, Prehn S (1993) A tetrameric complex of membrane proteins in the endoplasmic reticulum.   Eur J Biochem   214:375–381. doi:10.1111/j.1432-1033.1993.tb17933.x

70. Kelleher DJ, Kreibich G, Gilmore R (1992) Oligosaccharyltransferase activity is associated with a protein complex composed of ribophorins I and II and a 48 kd protein. Cell 69:55–65. doi:10.1016/0092-8674(92)90118-V

71. Kelleher DJ, Gilmore R (1997) DAD1, the defender against apoptotic cell death, is a subunit of the mammalian oligosaccharyltransferase. Proc Natl Acad Sci U S A 94:4994–4999. doi:10.1073/pnas.94.10.4994

72. Nilsson I, Kelleher DJ, Miao Y, Shao Y, Kreibich G, Gilmore R, von Heijne G, Johnson AE (2003) Photocross-linking of nascent chains to the STT3 subunit of the oligosaccharyltransferase complex. J Cell Biol 161:715–725. doi:10.1083/jcb.200301043

73. Evans EA, Gilmore R, Blobel G (1986) Purification of microsomal signal peptidase as a complex. Proc Natl Acad Sci U S A 83:581–585. doi:10.1073/pnas.83.3.581

74. Meyer HA, Grau H, Kraft R, Kostka S, Prehn S, Kalies KU, Hartmann E (2000) Mammalian Sec61 is associated with Sec62 and Sec63. J   Biol   Chem   275:14550–14557. doi:10.1074/jbc.275.19.14550

75. Tyedmers J, Lerner M, Bies C, Dudek J, Skowronek MH, Haas IG, Heim N, Nastainczyk W, Volkmer J, Zimmermann R (2000) Homologs of the yeast Sec complex subunits Sec62p and Sec63p are abundant

proteins in dog pancreas microsomes. Proc Natl Acad Sci U S A 97:7214–7219. doi:10.1073/pnas.97.13.7214

76. Böhni PC, Deshaies RJ, Schekman RW (1988) SEC11 is required for signal peptide processing and yeast cell growth. J Cell Biol 106:1035–1042. doi:10.1083/jcb.106.4.1035

77. Dempski RE Jr, Imperiali B (2002) Oligosaccharyl transferase: gatekeeper to the secretory pathway. Curr Opin Chem Biol 6:844–850. doi:10.1016/S1367-5931(02)00390-3

78. Mothes W, Heinrich SU, Graf R, Nilsson I, von Heijne G, Brunner J, Rapoport TA (1997) Molecular mechanism of membrane protein integration into the endoplasmic reticulum. Cell 89:523–533. doi:10.1016/S0092-8674(00)80234-2

79. Ng DT, Brown JD, Walter P (1996) Signal sequences specify the targeting route to the endoplasmic reticulum membrane. J Cell Biol 134:269–278. doi:10.1083/jcb.134.2.269

80. Hansen W, Garcia PD, Walter P (1986) In vitro protein translocation across the yeast endoplasmic reticulum: ATP-dependent posttranslational translocation of the prepro-alpha-factor. Cell 45:397–406. doi:10.1016/0092-8674(86)90325-9

81. Chirico WJ, Waters MG, Blobel G (1988) 70K heat shock related proteins stimulate protein translocation into microsomes. Nature 332:805–810. doi:10.1038/332805a0

82. Deshaies RJ, Sanders SL, Feldheim DA, Schekman R (1991) Assembly of yeast Sec proteins involved in translocation into the endoplasmic reticulum into a membrane-bound multisubunit complex. Nature 349:806–808. doi:10.1038/349806a0

83. Panzner S, Dreier L, Hartmann E, Kostka S, Rapoport TA (1995) Posttranslational protein transport in yeast reconstituted with a purified complex of Sec proteins and Kar2p. Cell 81:561–570. doi:10.1016/0092-8674(95)90077-2

84. Hanein D, Matlack KES, Jungnickel B, Plath K, Kalies KU, Miller KR, Rapoport TA, Akey CW (1996) Oligomeric rings of the Sec61p complex induced by ligands required for protein translocation. Cell 87:721–732. doi:10.1016/S0092-8674(00)81391-4

85. Rapoport TA (2007) Protein translocation across the eukaryotic endoplasmic reticulum and bacterial plasma membranes. Nature 450:663–669. doi:10.1038/nature06384

86. Johnson N, Powis K, High S (2013) Post-translational translocation into the endoplasmic reticulum. Biochim Biophys Acta 1833:2403–2409. doi:10.1016/j.bbamcr.2012.12.008

87. Stefanovic S, Hegde RS (2007) Identification of a targeting factor for posttranslational membrane protein insertion into the ER. Cell 128:1147–1159. doi:10.1016/j.cell.2007.01.036

88. Favaloro V, Spasic M, Schwappach B, Dobberstein B (2008) Distinct targeting pathways for the membrane insertion of tail-anchored (TA) proteins. J Cell Sci 121:1832–1840. doi:10.1242/jcs.020321

89. Mariappan M, Li X, Stefanovic S, Sharma A, Mateja A, Keenan RJ, Hegde RS (2010) A ribosome-associating factor chaperones tail-anchored membrane proteins. Nature 466:1120–1124. doi:10.1038/nature09296

90. Schuldiner M, Metz J, Schmid V, Denic V, Rakwalska M, Schmitt HD, Schwappach B, Weissman JS (2008) The GET complex mediates insertion of tail-anchored proteins into the ER membrane. Cell 134:634–645. doi:10.1016/j.cell.2008.06.025

91. Vilardi F, Lorenz H, Dobberstein B (2011) WRB is the receptor for TRC40/Asna1-mediated insertion of tail-anchored proteins into the ER membrane. J Cell Sci 124:1301–1307. doi:10.1242/jcs.084277

92. Yamamoto Y, Sakisaka T (2012) Molecular machinery for insertion of tail-anchored membrane proteins into the endoplasmic reticulum membrane in mammalian cells. Mol Cell 48:387–397. doi:10.1016/j.molcel.2012.08.028

93. Mariappan M, Mateja A, Dobosz M, Bove E, Hegde RS, Keenan RJ (2011) The mechanism of membrane-associated steps in tail-anchored protein insertion. Nature 477:61–66. doi:10.1038/nature10362

94. Walter P, Ron D (2011) The unfolded protein response: from stress pathway to homeostatic regulation. Science 334:1081–1086. doi:10.1126/science.1209038

95. Howell SH (2013) Endoplasmic reticulum stress responses in plants. Annu Rev Plant Biol 64:477–499. doi:10.1146/annurev-arplant-050312-120053

96. Barlowe C, Orci L, Yeung T, Hosobuchi M, Hamamoto S, Salama N, Rexach MF, Ravazzola M, Amherdt M, Schekman R (1994) COPII–a membrane coat formed by Sec proteins that drive vesicle budding from the endoplasmic reticulum. Cell 77:895–907. doi:10.1016/0092-8674(94)90138-4

97. Brandizzi F, Barlowe C (2013) Organization of the ER-Golgi interface for membrane traf-

fic control. Nat Rev Mol Cell Biol 14:382–392. doi:10.1038/nrm3588

98. Nakano A, Muramatsu M (1989) A novel GTP-binding protein, Sar1p, is involved in transport from the endoplasmic reticulum to the Golgi apparatus. J Cell Biol 109:2677–2691. doi:10.1083/jcb.109.6.2677

99. Yoshihisa T, Barlowe C, Schekman R (1993) Requirement for a GTPase-activating protein in vesicle budding from the endoplasmic reticulum. Science 259:1466–1468. doi:10.1126/science.8451644

100. Nakano A, Brada D, Schekman R (1988) A membrane glycoprotein, Sec12p, required for protein transport from the endoplasmic reticulum to the Golgi apparatus in yeast. J Cell Biol 107:851–863. doi:10.1083/jcb.107.3.851

101. Barlowe C, Schekman R (1993) SEC12 encodes a guanine-nucleotide- exchange factor essential for transport vesicle budding from the ER. Nature 365:347–349. doi:10.1038/365347a0

102. Goldberg J (1998) Structural basis for activation of ARF GTPase: mechanisms of guanine nucleotide exchange and GTP-myristoyl switching. Cell 95:237–248. doi:10.1016/S0092-8674(00)81754-7

103. Huang M, Weissman JT, Beraud-Dufour S, Luan P, Wang C, Chen W, Aridor M, Wilson IA, Balch WE (2001) Crystal structure of Sar1-GDP at 1.7 Å resolution and the role of the NH2 terminus in ER export. J Cell Biol 155:937–948. doi:10.1083/jcb.200106039

104. Lee MC, Orci L, Hamamoto S, Futai E, Ravazzola M, Schekman R (2005) Sar1p N-terminal helix initiates membrane curvature and completes the fission of a COPII vesicle. Cell 122:605–617. doi:10.1016/j.cell.2005.07.025

105. Matsuoka K, Orci L, Amherdt M, Bednarek SY, Hamamoto S, Schekman R, Yeung T (1998) COPII-coated vesicle formation reconstituted with purified coat proteins and chemically defined liposomes. Cell 93:263–275. doi:10.1016/S0092-8674(00)81577-9

106. Bi X, Corpina RA, Goldberg J (2002) Structure of the Sec23/24-Sar1 pre-budding complex of the COPII vesicle coat. Nature 419:271–277. doi:10.1038/nature01040

107. Stagg SM, Gürkan C, Fowler DM, LaPointe P, Foss TR, Potter CS, Carragher B, Balch WE (2006) Structure of the Sec13/31 COPII coat cage. Nature 439:234–238. doi:10.1038/nature04339

108. Fath S, Mancias JD, Bi X, Goldberg J (2007) Structure and organization of coat proteins in the COPII cage. Cell 129:1325–1336. doi:10.1016/j.cell.2007.05.036

109. Stagg SM, LaPointe P, Razvi A, Gürkan C, Potter CS, Carragher B, Balch WE (2008) Structural basis for cargo regulation of COPII coat assembly. Cell 134:474–484. doi:10.1016/j.cell.2008.06.024

110. Miller EA, Beilharz TH, Malkus PN, Lee MC, Hamamoto S, Orci L, Schekman R (2003) Multiple cargo binding sites on the COPII subunit Sec24p ensure capture of diverse membrane proteins into transport vesicles. Cell 114:497–509. doi:10.1016/S0092-8674(03)00609-3

111. Sato K, Nakano A (2005) Dissection of COPII subunit-cargo assembly and disassembly kinetics during Sar1p-GTP hydrolysis. Nat Struct Mol Biol 12:167–174. doi:10.1038/nsmb893

112. Bi X, Mancias JD, Goldberg J (2007) Insights into COPII coat nucleation from the structure of Sec23.Sar1 complexed with the active fragment of Sec31. Dev Cell 13:635–645. doi:10.1016/j.devcel.2007.10.006

113. Watson P, Townley AK, Koka P, Palmer KJ, Stephens DJ (2006) Sec16 defines endoplasmic reticulum exit sites and is required for secretory cargo export in mammalian cells. Traffic 7:1678–1687. doi:10.1111/j.1600-0854.2006.00493.x

114. Connerly PL, Esaki M, Montegna EA, Strongin DE, Levi S, Soderholm J, Glick BS (2005) Sec16 is a determinant of transitional ER organization. Curr Biol 15:1439–1447. doi:10.1016/j.cub.2005.06.065

115. Hughes H, Budnik A, Schmidt K, Palmer KJ, Mantell J, Noakes C, Johnson A, Carter DA, Verkade P, Watson P, Stephens DJ (2009) Organisation of human ER-exit sites: requirements for the localisation of Sec16 to transitional ER. J Cell Sci 122:2924–2934. doi:10.1242/jcs.044032

116. Whittle JR, Schwartz TU (2010) Structure of the Sec13-Sec16 edge element, a template for assembly of the COPII vesicle coat. J Cell Biol 190:347–361. doi:10.1083/jcb.201003092

117. Zanetti G, Pahuja KB, Studer S, Shim S, Schekman R (2011) COPII and the regulation of protein sorting in mammals. Nat Cell Biol 14:20–28. doi:10.1038/ncb2390

118. O'Kelly I, Butler MH, Zilberberg N, Goldstein SA (2002) Forward transport. 14-3-3 binding overcomes retention in endoplasmic reticulum by dibasic signals. Cell

111:577–588.   doi:10.1016/S0092-8674(02)01040-1

119. Nakamura T, Hayashi T, Nasu-Nishimura Y, Sakaue F, Morishita Y, Okabe T, Ohwada S, Matsuura K, Akiyama T (2008) PX-RICS mediates ER-to-Golgi transport of the N-cadherin/beta-catenin complex. Genes Dev 22:1244–1256. doi:10.1101/gad.1632308

120. Wang J, Hamblet NS, Mark S, Dickinson ME, Brinkman BC, Segil N, Fraser SE, Chen P, Wallingford JB, Wynshaw-Boris A (2006) Dishevelled genes mediate a conserved mammalian PCP pathway to regulate convergent extension during neurulation. Development 133:1767–1778. doi:10.1242/dev.02347

121. Simons M, Gault WJ, Gotthardt D, Rohatgi R, Klein TJ, Shao Y, Lee HJ, Wu AL, Fang Y, Satlin LM, Dow JT, Chen J, Zheng J, Boutros M, Mlodzik M (2009) Electrochemical cues regulate assembly of the Frizzled/Dishevelled complex at the plasma membrane during planar epithelial polarization. Nat Cell Biol 11:286–294. doi:10.1038/ncb1836

122. Malhotra V, Erlmann P (2015) The pathway of collagen secretion. Annu Rev Cell Dev Biol. doi:10.1146/annurev-cellbio-100913-013002

123. Townley AK, Feng Y, Schmidt K, Carter DA, Porter R, Verkade P, Stephens DJ (2008) Efficient coupling of Sec23–Sec24 to Sec13–Sec31 drives COPII-dependent collagen secretion and is essential for normal craniofacial development. J Cell Sci 121:3025–3034. doi:10.1242/jcs.031070

124. Sarmah S, Barrallo-Gimeno A, Melville DB, Topczewski J, Solnica-Krezel L, Knapik EW (2010) Sec24D dependent transport of extracellular matrix proteins is required for zebrafish skeletal morphogenesis. PLoS One 5, e10367. doi:10.1371/journal.pone.0010367

125. Boyadjiev SA, Kim SD, Hata A, Haldeman-Englert C, Zackai EH, Naydenov C, Hamamoto S, Schekman RW, Kim J (2011) Cranio-lenticulo-sutural dysplasia associated with defects in collagen secretion. Clin Genet 80:169–176.doi:10.1111/j.1399-0004.2010.01550.x

126. Venditti R, Scanu T, Santoro M, Di Tullio G, Spaar A, Gaibisso R, Beznoussenko GV, Mironov AA, Mironov A Jr, Zelante L, Piemontese MR, Notarangelo A, Malhotra V, Vertel BM, Wilson C, De Matteis MA (2012) Sedlin controls the ER export of procollagen by regulating the Sar1 cycle. Science 337:1668–1672. doi:10.1126/science.1224947

127. Nogueira C, Erlmann P, Villeneuve J, Santos AJ, Martínez-Alonso E, Martínez-Menárguez JÁ, Malhotra V (2014) SLY1 and Syntaxin 18 specify a distinct pathway for procollagen VII export from the endoplasmic reticulum. Elife 3, e02784. doi:10.7554/eLife.02784

128. Bacia K, Futai E, Prinz S, Meister A, Daum S, Glatte D, Briggs JA, Schekman R (2011) Multibudded tubules formed by COPII on artificial liposomes. Sci Rep 1:17. doi:10.1038/srep00017

129. Zanetti G, Prinz S, Daum S, Meister A, Schekman R, Bacia K, Briggs JA (2013) The structure of the COPII transport-vesicle coat assembled on membranes. Elife 2, e00951. doi:10.7554/eLife.00951

130. Bard F, Casano L, Mallabiabarrena A, Wallace E, Saito K, Kitayama H, Guizzunti G, Hu Y, Wendler F, Dasgupta R, Perrimon N, Malhotra V (2006) Functional genomics reveals genes involved in protein secretion and Golgi organization. Nature 439:604–607. doi:10.1038/nature04377

131. Lerner DW, McCoy D, Isabella AJ, Mahowald AP, Gerlach GF, Chaudhry TA, Horne-Badovinac S (2013) A Rab10-dependent mechanism for polarized basement membrane secretion during organ morphogenesis. Dev Cell 24:159–168. doi:10.1016/j.devcel.2012.12.005

132. Saito K, Chen M, Bard F, Chen S, Zhou H, Woodley D, Polischuk R, Schekman R, Malhotra V (2009) TANGO1 facilitates cargo loading at endoplasmic reticulum exit sites. Cell 136:891–902. doi:10.1016/j.cell.2008.12.025

133. Saito K, Yamashiro K, Ichikawa Y, Erlmann P, Kontani K, Malhotra V, Katada T (2011) cTAGE5 mediates collagen secretion through interaction with TANGO1 at endoplasmic reticulum exit sites. Mol Biol Cell 22:2301–2308. doi:10.1091/mbc.E11-02-0143

134. Saito K, Yamashiro K, Shimazu N, Tanabe T, Kontani K, Katada T (2014) Concentration of Sec12 at ER exit sites via interaction with cTAGE5 is required for collagen export. J Cell Biol 206:751–762. doi:10.1083/jcb.201312062

135. Wieland FT, Gleason ML, Serafini TA, Rothman JE (1987) The rate of bulk flow from the endoplasmic reticulum to the cell surface. Cell 50:289–300.doi:10.1016/0092-8674(87)90224-8

136. Denecke J, Botterman J, Deblaere R (1990) Protein secretion in plant cells can occur via a default pathway. Plant Cell 2:51–59. doi:10.1105/tpc.2.1.51

137. Phillipson BA, Pimpl P, daSilva LL, Crofts AJ, Taylor JP, Movafeghi A, Robinson DG, Denecke J (2001) Secretory bulk flow of soluble proteins is efficient and COPII depen-

dent. Plant Cell 13:2005–2020. doi:10.1105/TPC.010110

138. Thor F, Gautschi M, Geiger R, Helenius A (2009) Bulk flow revisited: transport of a soluble protein in the secretory pathway. Traffic 10:1819–1830. doi:10.1111/j.1600-0854.2009.00989.x

139. Kappeler F, Klopfenstein DR, Foguet M, Paccaud JP, Hauri HP (1997) The recycling of ERGIC-53 in the early secretory pathway. ERGIC-53 carries a cytosolic endoplasmic reticulum-exit determinant interacting with COPII. J Biol Chem 272:31801–31808. doi:10.1074/jbc.272.50.31801

140. Nishimura N, Balch WE (1997) A di-acidic signal required for selective export from the endoplasmic reticulum. Science 277:556–558. doi:10.1126/science.277.5325.556

141. Contreras I, Yang Y, Robinson DG, Aniento F (2004) Sorting signals in the cytosolic tail of plant p24 proteins involved in the interaction with the COPII coat. Plant Cell Physiol 45:1779–1786. doi:10.1093/pcp/pch200

142. Hanton SL, Renna L, Bortolotti LE, Chatre L, Stefano G, Brandizzi F (2005) Diacidic motifs influence the export of transmembrane proteins from the endoplasmic reticulum in plant cells. Plant Cell 17:3081–3093. doi:10.1105/tpc.105.034900

143. Hay JC, Chao DS, Kuo CS, Scheller RH (1997) Protein interactions regulating vesicle transport between the endoplasmic reticulum and Golgi apparatus in mammalian cells. Cell 89:149–158. doi:10.1016/S0092-8674(00)80191-9

144. Cao X, Ballew N, Barlowe C (1998) Initial docking of ER-derived vesicles requires Uso1p and Ypt1p but is independent of SNARE proteins. EMBO J 17:2156–2165. doi:10.1093/emboj/17.8.2156

145. Allan BB, Moyer BD, Balch WE (2000) Rab1 recruitment of p115 into a cis-SNARE complex: programming budding COPII vesicles for fusion. Science 289:444–448. doi:10.1126/science.289.5478.444

146. Moyer BD, Allan BB, Balch WE (2001) Rab1 interaction with a GM130 effector complex regulates COPII vesicle cis-Golgi tethering. Traffic 2:268–276. doi:10.1034/j.1600-0854.2001.1o007.x

147. Sacher M, Barrowman J, Wang W, Horecka J, Zhang Y, Pypaert M, Ferro-Novick S (2001) TRAPP I implicated in the specificity of tethering in ER-to-Golgi transport. Mol Cell 7:433–442. doi:10.1016/S1097-2765(01)00190-3

148. Shorter J, Beard MB, Seemann J, Dirac-Svejstrup AB, Warren G (2002) Sequential tethering of Golgins and catalysis of SNAREpin assembly by the vesicle-tethering protein p115. J Cell Biol 157:45–62. doi:10.1083/jcb.200112127

149. Cai Y, Chin HF, Lazarova D, Menon S, Fu C, Cai H, Sclafani A, Rodgers DW, De La Cruz EM, Ferro-Novick S, Reinisch KM (2008) The structural basis for activation of the Rab Ypt1p by the TRAPP membrane-tethering complexes. Cell 133:1202–1213. doi:10.1016/j.cell.2008.04.049

150. Wong M, Munro S (2014) Membrane trafficking. The specificity of vesicle traffic to the Golgi is encoded in the golgin coiled-coil proteins. Science 346:1256898. doi:10.1126/science.1256898

151. Cai H, Yu S, Menon S, Cai Y, Lazarova D, Fu C, Reinisch K, Hay JC, Ferro-Novick S (2007) TRAPPI tethers COPII vesicles by binding the coat subunit Sec23. Nature 445:941–944. doi:10.1038/nature05527

152. Lord C, Bhandari D, Menon S, Ghassemian M, Nycz D, Hay J, Ghosh P, Ferro-Novick S (2011) Sequential interactions with Sec23 control the direction of vesicle traffic. Nature 473:181–186. doi:10.1038/nature09969

153. Rowe T, Dascher C, Bannykh S, Plutner H, Balch WE (1998) Role of vesicle-associated syntaxin 5 in the assembly of pre-Golgi intermediates. Science 279:696–700. doi:10.1126/science.279.5351.696

154. Xu D, Joglekar AP, Williams AL, Hay JC (2000) Subunit structure of a mammalian ER/Golgi SNARE complex. J Biol Chem 275:39631–39639. doi:10.1074/jbc.M007684200

155. Lowe SL, Peter F, Subramaniam VN, Wong SH, Hong W (1997) A SNARE involved in protein transport through the Golgi apparatus. Nature 389:881–884. doi:10.1038/39923

156. Yamaguchi T, Dulubova I, Min SW, Chen X, Rizo J, Südhof TC (2002) Sly1 binds to Golgi and ER syntaxins via a conserved N-terminal peptide motif. Dev Cell 2:295–305. doi:10.1016/S1534-5807(02)00125-9

157. Söllner T, Whiteheart SW, Brunner M, Erdjument-Bromage H, Geromanos S, Tempst P, Rothman JE (1993) SNAP receptors implicated in vesicle targeting and fusion. Nature 362:318–324. doi:10.1038/362318a0

158. Sutton RB, Fasshauer D, Jahn R, Brunger AT (1998) Crystal structure of a SNARE complex involved in synaptic exocytosis at 2.4 A resolution. Nature 395:347–353. doi:10.1038/26412

159. Weber T, Zemelman BV, McNew JA, Westermann B, Gmachl M, Parlati F, Söllner

TH, Rothman JE (1998) SNAREpins: minimal machinery for membrane fusion. Cell 92:759–772. doi:10.1016/S0092-8674(00)81404-X

160. Parlati F, McNew JA, Fukuda R, Miller R, Söllner TH, Rothman JE (2000) Topological restriction of SNARE-dependent membrane fusion. Nature 407:194–198. doi:10.1038/35025076

161. Südhof TC, Rothman JE (2009) Membrane fusion: grappling with SNARE and SM proteins. Science 323:474–477. doi:10.1126/science.1161748

162. Nebenfuhr A, Gallagher LA, Dunahay TG, Frohlick JA, Mazurkiewicz AM, Meehl JB, Staehelin LA. (1999) Stop-and-go movements of plant Golgi stacks are mediated by the actomyosin system. Plant Physiol 121:1127–1142. doi:10.1104/pp.121.4.1127

163. Boevink P, Oparka K, Santa Cruz S, Martin B, Betteridge A, Hawes C (1998) Stacks on tracks: the plant Golgi apparatus traffics on an actin/ER network. Plant J 15:441–447. doi:10.1046/j.1365-313X.1998.00208.x

164. Avisar D, Prokhnevsky AI, Makarova KS, Koonin EV, Dolja VV (2008) Myosin XI-K is required for rapid trafficking of Golgi stacks, peroxisomes, and mitochondria in leaf cells of Nicotiana benthamiana. Plant Physiol 146:1098–1108. doi:10.1104/pp.107.113647

165. Peremyslov VV, Prokhnevsky AI, Avisar D, Dolja VV (2008) Two class XI myosins function in organelle trafficking and root hair development in Arabidopsis. Plant Physiol 146:1109–1116. doi:10.1104/pp.107.113654

166. Prokhnevsky AI, Peremyslov VV, Dolja VV (2008) Overlapping functions of the four class XI myosins in Arabidopsis growth, root hair elongation, and organelle motility. Proc Natl Acad Sci U S A 105:19744–19749. doi:10.1073/pnas.0810730105

167. Sparkes IA, Teanby NA, Hawes C (2008) Truncated myosin XI tail fusions inhibit peroxisome, Golgi, and mitochondrial movement in tobacco leaf epidermal cells: a genetic tool for the next generation. J Exp Bot 59:2499–2512. doi:10.1093/jxb/ern114

168. Sparkes IA, Ketelaar T, Ruijter NC, Hawes C (2009) Grab a Golgi: laser trapping of Golgi bodies reveals in vivo interactions with the endoplasmic reticulum. Traffic 10:567–571. doi:10.1111/j.1600-0854.2009.00891.x

169. Brandizzi F, Snapp EL, Roberts AG, Lippincott-Schwartz J, Hawes C (2002) Membrane protein transport between the endoplasmic reticulum and the Golgi in tobacco leaves is energy dependent but cytoskeleton independent: evidence from selective

photobleaching. Plant Cell 14:1293–1309. doi:10.1105/tpc.001586

170. Kang BH, Staehelin LA (2008) ER-to-Golgi transport by COPII vesicles in Arabidopsis involves a ribosome-excluding scaffold that is transferred with the vesicles to the Golgi matrix. Protoplasma 234:51–64. doi:10.1007/s00709-008-0015-6

171. daSilva LLP, Snapp EL, Denecke J, Lippincott-Schwartz J, Hawes C, Brandizzi F (2004) Endoplasmic reticulum export sites and Golgi bodies behave as single mobile secretory units in plant cells. Plant Cell 16, 1753–1771. doi:10.1105/tpc.022673

172. Stefano G, Renna L, Chatre L, Hanton SL, Moreau P, Hawes C, Brandizzi F (2006) In tobacco leaf epidermal cells, the integrity of protein export from the endoplasmic reticulum and of ER export sites depends on active COPI machinery. Plant J 46:95–110. doi:10.1111/j.1365-313X.2006.02675.x

173. Langhans M, Meckel T, Kress A, Lerich A, Robinson DG (2012) ERES (ER exit sites) and the "secretory unit concept". J Microsc 247:48–59. doi:10.1111/j.1365-2818.2011.03597.x

174. Lerich A, Hillmer S, Langhans M, Scheuring D, van Bentum P, Robinson DG (2012) ER import sites and their relationship to ER exit sites: a new model for bidirectional ER-Golgi transport in higher plants. Front Plant Sci 3:143. doi:10.3389/fpls.2012.00143

175. Pimpl P, Movafeghi A, Coughlan S, Denecke J, Hillmer S, Robinson DG (2000) In situ localization and in vitro induction of plant COPI-coated vesicles. Plant Cell 12:2219–2236. doi:10.1105/tpc.12.11.2219

176. Ritzenthaler C, Nebenführ A, Movafeghi A, Stussi-Garaud C, Behnia L, Pimpl P, Staehelin LA, Robinson DG (2002) Reevaluation of the effects of brefeldin A on plant cells using tobacco Bright Yellow 2 cells expressing Golgi-targeted green fluorescent protein and COPI antisera. Plant Cell 14:237–261. doi:10.1105/tpc.010237

177. Robinson DG, Herranz MC, Bubeck J, Pepperkok R, Ritzenthaler C (2007) Membrane dynamics in the early secretory pathway. Crit Rev Plant Sci 26:199–225. doi:10.1080/07352680701495820

178. Staehelin LA, Kang BH (2008) Nanoscale architecture of endoplasmic reticulum export sites and of Golgi membranes as determined by electron tomography. Plant Physiol 147:1454–1468

179. Robinson DG, Brandizzi F, Hawes C, Nakano A (2015) Vesicles versus tubes: is endoplasmic

reticulum-Golgi transport in plants fundamentally different from other eukaryotes? Plant Physiol 168:393–406. doi:10.1104/pp.15.00124

180. Malhotra V, Serafini T, Orci L, Shepherd JC, Rothman JE (1989) Purification of a novel class of coated vesicles mediating biosynthetic protein transport through the Golgi stack. Cell 58:329–336. doi:10.1016/0092-8674(89)90847-7

181. Serafini T, Stenbeck G, Brecht A, Lottspeich F, Orci L, Rothman JE, Wieland FT (1991) A coat subunit of Golgi-derived non-clathrin-coated vesicles with homology to the clathrin-coated vesicle coat protein beta-adaptin. Nature 349:215–220. doi:10.1038/349215a0

182. Waters MG, Serafini T, Rothman JE (1991) 'Coatomer': a cytosolic protein complex containing subunits of non-clathrin-coated Golgi transport vesicles. Nature 349:248–251. doi:10.1038/349248a0

183. Serafini T, Orci L, Amherdt M, Brunner M, Kahn RA, Rothman JE (1991) ADP-ribosylation factor is a subunit of the coat of Golgi-derived COP-coated vesicles: a novel role for a GTP-binding protein. Cell 67:239–253. doi:10.1016/0092-8674(91)90176-Y

184. Orci L, Palmer DJ, Ravazzola M, Perrelet A, Amherdt M, Rothman JE (1993) Budding from Golgi membranes requires the coatomer complex of non-clathrin coat proteins. Nature 362:648–652. doi:10.1038/362648a0

185. Rothman JE, Wieland FT (1996) Protein sorting by transport vesicles. Science 272:227–234. doi:10.1126/science.272.5259.227

186. Zink S, Wenzel D, Wurm CA, Schmitt HD (2009) A link between ER tethering and COP-I vesicle uncoating. Dev Cell 17:403–416. doi:10.1016/j.devcel.2009.07.012

187. Hong W (2005) SNAREs and traffic. Biochim Biophys Acta 1744:493–517. doi:10.1016/j.bbamcr.2005.03.014

188. Klumperman J (2011) Architecture of the mammalian Golgi. Cold Spring Harb Perspect Biol 3:a005181. doi:10.1101/cshperspect.a005181

189. Marsh BJ, Volkmann N, McIntosh JR, Howell KE (2004) Direct continuities between cisternae at different levels of the Golgi complex in glucose-stimulated mouse islet b cells. Proc Natl Acad Sci 101:5565–5570. doi:10.1073/pnas.0401242101

190. Trucco A, Polishchuk RS, Martella O, Di Pentima A, Fusella A, Di Giandomenico D, San Pietro E, Beznoussenko GV, Polishchuk EV, Baldassarre M, Buccione R, Geerts WJ, Koster AJ, Burger KN, Mironov AA, Luini A (2004) Secretory traffic triggers the formation of tubular continuities across Golgi sub-compartments. Nat Cell Biol 6:1071–1081. doi:10.1038/ncb1180

191. Griffiths G, Pfeiffer S, Simons K, Matlin K (1985) Exit of newly synthesized membrane proteins from the trans cisterna of the Golgi complex to the plasma membrane. J Cell Biol 101:949–964. doi:10.1083/jcb.101.3.949

192. Griffiths G, Simons K (1986) The trans Golgi network: sorting at the exit site of the Golgi complex. Science 234:438–443. doi:10.1126/science.2945253

193. Viotti C, Bubeck J, Stierhof YD, Krebs M, Langhans M, van den Berg W, van Dongen W, Richter S, Geldner N, Takano J, Jürgens G, de Vries SC, Robinson DG, Schumacher K (2010) Endocytic and secretory traffic in Arabidopsis merge in the trans-Golgi network/early endosome, an independent and highly dynamic organelle. Plant Cell 22:1344–1357. doi:10.1105/tpc.109.072637

194. Dettmer J, Hong-Hermesdorf A, Stierhof YD, Schumacher K (2006) Vacuolar H+-ATPase activity is required for endocytic and secretory trafficking in Arabidopsis. Plant Cell 18:715–730. doi:10.1105/tpc.105.037978

195. Lam SK, Siu CL, Hillmer S, Jang S, An G, Robinson DG, Jiang L (2007) Rice SCAMP1 defines clathrin-coated, trans-Golgi-located tubular-vesicular structures as an early endosome in tobacco BY-2 cells. Plant Cell 19:296–319. doi:10.1105/tpc.106.045708

196. Glick BS, Luini A (2011) Models for Golgi traffic: a critical assessment. Cold Spring Harb Perspect Biol 3:a005215. doi:10.1101/cshperspect.a005215

197. Rabouille C, Hui N, Hunte F, Kieckbusch R, Berger EG, Warren G, Nilsson T (1995) Mapping the distribution of Golgi enzymes involved in the construction of complex oligosaccharides. J Cell Sci 108:1617–1627

198. Guo Y, Sirkis DW, Schekman R (2014) Protein sorting at the trans-Golgi network. Annu Rev Cell Dev Biol 30:169–206. doi:10.1146/annurev-cellbio-100913-013012

199. Simmen T, Honing S, Icking A, Tikkanen R, Hunziker W (2002) AP-4 binds basolateral signals and participates in basolateral sorting in epithelial MDCK cells. Nat Cell Biol 4:154–159. doi:10.1038/ncb745

200. Gravotta D, Carvajal-Gonzalez JM, Mattera R, Deborde S, Banfelder JR, Bonifacino JS, Rodriguez-Boulan E (2012) The clathrin

adaptor AP-1A mediates basolateral polarity. Dev Cell 22:811–823. doi:10.1016/j.devcel.2012.02.004

201. Bonifacino JS (2014) Adaptor proteins involved in polarized sorting. J Cell Biol 204:7–17. doi:10.1083/jcb.201310021

202. Simons K, Ikonen E (1997) Functional rafts in cell membranes. Nature 387:569–572. doi:10.1038/42408

203. Bard F, Malhotra V (2006) The formation of TGN-to-plasma-membrane transport carriers. Annu Rev Cell Dev Biol 22:439–455. doi:10.1146/annurev.cellbio.21.012704. 133126

204. Wang CW, Hamamoto S, Orci L, Schekman R (2006) Exomer: a coat complex for transport of select membrane proteins from the trans-Golgi network to the plasma membrane in yeast. J Cell Biol 174:973–983. doi:10.1083/jcb.200605106

205. Sanchatjate S, Schekman R (2006) Chs5/6 complex: a multiprotein complex that interacts with and conveys chitin synthase III from the trans-Golgi network to the cell surface. Mol Biol Cell 17:4157–4166. doi:10.1091/mbc.E06-03-0210

206. Barfield RM, Fromme JC, Schekman R (2009) The exomer coat complex transports Fus1p to the plasma membrane via a novel plasma membrane sorting signal in yeast. Mol Biol Cell 20:4985–4996. doi:10.1091/mbc.E09-04-0324

207. Paczkowski JE, Richardson BC, Strassner AM, Fromme JC (2012) The exomer cargo adaptor structure reveals a novel GTPase-binding domain. EMBO J 31:4191–4203. doi:10.1038/emboj.2012.268

208. Starr TL, Pagant S, Wang CW, Schekman R (2012) Sorting signals that mediate traffic of chitin synthase III between the TGN/endosomes and to the plasma membrane in yeast. PLoS One 7, e46386. doi:10.1371/journal.pone.0046386

209. Santos B, Snyder M (2003) Specific protein targeting during cell differentiation: polarized localization of Fus1p during mating depends on Chs5p in *Saccharomyces cerevisiae*. Eukaryot Cell 2:821–825. doi:10.1128/EC.2.4.821-825.2003

210. Trautwein M, Schindler C, Gauss R, Dengjel J, Hartmann E, Spang A (2006) Arf1p, Chs5p and the ChAPs are required for export of specialized cargo from the Golgi. EMBO J 25:943–954. doi:10.1038/sj.emboj.7601007

211. Goud B, Salminen A, Walworth NC, Novick PJ (1988) A GTP-binding protein required for secretion rapidly associates with secretory vesicles and the plasma membrane in yeast. Cell 53:753–768. doi:10.1016/0092-8674(88)90093-1

212. Rutherford S, Moore I (2002) The Arabidopsis Rab GTPase family: another enigma variation. Curr Opin Plant Biol 5:518–528. doi:10.1016/S1369-5266(02)00307-2

213. Gendre D, Oh J, Boutté Y, Best JG, Samuels L, Nilsson R, Uemura T, Marchant A, Bennett MJ, Grebe M, Bhalerao RP (2011) Conserved Arabidopsis ECHIDNA protein mediates trans-Golgi-network trafficking and cell elongation. Proc Natl Acad Sci U S A 108:8048–8053. doi:10.1073/pnas.1018371108

214. Gendre D, McFarlane HE, Johnson E, Mouille G, Sjödin A, Oh J, Levesque-Tremblay G, Watanabe Y, Samuels L, Bhalerao RP (2013) Trans-Golgi network localized ECHIDNA/Ypt interacting protein complex is required for the secretion of cell wall polysaccharides in Arabidopsis. Plant Cell 25:2633–2646. doi:10.1105/tpc.113.112482

215. McFarlane HE, Döring A, Persson S (2014) The cell biology of cellulose synthesis. Annu Rev Plant Biol 65:69–94. doi:10.1146/annurev-arplant-050213-040240

216. Boutté Y, Jonsson K, McFarlane HE, Johnson E, Gendre D, Swarup R, Friml J, Samuels L, Robert S, Bhalerao RP (2014) ECHIDNA-mediated post-Golgi trafficking of auxin carriers for differential cell elongation. Proc Natl Acad Sci U S A 110:16259–16264. doi:10.1073/pnas.1309057110

217. Jürgens G (2005) Plant cytokinesis: Fission by fusion. Trends Cell Biol 15:277–283. doi:10.1016/j.tcb.2005.03.005

218. Müller S, Jürgens G (2015) Plant cytokinesis-No ring, no constriction but centrifugal construction of the partitioning membrane. Semin Cell Dev Biol. doi:10.1016/j.semcdb.2015.10.037

219. Lukowitz W, Mayer U, Jürgens G (1996) Cytokinesis in the Arabidopsis embryo involves the syntaxin-related KNOLLE gene product. Cell 84:61–71. doi:10.1016/S0092-8674(00)80993-9

220. Lauber MH, Waizenegger I, Steinmann T, Schwarz H, Mayer U, Hwang I, Lukowitz W, Jürgens G (1997) The Arabidopsis KNOLLE protein is a cytokinesis-specific syntaxin. J Cell Biol 139:1485–1493. doi:10.1083/jcb.139.6.1485

221. Assaad FF, Huet Y, Mayer U, Jürgens G (2001) The cytokinesis gene KEULE encodes

a Sec1 protein that binds the syntaxin KNOLLE. J Cell Biol 152:531–543. doi:10.1083/jcb.152.3.531

222. Heese M, Gansel X, Sticher L, Wick P, Grebe M, Granier F, Jurgens G (2001) Functional characterization of the KNOLLE-interacting t-SNARE AtSNAP33 and its role in plant cytokinesis. J Cell Biol 155:239–249. doi:10.1083/jcb.200107126

223. Hammer JA 3rd, Wu XS (2002) Rabs grab motors: defining the connections between Rab GTPases and motor proteins. Curr Opin Cell Biol 14:69–75. doi:10.1016/S0955-0674(01)00296-4

224. TerBush DR, Maurice T, Roth D, Novick P (1996) The Exocyst is a multiprotein com-plex required for exocytosis in *Saccharomyces cerevisiae*. EMBO J 15:6483–6494

225. Cai H, Reinisch K, Ferro-Novick S (2007) Coats, tethers, Rabs, and SNAREs work together to mediate the intracellular destina-tion of a transport vesicle. Dev Cell 12:671–682. doi:10.1016/j.devcel.2007.04.005

226. Zárský V, Kulich I, Fendrych M, Pečenková T (2013) Exocyst complexes multiple functions in plant cells secretory pathways. Curr Opin Plant Biol 16:726–733. doi:10.1016/j.pbi.2013.10.013

227. Borgonovo B, Ouwendijk J, Solimena M (2006) Biogenesis of secretory granules. Curr Opin Cell Biol 18:365–370. doi:10.1016/j.ceb.2006.06.010

Chapter 2

Unconventional Protein Secretion in Animal Cells

Fanny Ng and Bor Luen Tang

Abstract

All eukaryotic cells secrete a range of proteins in a constitutive or regulated manner through the conventional or canonical exocytic/secretory pathway characterized by vesicular traffic from the endoplasmic reticulum, through the Golgi apparatus, and towards the plasma membrane. However, a number of proteins are secreted in an unconventional manner, which are insensitive to inhibitors of conventional exocytosis and use a route that bypasses the Golgi apparatus. These include cytosolic proteins such as fibroblast growth factor 2 (FGF2) and interleukin-1β (IL-1β), and membrane proteins that are known to also traverse to the plasma membrane by a conventional process of exocytosis, such as α integrin and the cystic fibrosis transmembrane conductor (CFTR). Mechanisms underlying unconventional protein secretion (UPS) are actively being analyzed and deciphered, and these range from an unusual form of plasma membrane translocation to vesicular processes involving the generation of exosomes and other extracellular microvesicles. In this chapter, we provide an overview on what is currently known about UPS in animal cells.

Key words Animal cells, Autophagy, Exosomes, GRASP, Unconventional protein secretion (UPS)

1 Introduction: Protein Secretion—Conventional and Unconventional

Eukaryotic cells are characterized by an elaborate intracellular membrane system, and protein secretion or exocytosis is classically viewed as occurring by multiple rounds of sequential budding and fusion of membranous vesicles or carriers from the endoplasmic reticulum (ER), the Golgi apparatus, and the trans-Golgi network (TGN), which ultimately fuse with the plasma membrane [1]. All cells secrete a range of proteins in a constitutive manner, but for specialized cells with regulated exocytosis, another classical route to the plasma membrane exists for specific cargoes such as hormone and neurotransmitters. Proteins destined for secretion (or those to be transported to the plasma membrane) are first targeted to the ER by N-terminal or internal signal sequences as the nascent polypeptide emerges from the ribosome. Secretory proteins exit the ER via coat protein complex II (COPII) vesicles nucleated by the small GTPase Sar1 [2, 3]. Many secretory proteins acquire a carbohydrate group-based post-translational modification, namely core

Andrea Pompa and Francesca De Marchis (eds.), *Unconventional Protein Secretion: Methods and Protocols*, Methods in Molecular Biology, vol. 1459, DOI 10.1007/978-1-4939-3804-9_2, © Springer Science+Business Media New York 2016

N-linked glycosylation at the ER, and undergo subsequent modifications to their N-linked carbohydrate groups by glycosyl-transferases as they traverse through the Golgi/TGN. The conventional exocytic transport process at the Golgi requires another GTPase, the ADP ribosylation factor 1 (Arf1), which nucleates the coat protein I (COPI) complex. COPI-mediated transport is known to be inhibited by a host of compounds, including the fungal metabolite brefeldin A (BFA), which inhibits guanine nucleotide exchange of Arf1 [4]. Viotti provides a more detail discussion on the conventional mode of protein secretion in a separate chapter.

Amongst proteins that are secreted from cells, a small number were released from cells in ways that appear independent of the conventional or canonical secretory pathway or mechanisms. Prominent in this regard are proteins such as FGF1/2 [5], cytokines like IL-1β [6], the *Drosophila* α-integrin [7], the nuclear protein amphoterin/high motility group protein B1 (HMGB1) [8], extracellular matrix proteins like galectins [9], yeast heat-shock protein 150 (Hsp150) [10], the neuropathogenic protein α-synuclein [11, 12], and more recently the *Dictyostelium* and yeast acyl-CoA-binding protein [13–15] as well as the membrane protein cystic fibrosis transmembrane conductor (CFTR) [16]. Evidently, both soluble and membrane-bound proteins located at various cellular compartments can undergo unconventional protein secretion (UPS). There is no distinct commonality between these in terms of identity and function.

There appears to be different modes by which UPS can take place. At least three different transport modes are apparent, depending on the nature and cellular location of the cargoes involved. Firstly, for proteins that are absolutely cytosolic and are never enclosed by membranous vesicles (such as FGF2) secretion would require some specific membrane translocation processes that bring them across the plasma membrane [17]. Secondly, cytoplasmic proteins could become membrane encased prior to secretion, and these processes may involve the generation of exosomes or ectosomes [18–23]. In the third scenario, there are cargoes, both soluble and membrane bound, that could initially enter the canonical secretory pathway through ER translocation as they possess ER-targeting signals. However, these could be subsequently transported to the cell surface or be secreted in a manner that is independent of COPII-mediated ER budding, and bypassing the Golgi apparatus. Some modes of UPS have been more extensively investigated, and some aspects of UPS are better known than others. For example, while the mechanism of generation of exosomes by the endosomal sorting complexes required for transport (ESCRT) complexes has been extensively examined [24], the mode of unconventional secretion that is dependent on autophagy is less clear in mechanistic terms [25, 26].

UPS appears to occur in organisms across the entire eukaryotic domain. In this chapter, we provide a brief overview of what is currently known about this process in animal cells, and also draw results from some lower eukaryotes. In the paragraphs that follow, we first describe some representative cargoes undergoing unconventional secretion or exocytic cell surface transport, followed by a discussion on the mechanisms involved.

2   Proteins Known to Undergo Unconventional Protein Secretion

We outline in this section a few prominent examples of proteins that are unconventionally secreted by animal cells. They are rather loosely categorized as below, and listed in Table 1. This list is far from being exhaustive. The reader is referred to more extensive and dedicated reviews in the literature for other examples [27–29].

2.1   Growth Factors: Fibroblast Growth Factors 2

Fibroblast growth factor 2 (FGF2, also known as basic FGF (bFGF)) is a member of the heparin-binding FGF family, which are key regulators of proliferative and differentiation processes in a wide spectrum of

Table 1
Proteins known to be secreted by unconventional secretion

Protein	Cellular localization	Known mechanistic insight
Fibroblast growth factor 2 (FGF2)	Cytosolic	Direct plasma membrane crossing with unique mechanism dependent on phosphoinositides and extracellular heparin sulfate proteoglycans
Interleukin 1-β (IL1-β)	Cytosolic	Autophagy- and GRASP-dependent UPS, secretory lysosomes, exosomes
Acyl-CoA-binding protein (ACBP)		Autophagy- and GRASP-dependent UPS
Galectin	Extracellular matrix	
α-Integrin	Plasma membrane (and internal membranes)	dGRASP dependent
Cystic fibrosis transmembrane conductor (CFTR)	Plasma membrane (and internal membranes)	Autophagy- and GRASP-dependent UPS
α-Synuclein	Cytosolic	Exosomal secretion
γ-Synuclein	Cytosolic	Exosomal secretion
Tau	Cytosolic/microtubule associated	Exosomal secretion

See text for details

tissues. FGF2's known roles in tumor angiogenesis and would healing and as a key factor in maintenance of renewability in stem cell cultures are particularly prominent [30]. FGF2 has several isoforms of high and low molecular sizes, and the 18 kDa small isoform is known to be secreted extracellularly in an unconventional manner [31]. A related FGF family member, FGF1, is likewise unconventionally secreted, particularly under the condition of stress and starvation [32, 33]. FGF2 has no signal peptide or leader sequence and is expected to be exclusively cytoplasmic, but its non-cell-autonomous activity is of vital physiological and pathological importance. FGF2's unconventional secretion represents a unique pathway as secretion appears to occur via direct crossing of the plasma membrane [5, 34], which is discussed further below.

2.2 Cytokines: Interleukin-1β (IL-1β)

IL-1β belongs to the IL-1 family of pro- and anti-inflammatory cytokines [35], which unlike most other cytokines do not would have an ER-targeting signal peptide and thus have no access to the conventional secretory pathway. Many if not all members of the IL-1 family are secreted unconventionally, but IL-1β is the best studied in this regard [36]. IL-1β is synthesized as a pro-peptide in monocytes, which is proteolytically processed by activated caspase-1 of the inflammasome complex upon infection, injury, and other forms of stress [37]. Cleaved/mature IL-1β could then exit the cells if these are lysed during pyroptosis, a mode of lytic cell death driven by caspase-1 or caspase-11 [38]. Otherwise, its release is not inhibited by perturbation of the secretory pathway with drugs such as BFA or monensin [39]. Unconventional secretion of IL-1β could occur by a variety of mechanisms that include plasma membrane translocation, exosomes, or other forms of secretory micovesicles [18, 25, 28, 36]. Caspase-1 is also apparently a driver for the unconventional secretion of damage-associated molecular patterns (DAMPs) or alarmins, such as the nuclear HMGB1 [40]. Recent findings suggest that IL-1β secretion is dependent on autophagy and the Golgi reassembly stacking protein 55 (GRASP55) [41]. This "GRAPSP and autophagy-dependent" (GAD) pathway/mechanism [25] is further discussed below.

2.3 Non-Cell-Autonomous Modulator: Acyl-CoA-Binding Protein

The evolutionarily conserved acyl-CoA-binding protein (ACBP) [42, 43] has both a cell-autonomous activity of binding to medium- and long-chain acy-CoA esters [44], as well as non-cell-autonomous functions as a secreted protein. The *Dictyostelium AsbA* product is secreted as the sporulation factor spore differentiation factor 2 (SDF-2) [45], while the mammalian ACBP is the precursor of benzodiazepine-binding inhibitor (BDI) [46]. A series of recent studies have revealed that ACBP orthologues in *Dictyostelium* and yeast are unconventionally secreted in a GRASP and autophagy-dependent manner [13–15].

2.4 Extracellular Matrix Components: Galectin-1 and Integrin-α

The evolutionarily conserved galectin-1 is a nuclear/cytoplasmic β-galactoside-binding protein that has cell adhesion, immune suppression, and neuroprotective activities [47–49]. Lacking an ER-targeting signal peptide, it is nonetheless secreted into the extracellular matrix

[9, 50]. Integrin subunits are known to go through conventional exocytosis en route to the cell surface. However, during certain stages of *Drosophila* wing imaginal disc epithelia development nascent integrin-α subunits are transported to the specific plasma membrane domains in contact with the basal membrane via a mechanism that appears to bypass the Golgi, but is dependent on the *Drosophila* dGRASP [7, 51]. Whether integrins could be unconventionally secreted in mammalian cells has not yet been clearly demonstrated.

2.5 Membrane Proteins with Conventional Exocytosis: The Cystic Fibrosis Transmembrane Conductance Regulator

The multi-membrane-spanning cystic fibrosis transmembrane conductance regulator (CFTR) is the protein mutated in cystic fibrosis [52]. It is ER targeted and transported via conventional exocytosis to the cell surface, which is critical for its function as a chloride channel. Surface CFTR also undergoes endocytic recycling and its surface transport was shown to be dependent on the TGN SNARE syntaxin 16 [53, 54]. The most common and prominent class of CFTR mutations, ΔF508-CFTR, has a defect in ER exit and exocytosis. Both wild-type and mutant CFTR are however capable of undergoing some form of unconventional ER to cell surface transport [55, 56]. Such a mode of exocytosis is not inhibited by BFA or by silencing of COPI and COPII components, and the transported CFTR retains its ER core-glycosylated forms [56]. This unconventional surface trafficking process of core-glycosylated CFTR is dependent on GRASP55 and autophagy, and appears to be enhanced by stress. The discovery of an unconventional transport mode for integrin-α and CFTR (particularly ER-entrapped ΔF508-CFTR) suggests that even proteins that usually undergo conventional exocytosis could be engaged in unconventional exocytosis under certain conditions, and this funding has important implications.

2.6 Neuropathogenic Proteins: α-Synuclein and Tau

Many neurodegenerative diseases are characterized by intra- or extracellular accumulation of protein aggregates with a dominant etiological component. The small molecule α-synuclein encode by the *SNCA* gene has the propensity to form toxic aggregates, which is a major component of the pathological feature of Lewy bodies (LB) in brains of Parkinson's disease (PD) and other LB disease [57]. α-Synuclein mutations could enhance its propensity to aggregate, and give rise to the juvenile onset form of PD. α-Synuclein pathology could spread in a prion-like manner from neuron to neuron [58], and evidence has accumulated to suggest that α-synuclein could be unconventionally secreted by an exosome-based mechanism [11, 59, 60]. The microtubule-binding protein tau, whose pathological hyperphosphorylated form is found in intracellular fibrillary tangle in Alzheimer's disease and other taupathies [61], is likewise secreted in an unconventional manner [62, 63], likely through the generation of microvesicles.

3   Mechanisms of Unconventional Protein Secretion

In this section, we review what is currently known about the mechanisms underlying unconventional secretion in animal cells. It should be noted that some proteins have possibly more than a single mode of unconventional secretion (such as IL-1β) while some others could be secreted by both canonical and unconventional means (such as CFTR). A schematic summary of the UPS pathways is presented in Fig. 1.

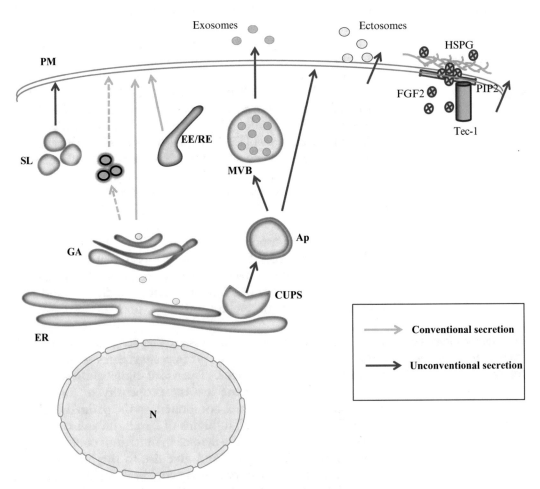

Fig. 1 Conventional versus unconventional secretion. A schematic diagram depicting paths and compartments known to be involved in UPS. The conventional secretory pathways are marked by *green arrows* while the unconventional pathways by *blue arrows*. For simplicity, the endocytic pathways are not depicted. FGF2 UPS requires phosphatidylinositol 4,5-bisphosphate (PIP2) and extracellular heparin sulfate proteoglycan (HSPG), and is regulated by the Tec-1 kinase. The underlying molecular components for other modes of UPS are less clear. *N* nucleus, *ER* endoplasmic reticulum, *CUPS* compartment for unconventional protein secretion, *Ap* autophagosome, *GA* Golgi apparatus, *MVB* multivesicular bodies, *EE/RE* early endosome/recycling endosome, *SL* secretory lysosomes, *PM* plasma membrane

3.1 Crossing the Plasma Membrane: The FGF2 Chronicle

Proteins that are absolutely cytosolic and have no access to any form of membranous vesicles would not be able to exit an intact cell unless there are ways to negotiate the plasma membrane (PM). In theory, peptides could penetrate a barrier of lipid bilayer via hydrophilic channels and pumps, such as the ATP-binding cassette (ABC) transporters and the mitochondrial translocon complexes. Membrane-penetrating peptides could interact with and perturb the structure of the phospholipid bilayer and form inverted micelles [64]. In most if not all of these mechanisms, however, the polypeptide is translocated in a denatured or misfolded form.

Work from the laboratory of Walter Nickle has shed light on a rather unique mode of plasma membrane translocation that is exhibited by FGF2 [5, 34], which appears to translocate directly through the PM in a fully folded form without relying on protein-conducting channels. FGF2 is able to interact with phosphoinositides at the inner leaflet of the PM [65]. Membrane-recruited FGF2 could oligomerize to form a membrane pore [66]. Phosphorylation of Tyr 81 of FGF2 by Tec kinase, a non-receptor tyrosine kinase which contains a pleckstrin homology (PH) domain that binds phosphoinositides, enhances the lipidic membrane pore formation [67]. FGF2 has a high affinity for heparin sulfate proteoglycans enriched at the outer leaflet of the PM bilayer [68], and the latter could literally extract FGF2 from their prior interaction with phosphoinositides, thus completing the translocation. At the moment it is yet uncertain whether this mode of PM translocation is utilized by other proteins. However, the HIV-TAT protein is known to require phosphoinositides for secretion [69] and a direct PM translocation mechanism has also been proposed for FGF1 and IL-1β.

3.2 Exosome, Ectosomes, and Other Microvesicles

A more common mode of unconventional secretion is one that utilizes some form of membranous vesicles. The unconventional secretion of IL-1β, for example, is perhaps largely through a vesicle-mediated mechanism. Amongst these, exosomes from multivesicular bodies (MVBs) are perhaps the best understood. These 40–100 nm endosome-derived vesicles [70–72] are formed in MVBs by an unusual luminal budding of vesicles from the limiting membrane of late endosomes. The mechanism involves the generation of intra-MVB vesicles by the ESCRT complexes [73, 74], and these vesicles are released as exosomes when MVBs move towards the cell periphery and fuse with the plasma membrane. Other than proteins, exosomes also contain RNA and DNA molecules, and these are potential mediators of intercellular communication.

Another possible mode of unconventional secretion of IL-1β involves the secretory lysosomes, or lytic granules [75, 76]. Lysosomes are traditionally viewed as lytic compartments for the terminal destruction of cellular materials. However, in some cases, modified lysosomes could undergo regulated secretion in response to intracellular Ca^{2+} elevation, for example resulting from the activation of P2X7 purinergic receptors on monocytes [77], which are ATP-gated ion channels [78]. When secretory lysosomes fuse with the plasma membrane, likely facilitated by a set of syntaxin 11-based [79] fusion machinery akin to that used by lytic granules [80], their contents could be released into the extracellular space. In human monocytes and dendritic cells, Ca^{2+} elevation triggers secretion of both IL-1β and the lysosomal hydrolase cathepsin D [81–83], which suggest that IL-1β could be co-released from the same compartment as a lysosomal marker.

Another possible mode of extracellular release of cytosolic materials is plasma membrane shedding of microvesicles. These plasma membrane-derived microvesicles have been given several different names, ranging from ectosomes to shedding microvesicles [23, 84]. These microvesicles are generally somewhat larger than the MVB-derived exosomes (100 nm to 1 μm in size), and are enriched in the inner leaflet phospholipid phosphatidylserine on their outer surface. The mechanism for microvesicle formation is yet unclear. Like the case for secretory lysosomes, there is evidence that P2X7 receptors expressed in the membrane of microvesicles may be involved in the regulation of IL-1β release [85, 86].

3.3 The GRASP and Autophagy-Dependent (GAD) Pathway of UPS

A particularly interesting recent development is emerging evidence for the involvement of GRASPs and the autophagy machinery in unconventional secretion. Both the unconventional secretion of the cytoplasmic ACBP [14, 15] and the membrane-bound CFTR [16] involve GRASP and autophagy. In fact, it was recently showed that the secretion of IL-1β [41] also requires these to be functional. GRASP orthologues act in Golgi cisternae stacking and Golgi ribbon formation [87–89] by forming oligomers through the N-terminal PDZ domains. Two paralogues exist in the mammalian genome GRASP65 and GRASP55; both are peripheral membrane proteins at the *cis-* and medial-*trans*-Golgi cisternae. Silencing of both GRASP proteins leads to disassembly of the Golgi stack [89].

The involvement of GRASPs in unconventional secretion may superficially appear paradoxical, as cargoes unconventionally secreted (including CFTR which also follows the canonical secretory route) could bypass the Golgi apparatus in reaching the plasma membrane. In that case why should the Golgi cisternae stacker GRASP be involved? It is possible that GRASPs' function in Golgi stack maintenance is unrelated to its role in unconventional secretion. It has been proposed that GRASPs are primarily membrane tethers, as GRASPs oligomerization through their PDZ motifs could bring opposing membranes in close proximity [90]. Thus, they could possibly act in membrane tethering of a specific subset of ER-derived vesicles with the PM. While this is an interesting hypothesis, the PM-targeting mechanism based on GRASP alone is not yet well defined. It is possible that canonical secretory pathway components such as Rabs and SNAREs are involved in this tethering and fusion step. Acb1secretion in *S. cerevisiae* requires the plasma membrane t-SNARE, Sso1p [14], and that in *P. pastoris* requires PM SNAREs of *Pichia* [15].

The apparent requirement for components of the autophagy pathway for the unconventional secretion of quite a range of cargo types is intriguing. Autophagy is an evolutionarily conserved process in which cytosolic materials and membranous organelles like the mitochondria are encased within a double-membrane autophagosome, which eventually fuses with the vacuole or lysosome for the degradation of its contents [91]. The roles of autophagy in cellular and systemic physiology as well as pathology have been extensively studied [92–97]. A particularly interesting point to note is that autophagy is typically induced under conditions of nutrient or growth factor starvation, or during various conditions of stress [98]. For the cases of ACBP, CFTR, and IL-1β, autophagy-dependent unconventional secretion of these proteins does indeed become apparent under conditions of stress. It is thus conceivable that during times of stress, the cell may resort to unconventional secretion of cytosolic proteins (such as IL-1β and ACBP) to elicit a non-cell-autonomous signal. For the case of proteins already targeted to the ER (such as CFTR), activation of UPS by stress could help relieve the accumulation of unfolded protein in the ER, or bypass of trafficking defects in the canonical secretory pathway to allow some degree of secretion to occur.

Exactly how autophagy leads to UPS is not yet clear, and several possibilities have been proposed. Interestingly, autophagy components that are important for unconventional secretion are largely those involved in the early stages of generation of autophagosome and for endosomal fusion, but not for the final lysosomal fusion. Secretion of the Acb-1 in yeast, for example, does not require the vacuole/lysosomal SNARE VAMP7p or the Rab Ypt7p that are important for vacuole fusion [14]. Possible intermediates for autophagy-mediated unconventional secretion may therefore be autophagosomes (specifically created or otherwise) that fuse with endosomes/MVBs to form amphisomes [99] that subsequently fuse with the PM [25, 26, 29], and not lysosomes. In the former case, exosomes are presumably generated from the intraluminal vesicles in the amphisome or MVB.

In yeast, a novel compartment for UPS, known as the compartment for unconventional protein secretion (CUPS), is induced by nutrient starvation and the GRASP orthologue Grh1p as well as autophagy proteins initiating autophagosome formation are recruited into these structures [100]. Whether the autophagosome generated by CUPS differs from other sites of autophagosomal origin is not yet clear, and structures analogous to CUPS have not yet been reported in mammalian cells. While amphisome-PM fusion may be akin to MVB-PM fusion, direct fusion between autophagosome and the PM has not been clearly documented. It is however conceivable that autophagosomes generated by CUPS-like structures carrying GRASPs could dock with the PM. What happens after this fusion is unclear. It should be noted that other than exosomes, microvesicles known as ectosomes could be generated by direct budding from the PM [101]. These microvesicles are known to be shed from the cilium and flagellum [102, 103]. At the moment the mechanism of ectosome generation remains unknown and any connection with autophagy remains speculative.

4   Studying Unconventional Protein Secretion in Animal and Yeast/Fungal Cells/Tissues

UPS, far from simply a curious set of biological phenomena, has important implications in health and disease. Its mechanism and functions have been rigorously tackled by workers in various research niches. A lot of the work is cargo centered, with important molecules such as FGF2 [34], IL-1β [36], and CFTR [16] being investigated as part of efforts to understand their basic biology. Another major aspect of the work pertains to investigations done on microvesicle-based secretion, which are of pathological interest [104, 105]. Biophysical and biochemical characterization of extracellular microvesicles is actively pursued in various contexts [106, 107]. With advances in high-throughput screening approaches, it is anticipated that more holistic analysis such as genome-wide siRNA screens will be performed to decipher components and pathways underlying the various modes of unconventional secretion. Genetic screens with model organisms such as S. cerevisiae should be ongoing, and are likely to reveal more mechanistic insights into the near future. Although unconventional protein secretion tends to bypass some, if not most, of the needs of the conventional secretory machinery, it is likely that some vesicular transport components responsible for conventional secretion are still involved. A bunch of Rab proteins, for example, are critically involved in various aspects of autophagy [108–110], and are therefore likely to influence UPS that rely on autophagy. UPS processes that rely on fusion of amphisomes, MVB, or other membranous carriers with the plasma membrane would likely need the participation of SNAREs. As proposed earlier, secretory lysosome may use a bunch of cell surface syntaxins or the atypical syntaxin 11 [79, 80] to facilitate plasma membrane fusion.

From a translational perspective, exosome and other extracellular microvesicles are being developed as specific sources for disease biomarkers [111] and as biomimetic drug delivery vehicles [112, 113]. Unconventional secretion appears to be a major contributor to the secretome of cancer cells and tissues [114], and some of the proteins that are unconventionally secreted could promote tumorigenesis and metastasis [115, 116]. UPS, particularly in the mode of exosomal release, has also been extensively implicated in neurodegenerative diseases [117, 118]. Causative agents of neurodegeneration such as prion protein [119] and tau [62, 63] could potentially spread from diseased neurons to healthy ones via UPS. On the other hand, discovery of ΔF508-CFTR's unconventional exocytosis to the cell surface opens up new therapeutic possibilities for cystic fibrosis [16] and potentially other diseases that result from impaired surface transport of mutated and misfolded proteins. A thorough understanding of both the modes and mechanism of UPS would therefore be of tremendous academic and clinical interest in the coming years.

In this collection, methods and approaches for investigating many of the proteins undergoing unconventional secretion shall be presented and discussed. Methods for preparation and analysis of the tissue secretome from tumor interstitial fluid (TIF) are presented by Gromov and colleagues. Amaral and co-workers examine unconventional transport of CFTR, Lacazette and colleagues discuss studies on FGF2, Shou and colleagues look at synuclein-γ in cancer cells, while Beer and colleagues examine the role of caspase-1 in unconventional secretion. Yeast and fungal genetics shall provide useful handles for dissecting mechanisms and identification of mechanistic components of UPS, and could be used for the preparation of molecules of biological interests via UPS. Schipper and colleagues present the use of UPS to express sugar-free heterologous proteins in the plant fungal pathogen *Ustilago maydis*. On UPS-related pathogenicity, MacLean and colleagues look at the hydrophilic acylated surface protein B of *Leishmania*, and Reynard and colleagues discuss the role of unconventional matrix protein VP40 secretion in Ebola pathogenicity. Shedding of microvesicles may underlie a large fraction of all UPS, and in their respective chapters, Hajj and colleagues discuss microvesicle-based UPS of the co-chaperone stress-inducible protein 1, while Rodrigues and co-workers examine extracellular vesicles from yeast.

Bellucci, Zhang, Goring, and Pocsfalvi discuss methods for preparation and isolation of exosome-like vesicles or secretome, derived from various plant materials, and chemical modulation of secretory pathway in protoplasts.

Acknowledgements

The authors are supported by NUS Graduate School for Integrative Sciences and Engineering. The authors declare no financial conflict of interest.

References

1. Bonifacino JS, Glick BS (2004) The mechanisms of vesicle budding and fusion. Cell 116:153–166

2. Tang BL, Wang Y, Ong YS, Hong W (2005) COPII and exit from the endoplasmic reticulum. Biochim Biophys Acta 1744:293–303

3. Jensen D, Schekman R (2011) COPII-mediated vesicle formation at a glance. J Cell Sci 124:1–4

4. Beck R, Rawet M, Ravet M, Wieland FT, Cassel D (2009) The COPI system: molecular mechanisms and function. FEBS Lett 583:2701–2709

5. Nickel W (2011) The unconventional secretory machinery of fibroblast growth factor 2. Traffic 12:799–805

6. Eder C (2009) Mechanisms of interleukin-1beta release. Immunobiology 214:543–553

7. Schotman H, Karhinen L, Rabouille C (2008) dGRASP-mediated noncanonical integrin secretion is required for Drosophila epithelial remodeling. Dev Cell 14:171–182

8. Gardella S, Andrei C, Ferrera D, Lotti LV, Torrisi MR, Bianchi ME, Rubartelli A (2002) The nuclear protein HMGB1 is secreted by monocytes via a non-classical, vesicle-mediated secretory pathway. EMBO Rep 3:995–1001

9. Seelenmeyer C, Wegehingel S, Tews I, Künzler M, Aebi M, Nickel W (2005) Cell surface counter receptors are essential components of the unconventional export machinery of galectin-1. J Cell Biol 171:373–381

10. Karhinen L, Bastos RN, Jokitalo E, Makarow M (2005) Endoplasmic reticulum exit of a secretory glycoprotein in the absence of sec24p family proteins in yeast. Traffic 6:562–574

11. Lee HJ, Patel S, Lee SJ (2005) Intravesicular localization and exocytosis of alpha-synuclein and its aggregates. J Neurosci 25:6016–6024

12. Ejlerskov P, Rasmussen I, Nielsen TT, Bergström AL, Tohyama Y, Jensen PH, Vilhardt F (2013) Tubulin polymerization-promoting protein (TPPP/p25α) promotes unconventional secretion of α-synuclein through exophagy by impairing autophagosome-lysosome fusion. J Biol Chem 288:17313–17335

13. Kinseth MA, Anjard C, Fuller D, Guizzunti G, Loomis WF, Malhotra V (2007) The Golgi-associated protein GRASP is required for unconventional protein secretion during development. Cell 130:524–534

14. Duran JM, Anjard C, Stefan C, Loomis WF, Malhotra V (2010) Unconventional secretion of Acb1 is mediated by autophagosomes. J Cell Biol 188:527–536

15. Manjithaya R, Anjard C, Loomis WF, Subramani S (2010) Unconventional secretion of Pichia pastoris Acb1 is dependent on GRASP protein, peroxisomal functions, and autophagosome formation. J Cell Biol 188:537–546

16. Gee HY, Noh SH, Tang BL, Kim KH, Lee MG (2011) Rescue of ΔF508-CFTR trafficking via a GRASP-dependent unconventional secretion pathway. Cell 146:746–760

17. Schäfer T, Zentgraf H, Zehe C, Brügger B, Bernhagen J, Nickel W (2004) Unconventional secretion of fibroblast growth factor 2 is mediated by direct translocation across the plasma membrane of mammalian cells. J Biol Chem 279:6244–6251

18. Nickel W, Rabouille C (2009) Mechanisms of regulated unconventional protein secretion. Nat Rev Mol Cell Biol 10:148–155

19. Mathivanan S, Ji H, Simpson RJ (2010) Exosomes: extracellular organelles important in intercellular communication. J Proteomics 73:1907–1920

20. Camussi G, Deregibus MC, Bruno S, Cantaluppi V, Biancone L (2010) Exosomes/microvesicles as a mechanism of cell-to-cell communication. Kidney Int 78:838–848

21. Sadallah S, Eken C, Schifferli JA (2011) Ectosomes as modulators of inflammation and immunity. Clin Exp Immunol 163:26–32

22. Record M, Subra C, Silvente-Poirot S, Poirot M (2011) Exosomes as intercellular signalosomes and pharmacological effectors. Biochem Pharmacol 81:1171–1182

23. Lee TH, D'Asti E, Magnus N, Al-Nedawi K, Meehan B, Rak J (2011) Microvesicles as mediators of intercellular communication in cancer--the emerging science of cellular 'debris'. Semin Immunopathol 33:455–467

24. Kowal J, Tkach M, Théry C (2014) Biogenesis and secretion of exosomes. Curr Opin Cell Biol 29:116–125

25. Chua CEL, Lim YS, Lee MG, Tang BL (2012) Non-classical membrane trafficking processes galore. J Cell Physiol 227:3722–3730

26. Jiang S, Dupont N, Castillo EF, Deretic V (2013) Secretory versus degradative autophagy: unconventional secretion of inflammatory mediators. J Innate Immun 5:471–479

27. Nickel W (2003) The mystery of nonclassical protein secretion. A current view on cargo proteins and potential export routes. Eur J Biochem 270:2109–2119

28. Nickel W, Seedorf M (2008) Unconventional mechanisms of protein transport to the cell

surface of eukaryotic cells. Annu Rev Cell Dev Biol 24:287–308

29. Manjithaya R, Subramani S (2011) Autophagy: a broad role in unconventional protein secretion? Trends Cell Biol 21:67–73

30. Dvorak P, Dvorakova D, Hampl A (2006) Fibroblast growth factor signaling in embryonic and cancer stem cells. FEBS Lett 580:2869–2874

31. Mignatti P, Morimoto T, Rifkin DB (1992) Basic fibroblast growth factor, a protein devoid of secretory signal sequence, is released by cells via a pathway independent of the endoplasmic reticulum-Golgi complex. J Cell Physiol 151:81–93

32. Jackson A, Friedman S, Zhan X, Engleka KA, Forough R, Maciag T (1992) Heat shock induces the release of fibroblast growth factor 1 from NIH 3T3 cells. Proc Natl Acad Sci U S A 89:10691–10695

33. Shin JT, Opalenik SR, Wehby JN, Mahesh VK, Jackson A, Tarantini F, Maciag T, Thompson JA (1996) Serum-starvation induces the extracellular appearance of FGF-1. Biochim Biophys Acta 1312:27–38

34. Steringer JP, Müller HM, Nickel W (2015) Unconventional secretion of fibroblast growth factor 2—a novel type of protein translocation across membranes? J Mol Biol 427:1202–1210

35. Garlanda C, Dinarello CA, Mantovani A (2013) The interleukin-1 family: back to the future. Immunity 39:1003–1018

36. Monteleone M, Stow JL, Schroder K (2015) Mechanisms of unconventional secretion of IL-1 family cytokines. Cytokine 74(2):213–218

37. Strowig T, Henao-Mejia J, Elinav E, Flavell R (2012) Inflammasomes in health and disease. Nature 481:278–286

38. Jorgensen I, Miao EA (2015) Pyroptotic cell death defends against intracellular pathogens. Immunol Rev 265:130–142

39. Rubartelli A, Cozzolino F, Talio M, Sitia R (1990) A novel secretory pathway for interleukin-1 beta, a protein lacking a signal sequence. EMBO J 9:1503–1510

40. Zhu H, Wang L, Ruan Y, Zhou L, Zhang D, Min Z, Xie J, Yu M, Gu J (2011) An efficient delivery of DAMPs on the cell surface by the unconventional secretion pathway. Biochem Biophys Res Commun 404:790–795

41. Dupont N, Jiang S, Pilli M, Ornatowski W, Bhattacharya D, Deretic V (2011) Autophagy-based unconventional secretory pathway for extracellular delivery of IL-1β. EMBO J 30:4701–4711

42. Knudsen J (1990) Acyl-CoA-binding protein (ACBP) and its relation to fatty acid-binding protein (FABP): an overview. Mol Cell Biochem 98:217–223

43. Faergeman NJ, Knudsen J (2002) Acyl-CoA binding protein is an essential protein in mammalian cell lines. Biochem J 368:679–682

44. Huang H, Atshaves BP, Frolov A, Kier AB, Schroeder F (2005) Acyl-coenzyme A binding protein expression alters liver fatty acyl-coenzyme A metabolism. Biochemistry 44:10282–10297

45. Cabral M, Anjard C, Malhotra V, Loomis WF, Kuspa A (2010) Unconventional secretion of AcbA in Dictyostelium discoideum through a vesicular intermediate. Eukaryot Cell 9:1009–1017

46. Chang JL, Tsai HJ (1996) Carp cDNA sequence encoding a putative diazepam-binding inhibitor/endozepine/acyl-CoA-binding protein. Biochim Biophys Acta 1298:9–11

47. Perillo NL, Marcus ME, Baum LG (1998) Galectins: versatile modulators of cell adhesion, cell proliferation, and cell death. J Mol Med 76:402–412

48. Welsh JW, Seyedin SN, Cortez MA, Maity A, Hahn SM (2014) Galectin-1 and immune suppression during radiotherapy. Clin Cancer Res 20:6230–6232

49. Wang J, Xia J, Zhang F, Shi Y, Wu Y, Pu H, Liou AKF, Leak RK, Yu X, Chen L, Chen J (2015) Galectin-1-secreting neural stem cells elicit long-term neuroprotection against ischemic brain injury. Sci Rep 5:9621

50. Seelenmeyer C, Stegmayer C, Nickel W (2008) Unconventional secretion of fibroblast growth factor 2 and galectin-1 does not require shedding of plasma membrane-derived vesicles. FEBS Lett 582:1362–1368

51. Schotman H, Karhinen L, Rabouille C (2009) Integrins mediate their unconventional, mechanical-stress-induced secretion via RhoA and PINCH in Drosophila. J Cell Sci 122:2662–2672

52. Quinton PM (1999) Physiological basis of cystic fibrosis: a historical perspective. Physiol Rev 79:S3–S22

53. Tang BL, Low DY, Lee SS, Tan AE, Hong W (1998) Molecular cloning and localization of human syntaxin 16, a member of the syntaxin family of SNARE proteins. Biochem Biophys Res Commun 242:673–679

54. Gee HY, Tang BL, Kim KH, Lee MG (2010) Syntaxin 16 binds to cystic fibrosis transmem-

brane conductance regulator and regulates its membrane trafficking in epithelial cells. J Biol Chem 285:35519–35527

55. Yoo JS, Moyer BD, Bannykh S, Yoo HM, Riordan JR, Balch WE (2002) Non-conventional trafficking of the cystic fibrosis transmembrane conductance regulator through the early secretory pathway. J Biol Chem 277:11401–11409

56. Tang BL, Gee HY, Lee MG (2011) The cystic fibrosis transmembrane conductance regulator's expanding SNARE interactome. Traffic 12:364–371

57. Kim WS, Kågedal K, Halliday GM (2014) Alpha-synuclein biology in Lewy body diseases. Alzheimer Res Ther 6:73

58. Recasens A, Dehay B (2014) Alpha-synuclein spreading in Parkinson's disease. Front Neuroanat 8:159

59. Emmanouilidou E, Melachroinou K, Roumeliotis T, Garbis SD, Ntzouni M, Margaritis LH, Stefanis L, Vekrellis K (2010) Cell-produced alpha-synuclein is secreted in a calcium-dependent manner by exosomes and impacts neuronal survival. J Neurosci 30:6838–6851

60. Danzer KM, Kranich LR, Ruf WP, Cagsal-Getkin O, Winslow AR, Zhu L, Vanderburg CR, McLean PJ (2012) Exosomal cell-to-cell transmission of alpha synuclein oligomers. Mol Neurodeg 7:42

61. Ballatore C, Lee VMY, Trojanowski JQ (2007) Tau-mediated neurodegeneration in Alzheimer's disease and related disorders. Nat Rev Neurosci 8:663–672

62. Saman S, Kim W, Raya M, Visnick Y, Miro S, Saman S, Jackson B, McKee AC, Alvarez VE, Lee NCY, Hall GF (2012) Exosome-associated tau is secreted in tauopathy models and is selectively phosphorylated in cerebrospinal fluid in early Alzheimer disease. J Biol Chem 287:3842–3849

63. Chai X, Dage JL, Citron M (2012) Constitutive secretion of tau protein by an unconventional mechanism. Neurobiol Dis 48:356–366

64. Di Pisa M, Chassaing G, Swiecicki JM (2015) Translocation mechanism(s) of cell-penetrating peptides: biophysical studies using artificial membrane bilayers. Biochemistry 54:194–207

65. Temmerman K, Ebert AD, Müller HM, Sinning I, Tews I, Nickel W (2008) A direct role for phosphatidylinositol-4,5-bisphosphate in unconventional secretion of fibroblast growth factor 2. Traffic 9:1204–1217

66. Steringer JP, Bleicken S, Andreas H, Zacherl S, Laussmann M, Temmerman K, Contreras FX, Bharat TAM, Lechner J, Müller HM, Briggs JAG, García-Sáez AJ, Nickel W (2012) Phosphatidylinositol 4,5-bisphosphate (PI(4,5)P2)-dependent oligomerization of fibroblast growth factor 2 (FGF2) triggers the formation of a lipidic membrane pore implicated in unconventional secretion. J Biol Chem 287:27659–27669

67. Ebert AD, Laussmann M, Wegehingel S, Kaderali L, Erfle H, Reichert J, Lechner J, Beer HD, Pepperkok R, Nickel W (2010) Tec-kinase-mediated phosphorylation of fibroblast growth factor 2 is essential for unconventional secretion. Traffic 11:813–826

68. Zehe C, Engling A, Wegehingel S, Schäfer T, Nickel W (2006) Cell-surface heparan sulfate proteoglycans are essential components of the unconventional export machinery of FGF-2. Proc Natl Acad Sci U S A 103:15479–15484

69. Rayne F, Debaisieux S, Yezid H, Lin YL, Mettling C, Konate K, Chazal N, Arold ST, Pugnière M, Sanchez F, Bonhoure A, Briant L, Loret E, Roy C, Beaumelle B (2010) Phosphatidylinositol-(4,5)-bisphosphate enables efficient secretion of HIV-1 Tat by infected T-cells. EMBO J 29:1348–1362

70. Février B, Raposo G (2004) Exosomes: endosomal-derived vesicles shipping extracellular messages. Curr Opin Cell Biol 16:415–421

71. Schorey JS, Bhatnagar S (2008) Exosome function: from tumor immunology to pathogen biology. Traffic 9:871–881

72. Simons M, Raposo G (2009) Exosomes—vesicular carriers for intercellular communication. Curr Opin Cell Biol 21:575–581

73. Babst M (2011) MVB vesicle formation: ESCRT-dependent, ESCRT-independent and everything in between. Curr Opin Cell Biol 23:452–457

74. Hurley JH (2010) The ESCRT complexes. Crit Rev Biochem Mol Biol 45:463–487

75. Blott EJ, Griffiths GM (2002) Secretory lysosomes. Nat Rev Mol Cell Biol 3:122–131

76. Holt OJ, Gallo F, Griffiths GM (2006) Regulating secretory lysosomes. J Biochem 140:7–12

77. Qu Y, Dubyak GR (2009) P2X7 receptors regulate multiple types of membrane trafficking responses and non-classical secretion pathways. Purinergic Signal 5:163–173

78. Bours MJL, Dagnelie PC, Giuliani AL, Wesselius A, Di Virgilio F (2011) P2 receptors and extracellular ATP: a novel homeostatic pathway in inflammation. Front Biosci (Schol Ed) 3:1443–1456

79. Tang BL, Low DY, Hong W (1998) Syntaxin 11: a member of the syntaxin family without a carboxyl terminal transmembrane domain. Biochem Biophys Res Commun 245:627–632

80. Halimani M, Pattu V, Marshall MR, Chang HF, Matti U, Jung M, Becherer U, Krause E, Hoth M, Schwarz EC, Rettig J (2014) Syntaxin11 serves as a t-SNARE for the fusion of lytic granules in human cytotoxic T lymphocytes. Eur J Immunol 44:573–584

81. Andrei C, Dazzi C, Lotti L, Torrisi MR, Chimini G, Rubartelli A (1999) The secretory route of the leaderless protein interleukin 1beta involves exocytosis of endolysosome-related vesicles. Mol Biol Cell 10:1463–1475

82. Andrei C, Margiocco P, Poggi A, Lotti LV, Torrisi MR, Rubartelli A (2004) Phospholipases C and A2 control lysosome-mediated IL-1 beta secretion: Implications for inflammatory processes. Proc Natl Acad Sci U S A 101:9745–9750

83. Gardella S, Andrei C, Lotti LV, Poggi A, Torrisi MR, Zocchi MR, Rubartelli A (2001) CD8(+) T lymphocytes induce polarized exocytosis of secretory lysosomes by dendritic cells with release of interleukin-1beta and cathepsin D. Blood 98:2152–2159

84. Cocucci E, Racchetti G, Meldolesi J (2009) Shedding microvesicles: artefacts no more. Trends Cell Biol 19:43–51

85. Bianco F, Pravettoni E, Colombo A, Schenk U, Möller T, Matteoli M, Verderio C (2005) Astrocyte-derived ATP induces vesicle shedding and IL-1 beta release from microglia. J Immunol 174:7268–7277

86. Pizzirani C, Ferrari D, Chiozzi P, Adinolfi E, Sandonà D, Savaglio E, Di Virgilio F (2007) Stimulation of P2 receptors causes release of IL-1beta-loaded microvesicles from human dendritic cells. Blood 109:3856–3864

87. Shorter J, Watson R, Giannakou ME, Clarke M, Warren G, Barr FA (1999) GRASP55, a second mammalian GRASP protein involved in the stacking of Golgi cisternae in a cell-free system. EMBO J 18:4949–4960

88. Feinstein TN, Linstedt AD (2008) GRASP55 regulates Golgi ribbon formation. Mol Biol Cell 19:2696–2707

89. Xiang Y, Wang Y (2010) GRASP55 and GRASP65 play complementary and essential roles in Golgi cisternal stacking. J Cell Biol 188:237–251

90. Giuliani F, Grieve A, Rabouille C (2011) Unconventional secretion: a stress on GRASP. Curr Opin Cell Biol 23:498–504

91. Yang Z, Klionsky DJ (2010) Eaten alive: a history of macroautophagy. Nat Cell Biol 12:814–822

92. Banerjee R, Beal MF, Thomas B (2010) Autophagy in neurodegenerative disorders: pathogenic roles and therapeutic implications. Trends Neurosci 33:541–549

93. Xilouri M, Stefanis L (2010) Autophagy in the central nervous system: implications for neurodegenerative disorders. CNS Neurol Disord Drug Targets 9:701–719

94. Beau I, Mehrpour M, Codogno P (2011) Autophagosomes and human diseases. Int J Biochem Cell Biol 43:460–464

95. Mathew R, White E (2011) Autophagy in tumorigenesis and energy metabolism: friend by day, foe by night. Curr Opin Genet Dev 21:113–119

96. Rosenfeldt MT, Ryan KM (2011) The multiple roles of autophagy in cancer. Carcinogenesis 32(7):955–963

97. Mariño G, Madeo F, Kroemer G (2011) Autophagy for tissue homeostasis and neuroprotection. Curr Opin Cell Biol 23:198–206

98. Kroemer G, Mariño G, Levine B (2010) Autophagy and the integrated stress response. Mol Cell 40:280–293

99. Fader CM, Colombo MI (2009) Autophagy and multivesicular bodies: two closely related partners. Cell Death Differ 16:70–78

100. Bruns C, McCaffery JM, Curwin AJ, Duran JM, Malhotra V (2011) Biogenesis of a novel compartment for autophagosome-mediated unconventional protein secretion. J Cell Biol 195:979–992

101. Cocucci E, Meldolesi J (2015) Ectosomes and exosomes: shedding the confusion between extracellular vesicles. Trends Cell Biol 25(6):364–372

102. Wood CR, Huang K, Diener DR, Rosenbaum JL (2013) The cilium secretes bioactive ectosomes. Curr Biol 23:906–911

103. Wood CR, Rosenbaum JL (2015) Ciliary ectosomes: transmissions from the cell's antenna. Trends Cell Biol 25:276–285

104. Hulsmans M, Holvoet P (2013) MicroRNA-containing microvesicles regulating inflammation in association with atherosclerotic disease. Cardiovasc Res 100:7–18

105. Candelario KM, Steindler DA (2014) The role of extracellular vesicles in the progression of neurodegenerative disease and cancer. Trends Mol Med 20:368–374

106. Raposo G, Stoorvogel W (2013) Extracellular vesicles: exosomes, microvesicles, and friends. J Cell Biol 200:373–383

107. Kastelowitz N, Yin H (2014) Exosomes and microvesicles: identification and targeting by particle size and lipid chemical probes. Chembiochem 15:923–928

108. Chua CEL, Gan BQ, Tang BL (2011) Involvement of members of the Rab family and related small GTPases in autophagosome formation and maturation. Cell Mol Life Sci 68:3349–3358

109. Szatmári Z, Sass M (2014) The autophagic roles of Rab small GTPases and their upstream regulators: a review. Autophagy 10:1154–1166

110. Ao X, Zou L, Wu Y (2014) Regulation of autophagy by the Rab GTPase network. Cell Death Differ 21:348–358

111. Julich H, Willms A, Lukacs-Kornek V, Kornek M (2014) Extracellular vesicle profiling and their use as potential disease specific biomarker. Front Immunol 5:413

112. Fleury A, Martinez MC, Le Lay S (2014) Extracellular vesicles as therapeutic tools in cardiovascular diseases. Front Immunol 5:370

113. Lässer C (2015) Exosomes in diagnostic and therapeutic applications: biomarker, vaccine and RNA interference delivery vehicle. Expert Opin Biol Ther 15:103–117

114. Villarreal L, Méndez O, Salvans C, Gregori J, Baselga J, Villanueva J (2013) Unconventional secretion is a major contributor of cancer cell line secretomes. Mol Cell Proteomics 12:1046–1060

115. Liu C, Qu L, Lian S, Tian Z, Zhao C, Meng L, Shou C (2014) Unconventional secretion of synuclein-γ promotes tumor cell invasion. FEBS J 281:5159–5171

116. Costa-Silva B et al (2015) Pancreatic cancer exosomes initiate pre-metastatic niche formation in the liver. Nat Cell Biol 17(6):816–826

117. Bellingham SA, Guo BB, Coleman BM, Hill AF (2012) Exosomes: vehicles for the transfer of toxic proteins associated with neurodegenerative diseases? Front Physiol 3:124

118. Schneider A, Simons M (2013) Exosomes: vesicular carriers for intercellular communication in neurodegenerative disorders. Cell Tissue Res 352:33–47

119. Fevrier B, Vilette D, Archer F, Loew D, Faigle W, Vidal M, Laude H, Raposo G (2004) Cells release prions in association with exosomes. Proc Natl Acad Sci U S A 101:9683–9688

Chapter 3

Unconventional Protein Secretion in Plants

Destiny J. Davis, Byung-Ho Kang, Angelo S. Heringer, Thomas E. Wilkop, and Georgia Drakakaki

Abstract

Unconventional protein secretion (UPS) describes secretion pathways that bypass one or several of the canonical secretion pit-stops on the way to the plasma membrane, and/or involve the secretion of leaderless proteins. So far, alternatives to conventional secretion were primarily observed and studied in yeast and animal cells. The sessile lifestyle of plants brings with it unique restraints on how they adapt to adverse conditions and environmental challenges. Recently, attention towards unconventional secretion pathways in plant cells has substantially increased, with the large number of leaderless proteins identified through proteomic studies. While UPS pathways in plants are certainly not yet exhaustively researched, an emerging notion is that induction of UPS pathways is correlated with pathogenesis and stress responses. Given the multitude UPS events observed, comprehensively organizing the routes proteins take to the apoplast in defined UPS categories is challenging. With the establishment of a larger collection of studied plant proteins taking these UPS pathways, a clearer picture of endomembrane trafficking as a whole will emerge. There are several novel enabling technologies, such as vesicle proteomics and chemical genomics, with great potential for dissecting secretion pathways, providing information about the cargo that travels along them and the conditions that induce them.

Key words Endomembrane trafficking, Protein secretion, Unconventional protein secretion (UPS), Leaderless proteins, Signal peptide, Chemical genomics, Vesicle proteomics

1 Introduction

Protein secretion is an essential process in all living organisms; it follows a highly regulated and dynamic sequence of events that culminates in the delivery of proteins to the extracellular space. Mammalian and yeast cell studies have firmly established that proteins are secreted via multiple pathways, which can be roughly classified into conventional and unconventional routes [1, 2]. In conventional secretion, proteins are trafficked through the endoplasmic reticulum (ER) to the Golgi apparatus, then further on to the *trans*-Golgi network, and eventually to the plasma membrane, where the protein is released into the extracellular space of animal cells or the apoplast of plant cells [2–4]. An N-terminal secretion

Andrea Pompa and Francesca De Marchis (eds.), *Unconventional Protein Secretion: Methods and Protocols*, Methods in Molecular Biology, vol. 1459, DOI 10.1007/978-1-4939-3804-9_3, © Springer Science+Business Media New York 2016

signal as part of the protein sequence determines the selective entrance into the endomembrane system. Upon or after entry into the ER, the signal peptide functioning as the secretion pathway entry is cleaved off [5]. In unconventional protein secretion (UPS) pathways, the majority of proteins do not feature a signal peptide. These proteins frequently bypass typical organelles involved in secretion, such as the Golgi apparatus [6–8]. Our current understanding of UPS is dominantly based on extensive studies in animal and yeast cells, with insights from plants only contributing to it recently. The hitherto limited attention towards UPS in plants is somewhat surprising, given the large number of leaderless proteins in the plant apoplast proteome [6, 9–11]. An increasing number of studies link pathogenesis and stress response to unconventional secretion pathways, offering an enticing research avenue towards understanding mechanisms potentially contributing to plant stress tolerance [12–19]. This review summarizes our current understanding UPS in higher plants.

2   Unconventional Protein Secretion in Animal and Yeast Cells

UPS includes vesicular and non-vesicular means of transport, both involving a variety of pathways and molecular interactions. In non-vesicular UPS, soluble cytosolic proteins are directly secreted and translocated across the plasma membrane. A primary example of non-vesicular UPS in mammalian cells is the fibroblast growth factor 2 (FGF2) which can directly translocate across the plasma membrane [20, 21]. Insertion of FGF2, starting on the cytosolic side of the plasma membrane, is mediated by phosphatidylinositol-4, 5-bisphosphate binding [22]. During this process, the correct conformation of the protein prior to insertion into the membrane is vital; this prerequisite potentially constitutes a quality control mechanism towards ensuring protein activity. At the extracellular side, heparan sulfate proteoglycans (HPsGs) facilitate the directional transport of FGF2 by a yet not fully understood mechanism [23].

Vesicular UPS involves a plethora of distinct vesicle types that bud off, travel between, or bypass secretion-related organelles in the cell. Specifically, this mode of secretion can encompass (a) bypassing the Golgi apparatus, (b) organelle fusion with the plasma membrane, (c) secretion via multi-vesicular bodies (MVBs), and (d) secretion via exosomes or intraluminal vesicles. While the majority of known unconventionally secreted proteins lack a signal sequence, some proteins bypass the Golgi apparatus despite featuring a signal peptide [24]. This particularly intriguing fact raises the question as to what induces a deviation in the selected pathway, i.e., the switch from conventional to unconventional secretions. Following insertion in the

ER, proteins are packaged into vesicles that can circumvent the Golgi apparatus during trafficking to the plasma membrane. In addition, proteins without a signal peptide can be packaged in vesicles, within MVBs or exosomes, and subsequently directed to the plasma membrane without involvement of the ER or the Golgi apparatus [25]. The secretion of acyl-coA-binding protein (ACB1) into the apoplast which is then converted to a spore differentiation factor in protists is a primary example for this behavior [25]. ACB1 does not feature a signal peptide and is therefore not translocated across the ER membrane into the lumen; its transport across the plasma membrane therefore must rely on an unconventional secretion mechanism. ACB1 secretion involves Golgi-associated proteins, even though the protein does not pass through the Golgi cisternae. In mammalian and yeast cells, the Golgi reassembly stacking protein (GRASP) plays a crucial role in both the secretion of proteins carrying a signal peptide bypassing the Golgi and the secretion of cytoplasmic proteins [26]. GRASP is localized at the periphery of the Golgi apparatus and, under special conditions, at the vesicle and plasma membranes. Studies in *Drosophila* have established that GRASP aids in plasma membrane tethering and/or attachment of vesicles destined for secretion [27]. Further studies have found that cells under tension show a marked increase in UPS, concurring with an upregulation of GRASP and the secretion of transmembrane proteins bypassing the Golgi apparatus [27, 28].

3   Conventional Secretion in Plants

Conventional secretion in plants begins at the rough ER. Secreted proteins are first translocated across the ER membrane by the attached ribosomes, utilizing the signal peptide in the protein sequences. Once the proteins have entered the ER lumen the signal peptides are cleaved off. These proteins are then transported through vesicles featuring specific coat proteins from the ER to the Golgi apparatus. Inside the Golgi, the proteins undergo additional folding and posttranslational modifications such as glycosylation [29, 30]. Secreted proteins are then packaged into *trans*-Golgi secretory vesicles, and following vesicle fusion with the plasma membrane delivered to the apoplast. Various fusion and docking-related proteins are involved along the secretory pathway from the ER to the plasma membrane, aiding in vesicle tethering, docking, and fusion with each successive target membrane compartment [4, 31, 32]. Vesicle fusion and the subsequent delivery of proteins to the apoplast share mechanistic elements between the conventional and unconventional pathways [7, 8].

4   UPS in Plants

With the insight that many secreted plant proteins do not feature a signal peptide, attention towards plant UPS has increased [6–8]. Similar to mammalian and yeast cells, plant cells secrete proteins in a variety of unconventional pathways. The major differences in plant cell UPS lie in the large variety of proteins and protein subunits involved and the unique secretion constraints within the cell. Given the sessile nature of plants, requiring in situ adaptation to their surroundings and environment, a higher degree of flexibility in their response might be required extending to the secretion pathways. On the basis of the large number of proteins lacking a signal peptide in plants, one is tempted to speculate that what is termed "unconventional protein secretion" may not be that unconventional after all. Two intriguing questions arising from this are the following: What are the triggering mechanisms for protein secretion along an unconventional route, and what is the evolutionary advantage?

5   Bypassing the Golgi

The Golgi apparatus is the hub of protein trafficking in the eukaryotic cell. Plant cells rely on the Golgi apparatus for a variety of vital processes, including the maintenance and buildup of the cell wall and packaging and glycosylation of proteins prior to trafficking to their destination [33, 34]. However, there are examples of proteins bypassing the Golgi apparatus on their pathway to the plasma membrane and therefore lack the posttranslational modifications occurring there. These alternative secretion pathways can be teased apart with the aid of chemicals disrupting specific aspects of the pathway. For example, brefeldin A (BFA) inhibits Golgi-mediated protein trafficking, but it does not affect the trafficking of proteins that bypass the Golgi [35, 36]. Two proteins secreted independently of the Golgi are mannitol dehydrogenase (MDH) and hygromycin phosphotransferase (HYG^R) (Table 1; Fig. 1). These two cytosolic proteins lack a signal sequence. Further, BFA treatment has no effect on the secretion of these proteins to the apoplast, indicating that they bypass the Golgi apparatus [17, 37]. It is hypothesized that these cytosolic proteins are directly translocated across the plasma membrane. Interestingly, even though the secretion of these proteins occurs independent of the Golgi apparatus, their secretion nonetheless requires the activity of several Golgi-localized proteins for membrane fusion, such as the synaptotagmin 2 (SYT2). Knockout mutations of the SYT2 gene in *Arabidopsis* prevent the secretion of HYG^R [37]. Instead, in the *syt2* mutant HYG^R is redirected to the prevacuolar compartment and then presumably degraded in the vacuole, preventing detoxification of hygromycin B in the plant [37] (Table 1; Fig. 1).

Table 1
Unconventionally secreted plant proteins

Plant UPS protein cargo	UPS pathway	Dominant associated (regulatory) partners	Localization of associated proteins	References
1. Mannitol dehydrogenase (MDH)	Golgi independent			[17]
2. Hygromycin phosphotransferase (HYG)	Golgi independent	Synaptotagmin 2	Golgi	[37]
3. Aleurain (ALEU)	Central vacuole fusion with PM	PBA1	Proteasome	[24]
4. Carboxypeptidase Y (CPY)	Central vacuole fusion with PM	PBA1	Proteasome	[24]
5. Aspartyl protease (AP)	Central vacuole fusion with PM	PBA1	Proteasome	[24]
6. Glucan synthase like 5 (PMR4/GSL5)	MVB (exosomes)	PEN1, SNAP33	PM	[16, 19, 41]
7. *Helianthus annuus* jacalin (Helja)	MVB-like			[42]
8. S-adenosylmethionine synthetase (SAMS2)	EXPO	Exo70E2	EXPO	[48]
9. Arabinogalactan glycosyltransferase (AG)	EXPO and uncharacterized endomembrane compartment		EXPO and small uncharacterized compartments	[49]
10. UDP-glucuronate epimerase 1 and 6 (GAE1, GAE6)	EXPO	Exo70E2	EXPO	[51]
11. Apyrase 3 (AP3)	EXPO	Exo70E2	EXPO	[51]

Listed are plant proteins with known unconventional secretion pathways. These include Golgi bypass, multivesicular body (MVB) or EXPO mediated, and central vacuole/plasma membrane fusion pathways. Associated partners and their localization are included

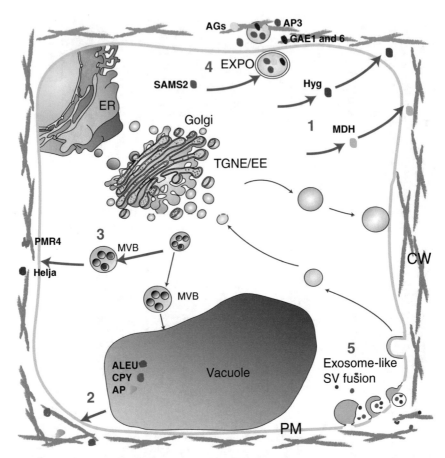

Fig. 1 Model of unconventional secretion pathways in plant cells. Pathways marked by *large red arrows* and *numbers* are unconventional secretion pathways. The identified proteins that travel along these pathways are labeled in *bold*, as indicated in Table 1. The main plant UPS pathways are (*1*) direct translocation of cytosolic proteins across the PM, (*2*) vacuolar fusion with the PM, (*3*) MVB fusion with the PM, (*4*) double-membrane EXPO releasing intraluminal vesicles into the apoplast, and (*5*) exosome-like secretory vesicle (SV) fusion with the PM. *Small blue arrows* denote canonical secretion and endocytosis pathways. *PM* plasma membrane, *CW* cell wall, *ER* endoplasmic reticulum, *TGNE/EE trans*-Golgi network/early endosome, *SV* secretory vesicle, *MVB* multivesicular body, *EXPO* exocyst-positive organelle

Mannitol dehydrogenase (MDH) is a mannitol catabolic enzyme highly expressed and secreted in response to pathogenic or stress signals, such as fungal elicitors or salicylic acid [17] (Table 1; Fig. 1). Following salicylic acid treatment, the leaderless MDH enzyme is highly abundant in the apoplast, but is not observed at the Golgi. Its apoplastic localization remains unaffected by BFA treatment, suggesting that the secretion of MDH is independent of the Golgi [17]. The secretion of pathogenic response proteins via unconventional routes raises interesting questions about the cellular dynamics during pathogen attack.

6   Organelle Fusion with the Plasma Membrane

In addition to the unconventional secretion pathways outlined thus far, fusion of organelles directly to the plasma membrane directs material to the apoplast. One prominent example of this behavior involves the fusion of the vacuole with the plasma membrane under attack from bacterial pathogens [18]. Proteins traveling to the apoplast along this route are typically trafficked first into the vacuole along the conventional secretion route (directed by a secretory and vacuolar sorting sequence), and then secreted to the apoplast unconventionally by membrane fusion between the tonoplast and the plasma membrane. Vacuoles accommodate degradative proteins and hydrolytic enzymes. Under pathogen attack, these proteins can be released upon tonoplast collapse or fusion with the plasma membrane. Following the vacuole's fusion with the plasma membrane, hydrolytic enzymes and antimicrobial proteins are released into the bacteria-occupied apoplast [18], with the consequence that both the bacteria and the plant cell die, due to a lack of protease discrimination. Programmed cell death (PCD) follows this pattern of tonoplast fusion or breakdown, in an attempt to control the extracellular proliferation of pathogens. PCD by tonoplast collapse involves caspase-1-like activity and is more closely associated with viral pathogenesis since the vacuolar enzymes are released directly into the cytoplasm [24]. Fusion of the vacuole with the plasma membrane also causes PCD, targeting the apoplast where bacterial pathogens aggregate.

The vacuole-plasma membrane fusion mechanism is likely coordinated by the degradation of some, yet unidentified, component of the proteasome complex involving the activity of a caspase-3-like subunit (PBA1) of the proteasome in plants. Inactivation of PBA1 inhibits the fusion between the tonoplast and the plasma membrane and therefore prevents the release of vacuolar proteins into the apoplast [24]. The dependence on the proteasome for membrane fusion between the tonoplast and the plasma membrane in bacteria-infected cells indicates a negative regulation of membrane fusion that is disabled upon infection. Support for this hypothesis is provided by the presence of hydrolytic and other vacuole-localized proteases, such as aleurain, carboxypeptidase Y, and aspartyl protease, in the extracellular fluid of *Arabidopsis* upon bacterial pathogen infection [24] (Table 1; Fig. 1).

7   MVB Fusion with the Plasma Membrane

While multivesicular bodies (MVBs), in plants also referred to as prevacuolar compartments, are considered most closely associated with cargo delivery to the vacuole, there is also evidence that MVBs are involved in unconventional secretion. MVBs can release their intraluminal vesicles (or exosomes) into the apoplast following fusion with the plasma membrane. MVB fusion with the plasma membrane is observed during pathogen attack and is also involved in the delivery of lectins. Following fungal pathogen attack, plasma membrane proteins involved in papillae formation and vesicle fusion are found both in MVBs and at the location of the infection. These proteins include the stress-induced callose synthase PMR4/GSL5 (Table 1; Fig. 1), the GTPase ARA6, and the SNARE proteins PEN1 and SNAP33 [16, 38–40]. It is thought that the membrane components present at the site of the fungal infection accompanying papillae formation are contributed by MVB fusion and intraluminal vesicle release [38]. The presence of plasma membrane proteins like PMR4 in MVBs during fungal attack suggests that MVBs are also involved in the recycling of certain plasma membrane proteins to the site of the infection [16, 19, 41]. BFA treatment substantially inhibits this mechanism of secretion, which is not surprising, given that MVBs must first endocytose the plasma membrane-localized PMR4 (and/or other callose deposition-related proteins) in order to deliver it to the site of the infection. However, the lack of complete susceptibility to fungal pathogen attack following BFA treatment suggests that other non-plasma membrane-localized proteins can be secreted by MVBs, including cytoplasmic proteins following invagination and cytosol capture [16].

In addition to cell defense-related secretion, MVBs are also involved in the secretion of lectins such as the jacalin-related lectin protein Helja (Table 1; Fig. 1). Helja is involved in a variety of processes including storage, stress defense, and plant development; it has been observed by immunolocalization in apoplastic fluid [42]. Given its lack of a signal peptide, its presence in the apoplast suggests an unconventional secretion route. In a series of merolectins sharing a lack of the signal peptide whose apoplastic location is predicated from data mining, Helja is the first lectin experimentally identified in the apoplast. Helja has the potential to function as a probe in bioinformatics analysis towards identification of unconventionally secreted lectins with a shared sequence [42, 43].

8 Exocyst-Positive Organelle (EXPO)

The yeast and mammalian exocyst complex has been extensively studied [44–46], whereas plant exocyst proteins have only recently gained attention. Exocyst complexes are composed of eight subunits that are responsible for the coordination of post-Golgi vesicle fusion with the plasma membrane [47]. Immunolabeling studies showed the exocyst subunit Exo70E2 at discrete punctate outside the plasma membrane, as well as at the unique double-membraned *exo*cyst-*p*ositive *o*rganelle (EXPO), and further showed that its localization pattern is unperturbed by trafficking inhibitors such as BFA and wortmannin [13, 48, 49]. EXPO is unique among trafficking compartments; it fuses with the plasma membrane to release cargo-containing vesicles into the apoplast [50]. To date, several proteins localized in EXPO have been identified [48, 49, 51–53]. One cargo protein of this new compartment is S-adenosylmethionine synthetase 2 (SAMS2), a leaderless cytosolic protein involved in lignin methylation [48] (Table 1; Fig. 1). Although SAMS2 has not been experimentally identified in the apoplast, proteomic analysis indicates its presence in *Arabidopsis* secondary cell walls. Colocalization studies with Exo70E2 in *Arabidopsis* protoplasts suggest that SAMS2 is likely engulfed by EXPO membranes prior to its release into the apoplast via exosomes [48]. Additionally plant enzymes involved in glycosylation, nucleotide sugar metabolism, and epimerization have been identified in the EXPO. For example, arabinogalactan glycosyltransferase localizes at the EXPO when transiently expressed in *N. benthamiana* leaf epidermal cells [49] (Table 1; Fig. 1). The presence of glycosylation enzymes in these small compartments (EXPO), beyond their presence in the ER and Golgi, suggests that there might be other avenues for the glycosylation of UPS proteins. Since the initial discovery of AGs in the EXPO, additional proteins under transient expression have been found in this compartment: two UDP-glucuronate epimerases and an apyrase colocalized with the AGs and Exo70E2 [51] (Table 1; Fig. 1).

One distinguishing aspect of the plant exocyst, most notably in the Exo70 subunits, is the large expansion of the subunits making up the complex. There are 23 paralogs of Exo70 in *Arabidopsis* and 47 in rice, which stands in strong contrast to a single copy in animal and yeast cells [54, 55]. Interestingly, yeast and animal exocyst subunit homologs expressed in plants do not induce the formation of EXPO, whereas plant subunits expressed in animal cells do [52]. Taken together, these observations support the hypotheses that plant exocyst subunits have additional or altered roles compared to their homologs in animal and yeast cells, potentially reflecting the unique plant-specific secretion challenges.

Given that autophagosomes and EXPO exhibit similar morphological characteristics, specifically the double membrane and the fact that an exocyst subunit (Exo70B1) localizes in autophagosomes, it has been hypothesized that autophagosomes and EXPO are the same cellular compartment [53]. This hypothesis is currently strongly contested [50, 52]. Electron microscopy immunolabeling studies indicate that the two compartments are distinct; autophagosomes fuse with the lytic vacuole and EXPO fuses exclusively with the plasma membrane [48]. Further support towards their distinct identities comes from the lack of EXPO induction upon sucrose starvation, a situation which generally induces the formation of autophagosomes, and the lack of colocalization with autophagosome markers [50]. However, there remains lingering doubt about the distinctness of EXPO from autophagosomes, given their morphological resemblance and common process of formation involving cytosol engulfment and membrane closure [48, 53].

9   Exosome-Like Carriers Produced at the Plasma Membrane by Secretory Vesicle Fusion

In plants, turgor pressure exerts forces on secretory vesicles (SVs) in the cytosol and on the plasma membrane. These forces can potentially play a role in defining the membrane curvature and the formation of exosome-like vesicles. Transmission electron microscopy (TEM) and freeze-fracture electron microscopy micrographs have shown horseshoe-shaped fusion intermediates of SVs in tobacco root cap cells, sycamore maple suspension culture cells, and developing pine tree xylem cells [56, 57]. The squeezed SVs (Fig. 2a) appear to fuse multiple times, forming exosome-like vesicles, with molecules inside the cytosolic volume, as well as molecules associated with the inner membrane, being delivered to the apoplast.

In mucilage-secreting root cap cells, large Golgi-derived SVs are abundant. Their fusion with the plasma membrane is easily identified in TEM micrographs, probably due to their large size and hence longer content discharging period compared to other cell types. Cytoplasm was captured during fusion of SVs in alfalfa root cap cells (Fig. 2b). These observations lead to the hypothesis that the exosome-like structures form when SVs are squeezed by turgor pressure and merge with each other before they complete the release of their mucilage cargo (Fig. 2c).

Another possibility is the formation of exosome-like structures from secretory vesicle clusters. Mobile compartments labeled by the secretory carrier membrane protein 2, carrying pectin cell wall polysaccharides, were identified in Bright-Yellow-2 tobacco cells [58]. They consist of vesicles and connecting tubules and were thus named secretory vesicle clusters (SVCs). These clusters correspond to the SV-rich-type *trans*-Golgi network compartments that eventually fragment to produce SVs in *Arabidopsis* [59, 60]. Interestingly, it was shown that SVCs fuse with the plasma membrane prior to their fragmentation (Fig. 2d). Since SVCs contain many SVs, of which each can fuse with the plasma membrane, the SVC membrane and the cytosol within the fusion sites can possibly form an exosome-like compartment.

EXPOs have been observed to generate exosome-like apoplastic vesicles in *Arabidopsis* and BY-2 cells [66]. The SV-derived exosome-like carriers differ from EXPOs in two major ways: (a) the double-membrane envelope of EXPOs is complete prior to their contact with the plasma membrane, while SVs fuse with the plasma membrane prior to packaging molecules associated with their cytosolic surface into the apoplastic vesicles; (b) EXPOs are larger (>200 nm) than SVs (<100 nm).

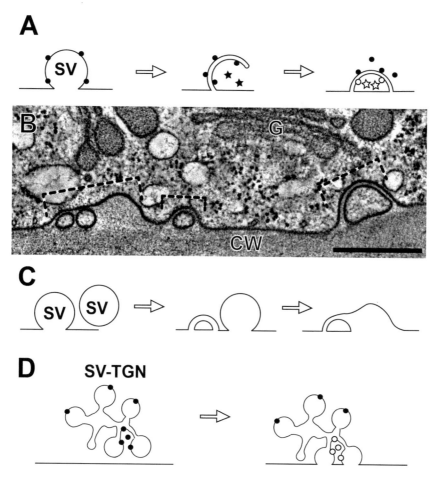

Fig. 2 Models depicting how secretory vesicle (SV) fusions can produce exosome-like carriers at the plasma membrane. (**a**) After an SV makes contact with the plasma membrane, the SV is compressed by turgor pressure. The in-folded PM originates from the SV and fuses with the PM to isolate membrane-bound molecules (*black circles*) and cytosolic molecules (*black stars*) in a membrane-bound compartment (*white circles* and *stars*). (**b**) A single image from an alfalfa root cap cell electron tomogram, showing three exosome-like carriers (*brackets*). The SV-1 has two exosome-like carriers that have potentially formed by the mechanisms illustrated in panels (**a**) and (**c**). (**c**) SVs grow at the plasma membrane by lateral fusion. (**d**) An SV-rich TGN (SV-TGN) fuses with the plasma membrane prior to fragmentation, enclosing cytosolic volumes and some of its membrane proteins (*white circles*) in an exosome-like carrier. *CW* cell wall, *SV* secretory vesicle, *TGN trans*-Golgi network, *G* Golgi apparatus. Scale bar: 500 nm

The proposed mechanisms for exosome-like carrier formation from SVs share a common feature; SVs connect multiple times to the plasma membrane prior to their complete absorption. Verifying the proposed novel mode of secretion would require characterization of the intermediate structures of the fusion processes and analysis of the cargo molecules inside the exosome-like carriers. The former could be achieved by skillful electron/immuno-electron tomography in combination with serial sectioning. The fast-paced SV fusion requires high-pressure freezing of samples to ensure accurate feature preservation [61]. The fusion intermediates have complex architectures whose interpretation from a single electron micrograph is challenging. Three-dimensional imaging involving serial sectioning and electron tomography in combination with immunolabeling is required for unambiguous determination of membrane topology and cargo in the transitional structures. Increasing widespread availability and adaptation of these advanced TEM techniques will probably contribute to the identification of additional vesicle fusion events producing exosome-like carriers.

10   Emerging Technologies for Elucidating Unconventional Secretion

10.1   Chemical Inhibition of Trafficking Pathways

A major strategy in dissecting the various trafficking pathways is their time point-specific disruption with small molecules [62]. The effectiveness of this method relies on the specificity of the chemical for the targeted event. Chemical genomics, in which large chemical libraries are screened for their activity on specific proteins and cellular pathways, is a way to identify suitable chemical/target combinations. Similar to forward genetics, in which a gene is mutated causing an aberrant phenotype, chemicals can achieve the same goal while offering several additional levels of experimental control while circumventing lethality issues: dose dependence, timing, and duration. The reversible nature of the cellular effect upon chemical degradation or removal can further provide information about the recovery mechanism and its kinetics [62–64].

Chemical genomics has identified a number of small molecules affecting endomembrane trafficking, among which BFA is probably the most well-known and most frequently used one. It disrupts Golgi-dependent secretion pathways in the cell and induces Golgi-derived vesicle aggregation into "BFA bodies" [35, 36, 65]. BFA is particularly useful in UPS studies, since a common characteristic of most UPS pathways is Golgi independence and therefore BFA treatment tolerance. Wortmannin is another agent used to dissect trafficking pathways, inducing homotypic MVB fusion in plant cells by targeting endosomal membrane lipids [66]. In recent screens, a new set of promising compounds affecting endomembrane trafficking in plants were identified [64, 67–69]. Endosidin 3 can distinguish between antagonistic signaling pathways controlling cell shape formation [67]. Endosidin 7, a cytokinesis inhibitor, differentially affects the arrival and fusion of cell plate vesicles during cytokinesis, including the delivery of two closely related populations of Rab GTPases at the cell plate [70, 71]. Endosidin 8 affects an early secretory pathway essential for basal polarity establishment in *Arabidopsis* [68]. C834 distinguishes a Golgi-independent route of tonoplast protein trafficking [72]. Chemical inhibition of trafficking pathways has great potential in our quest towards understanding unconventional pathways in the highly complex endomembrane trafficking network. Plants with their sessile lifestyle and hence strong coordination of trafficking events during adaptation to the local environment are prime study targets for this approach.

10.2 Subcellular Proteomics

A direct approach towards identification of cargo delivered through UPS is the isolation of the compartments involved and proteomic analysis of their contents. Analyzing the proteomes of distinct organelle and vesicle populations *en route* can dissect protein transport routes and potentially contribute to an understanding of UPS induction. To date only a limited number of subcellular vesicle proteomes have been analyzed. These include the isolation and characterization of *trans*-Golgi vesicles marked by the syntaxin SYP61 and VHA-a1, selected endosomal compartments, and the Golgi apparatus [73–79].

Proteomic analysis of SYP61 vesicles revealed the presence of cellulose synthase proteins and the SYP121 SNARE complex, which has an ascribed role in fungal infection response [80]. Taken together, this suggests the involvement of the SYP61 compartment in CESA recycling and biotic stress responses [73]. A comparison of the Golgi, *trans*-Golgi vesicle, and apoplastic proteomes can identify cargo that follows a Golgi and TGN-independent UPS secretion route. This approach can be extended towards the cargo characterization of specific UPS compartments, such as EXPO, and elucidate their biological role.

11   Conclusion and Perspectives

UPS is a nascent research field in plant trafficking research, addressing how the outstanding number of leaderless proteins in plants reach their extracellular destination. Advances in this area will provide many novel insights into plant growth, development, and pathogen and stress response. Given the variety of UPS pathways, each with its various permutations, it is likely that there are still undiscovered and unobserved UPS pathways in plants. Much remains to be understood regarding the various conditions that induce UPS and the apparent benefit these pathways offer the plant over traditional secretion.

So far, the majority of known examples of unconventionally secreted proteins in plants are related to stress and/or pathogenesis responses. The advantage of this secretion pathway over the canonical trafficking route in response to stress is yet unknown. One could speculate that under stress these specialized types of secretion might act as a broad defense strategy allowing for a more flexible, adaptive, and less specific response to environmental stresses. Knowledge of different secretion pathways, their regulation, and biological role with respect to their cargo is vital for a comprehensive understanding of plant cell interactions and responses to different stimuli.

Acknowledgements

This work was supported by the NSF-IOS-1258135 to G.D. and the Area of Excellence grant (AoE/M-05/12) from the Research Grants Council of Hong Kong to B.K.

References

1. Nickel W (2010) Pathways of unconventional protein secretion. Curr Opin Biotechnol 21:621–626. doi:10.1016/j.copbio.2010.06.004

2. Lee MCS, Miller EA, Goldberg J et al (2004) Bi-directional protein transport between the ER and Golgi. Annu Rev Cell Dev Biol 20:87–123. doi:10.1146/annurev.cellbio.20.010403.105307

3. Park M, Jürgens G (2011) Membrane traffic and fusion at post-Golgi compartments. Front Plant Sci 2:111. doi:10.3389/fpls.2011.00111

4. Alberts B, Johnson A, Lewis J et al (2008) Molecular biology of the cell, 5th edn. Garland Science, New York, NY

5. Blobel G, Dobberstein B (1975) Transfer of proteins across membranes. I. Presence of pro-teolytically processed and unprocessed nascent immunoglobulin light chains on membrane-bound ribosomes of murine myeloma. J Cell Biol 67:835–851

6. Ding Y, Wang J, Wang J et al (2012) Unconventional protein secretion. Trends Plant Sci 17:606–615. doi:10.1016/j.tplants.2012.06.004

7. Drakakaki G, Dandekar A (2013) Protein secretion: how many secretory routes does a plant cell have? Plant Sci 203–204:74–78. doi:10.1016/j.plantsci.2012.12.017

8. Ding Y, Robinson DG, Jiang L (2014) Unconventional protein secretion (UPS) path-ways in plants. Curr Opin Cell Biol 29:107–115. doi:10.1016/j.ceb.2014.05.008

9. Agrawal GK, Jwa N-S, Lebrun M-H et al (2010) Plant secretome: unlocking secrets of the secreted proteins. Proteomics 10:799–827. doi:10.1002/pmic.200900514

10. Krause C, Richter S, Knöll C, Jürgens G (2013) Plant secretome – from cellular process to biologi-cal activity. Biochim Biophys Acta 1834:2429–2441. doi:10.1016/j.bbapap.2013.03.024

11. Albenne C, Canut H, Jamet E (2013) Plant cell wall proteomics: the leadership of Arabidopsis thaliana. Front Plant Sci 4:111. doi:10.3389/fpls.2013.00111

12. Stegmann M, Anderson RG, Westphal L et al (2013) The exocyst subunit Exo70B1 is involved in the immune response of Arabidopsis thaliana to different pathogens and cell death.

Plant Signal Behav 8, e27421. doi:10.4161/psb.27421

13. Chen X, Ebbole DJ, Wang Z (2015) The exo-cyst complex: delivery hub for morphogenesis and pathogenesis in filamentous fungi. Curr Opin Plant Biol 28:48–54. doi:10.1016/j.pbi.2015.09.003

14. Micali CO, Neumann U, Grunewald D et al (2011) Biogenesis of a specialized plant-fungal interface during host cell internalization of Golovinomyces orontii haustoria. Cell Microbiol 13:210–226. doi:10.1111/j.1462-5822.2010.01530.x

15. Bednarek P, Kwon C, Schulze-Lefert P (2010) Not a peripheral issue: secretion in plant-microbe interactions. Curr Opin Plant Biol 13:378–387. doi:10.1016/j.pbi.2010.05.002

16. Meyer D, Pajonk S, Micali C et al (2009) Extracellular transport and integration of plant secretory proteins into pathogen-induced cell wall compartments. Plant J 57:986–999. doi:10.1111/j.1365-313X.2008.03743.x

17. Cheng F, Zamski E, Guo W et al (2009) Salicylic acid stimulates secretion of the normally sym-plastic enzyme mannitol dehydrogenase: a possible defense against mannitol-secreting fungal pathogens. Planta 230:1093–1103. doi:10.1007/s00425-009-1006-3

18. Hatsugai N, Hara-Nishimura I (2010) Two vacuole-mediated defense strategies in plants. Plant Signal Behav 5:1568–1570. doi:10.4161/psb.5.12.13319

19. An Q, Hückelhoven R, Kogel K-H, van Bel AJE (2006) Multivesicular bodies participate in a cell wall-associated defence response in bar-ley leaves attacked by the pathogenic powdery mildew fungus. Cell Microbiol 8:1009–1019. doi:10.1111/j.1462-5822.2006.00683.x

20. Torrado LC, Temmerman K, Müller H-M et al (2009) An intrinsic quality-control mechanism ensures unconventional secretion of fibroblast growth factor 2 in a folded conformation. J Cell Sci 122:3322–3329. doi:10.1242/jcs.049791

21. La Venuta G, Zeitler M, Steringer JP et al (2015) The startling properties of fibroblast growth factor 2: how to exit mammalian cells without a signal peptide at hand? J Biol Chem. doi:10.1074/jbc.R115.689257

22. Temmerman K, Ebert AD, Müller H-M et al (2008) A direct role for phosphatidylinositol-4,5-bisphosphate in unconventional secretion of fibroblast growth factor 2. Traffic 9:1204–1217. doi:10.1111/j.1600-0854.2008.00749.x

23. Nickel W (2011) The unconventional secretory machinery of fibroblast growth factor 2. Traffic 12:799–805. doi:10.1111/j.1600-0854.2011.01187.x

24. Hatsugai N, Iwasaki S, Tamura K et al (2009) A novel membrane fusion-mediated plant immunity against bacterial pathogens. Genes Dev 23:2496–2506. doi:10.1101/gad.1825209

25. Cabral M, Anjard C, Malhotra V et al (2010) Unconventional secretion of AcbA in Dictyostelium discoideum through a vesicular intermediate. Eukaryot Cell 9:1009–1017. doi:10.1128/EC.00337-09

26. Kinseth MA, Anjard C, Fuller D et al (2007) The Golgi-associated protein GRASP is required for unconventional protein secretion during development. Cell 130:524–534. doi:10.1016/j.cell.2007.06.029

27. Schotman H, Karhinen L, Rabouille C (2008) dGRASP-mediated noncanonical integrin secretion is required for Drosophila epithelial remodeling. Dev Cell 14:171–182. doi:10.1016/j.devcel.2007.12.006

28. Giuliani F, Grieve A, Rabouille C (2011) Unconventional secretion: a stress on GRASP. Curr Opin Cell Biol 23:498–504. doi:10.1016/j.ceb.2011.04.005

29. Dupree P, Sherrier DJ (1998) The plant Golgi apparatus. Biochim Biophys Acta 1404:259–270

30. Schoberer J, Strasser R (2011) Subcompartmental organization of Golgi-resident N-glycan processing enzymes in plants. Mol Plant 4:220–228. doi:10.1093/mp/ssq082

31. Hwang I, Robinson DG (2009) Transport vesicle formation in plant cells. Curr Opin Plant Biol 12:660–669. doi:10.1016/j.pbi.2009.09.012

32. Surpin M, Raikhel N (2004) Traffic jams affect plant development and signal transduction. Nat Rev Mol Cell Biol 5:100–109. doi:10.1038/nrm1311

33. Faso C, Boulaflous A, Brandizzi F (2009) The plant Golgi apparatus: last 10 years of answered and open questions. FEBS Lett 583:3752–3757. doi:10.1016/j.febslet.2009.09.046

34. Worden N, Park E, Drakakaki G (2012) Trans-Golgi network: an intersection of trafficking cell wall components. J Integr Plant Biol 54:875–886. doi:10.1111/j.1744-7909.2012.01179.x

35. Staehelin LA, Driouich A (1997) Brefeldin A effects in plants (are different Golgi responses caused by different sites of action?). Plant Physiol 114:401–403

36. Nebenführ A, Ritzenthaler C, Robinson DG (2002) Brefeldin A: deciphering an enigmatic inhibitor of secretion. Plant Physiol 130:1102–1108. doi:10.1104/pp.011569

37. Zhang H, Zhang L, Gao B et al (2011) Golgi apparatus-localized synaptotagmin 2 is required for unconventional secretion in Arabidopsis. PLoS One 6, e26477. doi:10.1371/journal.pone.0026477

38. Nielsen ME, Feechan A, Böhlenius H et al (2012) Arabidopsis ARF-GTP exchange factor, GNOM, mediates transport required for innate immunity and focal accumulation of syntaxin PEN1. Proc Natl Acad Sci U S A 109:11443–11448. doi:10.1073/pnas.1117596109

39. Ueda T, Yamaguchi M, Uchimiya H, Nakano A (2001) Ara6, a plant-unique novel type Rab GTPase, functions in the endocytic pathway of Arabidopsis thaliana. EMBO J 20:4730–4741. doi:10.1093/emboj/20.17.4730

40. Nielsen ME, Thordal-Christensen H (2013) Transcytosis shuts the door for an unwanted guest. Trends Plant Sci 18:611–616. doi:10.1016/j.tplants.2013.06.002

41. An Q, Ehlers K, Kogel K-H et al (2006) Multivesicular compartments proliferate in susceptible and resistant MLA12-barley leaves in response to infection by the biotrophic powdery mildew fungus. New Phytol 172:563–576. doi:10.1111/j.1469-8137.2006.01844.x

42. Pinedo M, Orts F, Carvalho AO et al (2015) Molecular characterization of Helja, an extracellular jacalin-related protein from Helianthus annuus: Insights into the relationship of this protein with unconventionally secreted lectins. J Plant Physiol 183:144–153. doi:10.1016/j.jplph.2015.06.004

43. Pinedo M, Regente M, Elizalde M et al (2012) Extracellular sunflower proteins: evidence on non-classical secretion of a jacalin-related lectin. Protein Pept Lett 19:270–276

44. TerBush DR, Maurice T, Roth D, Novick P (1996) The exocyst is a multiprotein complex required for exocytosis in Saccharomyces cerevisiae. EMBO J 15:6483–6494

45. Murthy M, Garza D, Scheller RH, Schwarz TL (2003) Mutations in the exocyst component Sec5 disrupt neuronal membrane traffic, but neurotransmitter release persists. Neuron 37:433–447

46. Ting AE, Hazuka CD, Hsu SC et al (1995) rSec6 and rSec8, mammalian homologs of yeast proteins essential for secretion. Proc Natl Acad Sci U S A 92:9613–9617

47. Hála M, Cole R, Synek L et al (2008) An exocyst complex functions in plant cell growth in Arabidopsis and tobacco. Plant Cell 20:1330–1345. doi:10.1105/tpc.108.059105

48. Wang J, Ding Y, Wang J et al (2010) EXPO, an exocyst-positive organelle distinct from multivesicular endosomes and autophagosomes, mediates cytosol to cell wall exocytosis in Arabidopsis and tobacco cells. Plant Cell 22:4009–4030. doi:10.1105/tpc.110.080697

49. Poulsen CP, Dilokpimol A, Mouille G et al (2014) Arabinogalactan glycosyltransferases target to a unique subcellular compartment that may function in unconventional secretion in plants. Traffic 15:1219–1234. doi:10.1111/tra.12203

50. Lin Y, Ding Y, Wang J et al (2015) Exocyst-positive organelles and autophagosomes are distinct organelles in plants. Plant Physiol 169:1917–1932. doi:10.1104/pp.15.00953

51. Poulsen CP, Dilokpimol A, Geshi N (2015) Arabinogalactan biosynthesis: Implication of AtGALT29A enzyme activity regulated by phosphorylation and co-localized enzymes for nucleotide sugar metabolism in the compartments outside of the Golgi apparatus. Plant Signal Behav 10, e984524. doi:10.4161/15592324.2014.984524

52. Ding Y, Wang J, Chun Lai JH et al (2014) Exo70E2 is essential for exocyst subunit recruitment and EXPO formation in both plants and animals. Mol Biol Cell 25:412–426. doi:10.1091/mbc.E13-10-0586

53. Kulich I, Pečenková T, Sekereš J et al (2013) Arabidopsis exocyst subcomplex containing subunit EXO70B1 is involved in autophagy-related transport to the vacuole. Traffic 14:1155–1165. doi:10.1111/tra.12101

54. Vukašinović N, Cvrčková F, Eliáš M et al (2014) Dissecting a hidden gene duplication: the Arabidopsis thaliana SEC10 locus. PLoS One 9, e94077. doi:10.1371/journal.pone.0094077

55. Cvrčková F, Grunt M, Bezvoda R et al (2012) Evolution of the land plant exocyst complexes. Front Plant Sci 3:159. doi:10.3389/fpls.2012.00159

56. Staehelin LA, Giddings TH, Kiss JZ, Sack FD (1990) Macromolecular differentiation of Golgi stacks in root tips of Arabidopsis and Nicotiana seedlings as visualized in high pressure frozen and freeze-substituted samples. Protoplasma 157:75–91

57. Samuels AL, Rensing KH, Douglas CJ et al (2002) Cellular machinery of wood production: differentiation of secondary xylem in Pinus contorta var. latifolia. Planta 216:72–82. doi:10.1007/s00425-002-0884-4

58. Toyooka K, Goto Y, Asatsuma S et al (2009) A mobile secretory vesicle cluster involved in mass transport from the Golgi to the plant cell exterior. Plant Cell 21:1212–1229. doi:10.1105/tpc.108.058933

59. Kang B, Nielsen E, Lai M et al (2011) Electron tomography of RabA4b- and PI-4 K β 1-labeled trans Golgi network compartments in Arabidopsis. Traffic 12:313–329. doi:10.1111/j.1600-0854.2010.01146.x

60. Staehelin LA, Kang B-H (2008) Nanoscale architecture of endoplasmic reticulum export sites and of Golgi membranes as determined by electron tomography. Plant Physiol 147:1454–1468. doi:10.1104/pp.108.120618

61. Kang B-H (2010) Electron microscopy and high-pressure freezing of Arabidopsis. Methods Cell Biol 96:259–283. doi:10.1016/s0091-679x(10)96012-3

62. Drakakaki G, Robert S, Raikhel NV, Hicks GR (2009) Chemical dissection of endosomal pathways. Plant Signal Behav 4(1):57–62. doi:10.1073/pnas.0711650.www.landesbioscience.com

63. Worden N, Girke T, Drakakaki G (2014) Endomembrane dissection using chemically induced bioactive clusters. Methods Mol Biol 1056:159–168. doi:10.1007/978-1-62703-592-7_16

64. Worden N, Wilkop TE, Esteve VE et al (2015) CESA trafficking inhibitor inhibits cellulose deposition and interferes with the trafficking of cellulose synthase complexes and their associated proteins KORRIGAN1 and POM2/cellulose synthase interactive protein 1. Plant Physiol 167:381–393. doi:10.1104/pp.114.249003

65. Klausner RD, Donaldson JG, Lippincott-Schwartz J (1992) Brefeldin A: insights into the control of membrane traffic and organelle structure. J Cell Biol 116:1071–1080

66. Wang J, Cai Y, Miao Y et al (2009) Wortmannin induces homotypic fusion of plant prevacuolar compartments. J Exp Bot 60:3075–3083. doi:10.1093/jxb/erp136

67. Drakakaki G, Robert S, Szatmari A-M et al (2011) Clusters of bioactive compounds target dynamic endomembrane networks in vivo. Proc Natl Acad Sci U S A 108:17850–17855. doi:10.1073/pnas.1108581108

68. Doyle SM, Haeger A, Vain T et al (2015) An early secretory pathway mediated by GNOM-LIKE 1 and GNOM is essential for basal polarity establishment in Arabidopsis thaliana. Proc Natl Acad Sci U S A 112:E806–E815. doi:10.1073/pnas.1424856112

69. Robert S, Chary SN, Drakakaki G et al (2008) Endosidin1 defines a compartment involved in endocytosis of the brassinosteroid receptor BRI1 and the auxin transporters PIN2 and

AUX1. Proc Natl Acad Sci U S A 105:8464–8469. doi:10.1073/pnas.0711650105

70. Davis DJ, McDowell S, Drakakaki G (2014) The RAB GTPases RABA5d and RABA1e localize to the cell plate and display distinct patterns upon exposure to the cytokinesis inhibitor ES7. Plant Signal Behav 10(3):e984520

71. Park E, Díaz-moreno SM, Davis DJ et al (2014) Endosidin 7 specifically arrests late cytokinesis and inhibits callose biosynthesis, revealing distinct trafficking events during cell plate maturation. Plant Physiol 165:1019–1034. doi:10.1104/pp.114.241497

72. Rivera-Serrano EE, Rodriguez-Welsh MF, Hicks GR, Rojas-Pierce M (2012) A small molecule inhibitor partitions two distinct pathways for trafficking of tonoplast intrinsic proteins in Arabidopsis. PLoS One 7, e44735. doi:10.1371/journal.pone.0044735

73. Drakakaki G, van de Ven W, Pan S et al (2012) Isolation and proteomic analysis of the SYP61 compartment reveal its role in exocytic trafficking in Arabidopsis. Cell Res 22:413–424. doi:10.1038/cr.2011.129

74. Park E, Drakakaki G (2014) Proteomics of endosomal compartments from plants case study: isolation of trans-Golgi network vesicles. Methods Mol Biol 1209:179–187. doi:10.1007/978-1-4939-1420-3_14

75. Nikolovski N, Rubtsov D, Segura MP et al (2012) Putative glycosyltransferases and other plant Golgi apparatus proteins are revealed by LOPIT proteomics. Plant Physiol 160:1037–1051. doi:10.1104/pp.112.204263

76. Groen AJ, Sancho-Andrés G, Breckels LM et al (2014) Identification of trans-Golgi network proteins in Arabidopsis thaliana root tissue. J Proteome Res 13:763–776. doi:10.1021/pr4008464

77. Heard E, Martienssen RA (2014) Transgenerational epigenetic inheritance: myths and mechanisms. Cell 157:95–109. doi:10.1016/j.cell.2014.02.045

78. Parsons HT, Drakakaki G, Heazlewood JL (2012) Proteomic dissection of the Arabidopsis Golgi and trans-Golgi network. Front Plant Sci 3:298. doi:10.3389/fpls.2012.00298

79. Parsons HT, Weinberg CS, Macdonald LJ et al (2013) Golgi enrichment and proteomic analysis of developing Pinus radiata xylem by free-flow electrophoresis. PLoS One 8, e84669. doi:10.1371/journal.pone.0084669

80. Rodrigues ML, Nosanchuk JD, Schrank A et al (2011) Vesicular transport systems in fungi. Future Microbiol 6:1371–1381. doi:10.2217/fmb.11.112

Part II

UPS Contemplates Multidisciplinary Approaches for Plants

Chemical Secretory Pathway Modulation in Plant Protoplasts

Francesca De Marchis, Andrea Pompa, and Michele Bellucci

Abstract

The classical Golgi pathway is not the only mechanism for vacuolar protein transport in plants because alternative transport mechanisms have been described. The existence of these alternative pathways can be demonstrated using several chemicals and here we describe the use of brefeldin A (BFA), endo-β-N-acetylglucosaminidase H (Endo-H), and tunicamycin, on isolated tobacco leaf protoplasts. Two main methods are illustrated in this chapter, protoplast pulse-chase followed by protein immunoprecipitation, and protoplast immunofluorescence.

Key words BFA, Endo-H, Golgi-mediated protein traffic, Pulse-chase, Tobacco, Tunicamycin

1 Introduction

The secretory pathway in eukaryotic cells is an important process that regulates key functions of cell biology by synthesizing and exporting proteins, complex carbohydrates, and lipids from the endoplasmic reticulum (ER) to their target compartments [1]. In plant cells, one of the most studied traffic routes in the secretory pathway is the transfer of proteins from the ER to the vacuole, and vacuolar proteins often reach the vacuole by endomembrane progression through the Golgi using secretory vesicles or direct tubular connections [2–4]. However, this classical Golgi pathway is not the only mechanism for vacuolar protein transport because alternative transport mechanisms are usually activated in particular cellular situations [5, 6]. The formation of protein aggregates is a key factor that appears to characterize these alternative pathways, but soluble or tonoplast membrane proteins can also be directly delivered to the vacuole [7]. Recently, a Golgi-independent vacuolar deliver of a soluble protein, the human lysosomal α-mannosidase, has been revealed in transgenic tobacco cells [8]. The authors have demonstrated, by incubating transgenic leaf protoplasts with both BFA and Endo-H, that the transport of this glycoprotein to the

Andrea Pompa and Francesca De Marchis (eds.), *Unconventional Protein Secretion: Methods and Protocols*, Methods in Molecular Biology, vol. 1459, DOI 10.1007/978-1-4939-3804-9_4, © Springer Science+Business Media New York 2016

vacuole does not involve the Golgi compartment. Moreover, to investigate if human α-mannosidase deliver to the vacuole involves its *N*-linked glycans, they have incubated transformed protoplasts with tunicamycin. Here we describe the protocols that enable protoplast isolation and the modulation of plant secretory pathway using these three chemicals.

2   Materials

2.1   Seed Sterilization and Plant Culture Media

1. Sterile water (*see* **Note 1**).

2. Ethanol 70%.

3. Disinfectant seed solution: 50% Commercial bleach (which contains about 5% sodium hypochlorite) and 0.05% Tween-20.

4. Murashige and Skoog medium (MS), stock solution 10× [9]: Dissolve the ready-to-use powder (commercially available) in 1 L of water and adjust pH between 2.0 and 3.0. Store at 4 °C.

5. MS medium for plant growth in sterile culture (1 L): 100 mL MS stock solution 10×, sucrose 30 g, pH to 5.8. Supplement the liquid medium with 6–8 g of agar and autoclave.

6. Petri dishes (diameter 9 cm) and sterile containers for plant growth.

2.2   Solutions and Media for Protoplast Isolation

1. α-Naphthaleneacetic acid (NAA) 1 mg/mL stock solution: Dissolve 50 mg NAA in 5 mL absolute ethanol with continuous stirring and heating the solution to 40–50 °C. When the solution is clear, add dropwise water to 50 mL. Filter sterilize and store at –20 °C.

2. 6-Benzylaminopurine (BA) 1 mg/mL stock solution: Dissolve as above, but help BA resuspension by adding a few drops of 1 N NaOH. Filter sterilize and store at –20 °C.

3. K3 medium (1 L): 3.78 g Gamborg's B-5 Basal Medium with Minimal Organics [10], 5.1 mM $CaCl_2 \cdot 2H_2O$ (750 mg), 3.12 mM NH_4NO_3 (250 mg), 0.4 M sucrose (136.2 g), 1.67 mM xylose (250 mg), BA 1 mg, NAA 1 mg. Bring to pH 5.5 with a few drops of 1 M KOH. Filter sterilize and store at –20 °C.

4. 10× Enzyme mix: Macerozyme Onozuka R-10 5%, Cellulase Onozuka R-10 10%. Dissolve in K3 medium, stir vigorously for 30 min or more, and collect in 50 mL conical tubes. Spin tubes at $10,000 \times g$ for 15 min to precipitate insoluble materials. Filter sterilize, aliquot, and store at –20 °C (–70 °C for very-long-term storage). Avoid more than two freeze-thaw cycles.

5. W5 medium (1 L): 152 mM NaCl (9 g), 5 mM KCl (0.37 g), 125 mM CaCl$_2$.2H$_2$O (18.37 g), 5 mM glucose (0.9 g). Filter sterilize and store at –20 °C.

6. Fluorescein diacetate (FDA) stock solution: Dissolve 5 mg FDA in 1 mL acetone. Store at –20 °C.

7. K3-FDA solution: Add 10 μL of FDA stock solution to 5 mL K3.

2.3 Protoplast Radiolabeling, Protein Immunoprecipitation, and Endo-H Treatment

1. Bovine serum albumin (BSA) stock solution: Solubilize 4 mg BSA in water and store at –20 °C.

2. Protein labeling mix containing both 35S-methionine and 35S-cysteine (Pro-Mix, GE Healthcare, Little Chalfont, Buckinghamshire, UK).

3. Unlabeled Met/Cys 10× stock solution: Solubilize, in the same tube, in K3 solution first 6.05 mg/mL Cys and then 15 mg/mL Met. Filter sterilize, prepare several aliquots in 1.5 mL tubes, and store at –20 °C (*see* **Note 2**).

4. Protoplast homogenization buffer: 150 mM Tris–HCl, pH 7.5, 150 mM NaCl, 1.5 mM EDTA, 1.5 % Triton X-100. Store at –20 °C in small aliquots. Before use, add immediately 1.5 mM phenylmethylsulfonyl fluoride (PMSF) and a mixture of broad-spectrum protease inhibitors (commercially available from several companies).

5. NET-gel buffer: 50 mM Tris–HCl, pH 7.5, 150 mM NaCl, 1 mM EDTA, 0.1 % Igepal, 0.25 % gelatin from porcine skin, 0.02 % NaN$_3$.

6. NET-buffer: 50 mM Tris–HCl, pH 7.5, 150 mM NaCl; 1 mM EDTA, 0.1 % Igepal, 0.02 % NaN$_3$.

7. Protein A-Sepharose (PAS) 10 % suspension: Work at room temperature. Swell about 1 g Protein A-Sepharose in 40 mL NET-buffer for 3 h with occasional agitation, in a 50 mL conical tube. Centrifuge at 300 × *g* for 20 min to precipitate the PAS or wait for the suspension sediments without centrifuging. Eliminate the supernatant, add 40 mL of 1 M Tris–HCl, pH 7.5, and let to stand for 1 h with occasional agitation. Centrifuge as above, discard the supernatant, and wash the pellet with 40 mL NET-buffer for 5 min with occasional agitation. Centrifuge again and remove the supernatant. Measure the volume occupied by the beads (about 3.5 mL) and add 9 volumes of NET-buffer. PAS can be stored at 4 °C for many months.

8. Denaturation buffer 6×: 120 mM Tris–HCl, pH 8.6, 6 % SDS, 2 % β-mercaptoethanol, 50 % glycerol.

9. Bromophenol blue (BPB) 6×: 0.1 % BPB, 10 % glycerol.

10. Loading buffer 2× for SDS-PAGE: Make a denaturation buffer 6×/BPB 6×/water (1:1:1) solution (*see* **Note 3**).

11. Amplify™ fluorography reagent (GE Healthcare, Little Chalfont, Buckinghamshire, UK).

12. Release buffer: 100 mM Tris–HCl, pH 8.0, 1% β-mercaptoethanol, 0.5% SDS.

2.4 Immuno-fluorescence

1. Polylysine-coated slides.

2. MaCa buffer: 0.5 M mannitol, 20 mM $CaCl_2$, 0.1% MES, pH 5.7.

3. Phosphate/mannitol buffer (~100 mL): 0.11 M Na_2HPO_4, 0.09 M $NaH_2PO_4 \cdot H_2O$, 1 M mannitol. Solubilize 13.8 g $NaH_2PO_4 \cdot H_2O$ in water and stir vigorously. After dissolving the phosphate salts, bring the solution to 500 mL and autoclave (Solution A). Do it again in exactly the same way but adding this time 14.2 g Na_2HPO_4 (Solution B). Mix together 45 mL Solution A and 55 mL solution B, dissolve in this phosphate buffer 18.2 g mannitol, and adjust the pH to 5.8 (try to keep the volume as much as possible close to 100 mL).

4. Fixative buffer with 4% paraformaldehyde for immunofluorescence (50 mL): Heat 15 mL water to 60 °C in a glass beaker and, under a fume hood, add 2 g paraformaldehyde with continuous stirring over a hot-plate magnetic stirrer to maintain the temperature at 60 °C (*see* **Note 4**). Cover the beaker to minimize evaporation loss (*see* **Note 5**). Then, slowly add 25 mL phosphate/mannitol buffer pH 5.8, bring to 50 mL with water, and allow the mixture to cool to room temperature. The pH of the fixative buffer must be ≤7.4 (adjust with drops of 1 N HCl if necessary). Filter the solution through a 0.45 μm membrane filter to remove any particulate matter, aliquot, and store at −20 °C. Do not refreeze thawed aliquots but keep them at 4 °C.

5. Gelatin stock solution 2%: Dissolve 4 g gelatin (from porcine skin) in 196 mL water. Autoclave and then add 0.02% NaN_3. Store at 4 °C. Before use, let gelatin solution completely liquefy at 37 °C, or, alternatively, put the solution in the microwave for few seconds.

6. TSW buffer: 10 mM Tris–HCl, pH 7.4, 0.9% NaCl, 0.25% gelatin, 0.02% SDS, 0.1% Triton X-100.

7. Mounting medium optimized for fluorescent samples.

2.5 Chemicals for Secretory Pathway Modulation

1. BFA stock solution: Solubilize 2 mg/mL in ethanol and store at −20 °C.

2. Tunicamycin (very toxic) stock solution: Dissolve 5 mg tunicamycin in 1 mL 10 mM NaOH and store at −20 °C. Do not refreeze thawed aliquots but keep them at 4 °C.

3. Endo-H: Commercially available in solution.

3 Methods

The methods described in Subheadings 3.1 and 3.2 must be carried out under sterile conditions; therefore handling plant tissues and culture media should be done in a laminar flow hood which provides an aseptic work area.

3.1 Plant Growth

1. Sterilize tobacco (*Nicotiana tabacum* cv. Petit Havana) seeds: Place seeds (~200) in a 50 mL conical sterile tube with 20 mL 70% ethanol and incubate with vigorous shaking for 1 min. Aspirate ethanol with a sterile glass Pasteur pipette (*see* **Note 6**), add 40 mL solution of 50% commercial bleach and 0.05% Tween-20, and incubate with shaking for 10 min. Aspirate solution and rinse seeds five times with sterile water (each time completely removing the liquid). Place seeds onto MS medium in culture plates for germination.

2. Culture in a chamber with 16–8-h (light-dark) photoperiod at 24 °C under 60 $\mu E/m^{-1}/s^{-2}$ light intensity. About 2 weeks after germination, transfer seedlings to a sterile container (e.g., glass pots) filled with MS medium. Young fully expanded leaves should be used for protoplast isolation (*see* **Note 7**).

3.2 Protoplasts from Leaves

1. Dilute the 10× enzyme mix in K3 buffer and keep it at room temperature. Add 10 mL 1× enzyme mix to each sterile 9 cm Petri dish.

2. Use young fully expanded leaves from tobacco plants grown under aseptic conditions on MS medium. Place the leaf on a sterile surface (e.g., an empty culture plate) and, with a blade, gently cut its abaxial (lower) surface every 1–2 mm, taking care not to cut through the whole leaf. Operate quickly to avoid excessive leaf drying. Remove the tip of the leaf if curled.

3. Remove the midrib (optional) and place the leaf in the Petri dish with the abaxial side in contact with the 1× enzyme mix, without wetting the adaxial (upper) side. Try to pack the dishes with as many whole leaves and leaf fragments as possible to fill in the gaps. Leave plates overnight in the dark at 25 °C. One plate should give about $1–2 \times 10^6$ protoplasts.

4. Remove completely the digestion mix with a sterile plastic Pasteur pipette.

5. Add up to 10 mL K3 medium, dropwise, to the leaves. Let the drops fall from a distance of 10–15 cm above the leaves to facilitate the protoplast's release. Shake the plates gently for 5–10 min to release the protoplasts. Recover the protoplasts with a sterile plastic Pasteur pipette (alternatively, use a sterile plastic 10 mL pipette and break its tip).

6. Repeat **step 5** by adding 3–5 mL of K3 medium.

7. Collect together the two protoplast aliquots and filter the protoplast suspension through a sterile 85 (100) μm-mesh nylon filter, previously wetted (bottom side) in K3 medium. Collect the filtrate in a sterile beaker.

8. Transfer the protoplasts with a sterile plastic Pasteur pipette into a 50 mL conical sterile tube. Otherwise, pour them gently from the beaker into the tube. One 50 mL tube can be filled with ≤30 mL protoplast suspension.

9. Centrifuge tubes for 20 min at 60×g with a swinging bucket rotor with the centrifuge brake set to off. Vital protoplasts float, and conversely dead cells are pelleted.

10. With a sterile glass Pasteur pipette form a window through the protoplast layer by pushing the cells from the center to the sides, and then carefully suck away the K3 medium leaving only 8–10 mL in the tube.

11. Add carefully 4 volumes W5 medium. First, add 2–3 mL of medium dropwise on the internal side of the tube, which is inclined by 45°, and mix by gently inverting the tube 2–3 times. Repeat the dropwise addition of W5 medium up to four times, and then the residual part of the medium can be added with a constant flow. Protoplasts sink in W5; therefore centrifuge for 10 min at 60×g and remove supernatant.

12. Resuspend the pellet very gently with 25 mL W5 (add 3–5 mL, mix gently by inverting the tube, and repeat until all the W5 is added), centrifuge again as above, and remove carefully the supernatant with sterile plastic 10 mL pipette.

13. Resuspend the protoplasts in 10 mL W5 medium and incubate for 30 min in the dark at 25 °C.

14. Count the viable protoplasts: Dilute 50 μL of protoplast suspension (always use cut pipette tips) into 450 μL K3-FDA, freshly prepared. Incubate for 5 min in the dark at 25 °C.

    Count fluorescent protoplasts under UV light using a microscope slide grid (*see* **Note 8**).

15. Centrifuge for 10 min at 60 x g, remove supernatant, and resuspend the protoplasts 1×10^6 mL^{-1} in K3 medium (Fig. 1).

3.3 Evaluation of Golgi-Mediated Protein Traffic by BFA Treatment

The fungal metabolite BFA can be used to verify if a secretory protein bypass the Golgi complex. BFA, inhibiting the formation of active ADP ribosylation factor (ARF) and thus the recruitment of preassembled coat protein complex I (COPI) coat on the Golgi membrane, prevents the bidirectional membrane trafficking between the endoplasmic reticulum (ER) and the Golgi [1].

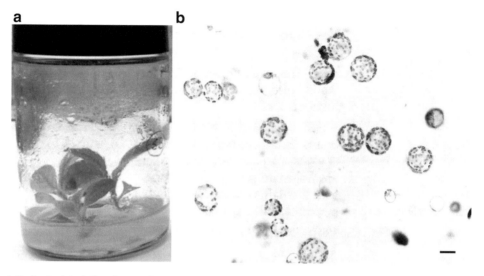

Fig. 1 Protoplast isolation from tobacco plants. (**a**) Glass pot containing a sterile culture of seed-derived tobacco plantlets. (**b**) Protoplasts are isolated from young fully expanded leaves and resuspended in K3 medium. Bar = 70 μm

3.3.1 Protoplast Radiolabeling (Pulse-Chase)

1. Pulse: Incubate 1×10^6 mL^{-1} protoplasts for 1 h (*see* **Note 9**) with 100 μCi/mL of a mixture of ^{35}S-methionine and ^{35}S-cysteine at 25 °C (*see* **Note 10**) in the dark in K3 medium supplemented with 150 μg/mL of BSA. Use 15 mL conical sterile tube to incubate the protoplasts.

2. Chase: Add to the protoplast solution 1/10 of the volume of unlabeled methionine and cysteine (e.g., 130 μL to 1.2 mL) from a 10× concentrated stock. The final concentration of unlabeled methionine and cysteine should be 10 mM and 5 mM, respectively.

3. For BFA treatment, preincubate 1×10^6 mL^{-1} protoplasts at 25 °C in the dark with 10 μg/mL BFA (from a 2 mg/mL stock solution) for 45 min before adding the mixture of radioactive amino acids. Maintain BFA throughout the pulse-chase experiment at the same conditions. Preincubate also an untreated aliquot as control, adding a volume of ethanol identical to the BFA added volume.

4. Stir gently to obtain a homogeneous solution, since the protoplasts tend to float. Then recover (chase 0) an aliquot of 0.3×10^6 protoplasts, always using cut pipette tips. Put the 15 mL conical tube with the other protoplasts at 25 °C in the dark.

5. Add 3 volumes of ice-cold W5 medium to the aliquot of protoplasts and centrifuge for 5 min at $50 \times g$ at 4 °C. Recover the supernatant containing the secreted proteins leaving about

50 μL above the pellet to cover the protoplasts. (Optional: Wash again with W5 if you want to get rid of all secreted proteins.)

6. Freeze the supernatant and protoplasts in separated tubes in liquid nitrogen and stores at −80 °C until ready for homogenization.

7. Repeat **steps** 5–6 to recover samples at the desired chase time points (routinely, they are 1, 2, 4, 8, 16, and 24 h).

3.3.2 Immuno-
precipitation
of Radioactive Proteins

1. Homogenate protoplasts and the corresponding supernatants with 2 volumes of ice-cold protoplast homogenization buffer, performing all steps at 4 °C.

2. Vortex briefly the samples and then add NET-gel buffer up to 1 mL.

3. Centrifuge twice for 4 min at $12,000 \times g$ at 4 °C and transfer the supernatant to a new tube each time.

4. Add to the supernatant 1 μL of an antiserum or antibody (not diluted) raised against the protein of interest.

5. Incubate the samples on ice for 90 min.

6. Add 100 μL of a 10% suspension of PAS, always using cut pipette tips.

7. After 90 min under gentle agitation, centrifuge the samples at $300 \times g$ for 3 min at 4 °C and discard supernatant. Wash the beads three times resuspending them with 1 mL of NET-gel buffer. Centrifuge as above and discard supernatant after each wash.

8. Add 50 μL of loading buffer 2× to the beads, denature for 5 min at 90 °C, spin briefly, and load the solution on a standard SDS-PAGE.

9. After electrophoresis, treat the gel with Amplify™ fluorography reagent according to the manufacturer's protocol, dry it with a gel-dryer, and visualize radioactive polypeptides by standard fluorography (Fig. 2a).

3.3.3 Immuno-
fluorescence of BFA-
Treated Protoplasts

1. BFA-treated protoplasts can be subjected to immunofluorescence to visualize the BFA effect on protein localization. The whole immunofluorescence procedure can be carried out at room temperature. An untreated protoplast aliquot must be subjected to immunofluorescence as control.

2. Assemble a humidity chamber to avoid slides drying out during the whole immunofluorescence procedure. According to the number of slides used in the experiment, choose an appropriate container (glass staining dish or plastic box). Cover the bottom of the container with a wet paper towel and use a

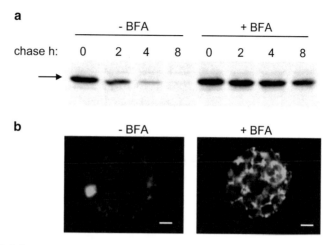

Fig. 2 Cellular transport of a protein of interest (PI) is affected by BFA. (**a**) Tobacco protoplasts were treated with BFA (+BFA) or untreated (−BFA). Then, protoplasts were pulse labeled for 1 h with a mixture of radioactive sulfur amino acids and chased for the indicated periods of time. After each chase point, protoplasts were homogenated, immunoprecipitated with a specific antibody, and analyzed by SDS-PAGE and fluorography. *Arrow* indicates the signal corresponding to the PI. In the absence of BFA, PI is transported to its cellular localization and its corresponding signal decreases. In the presence of BFA, PI transport is interrupted with the corresponding increase of the PI half-life. (**b**) Aliquots of BFA-treated or untreated protoplasts from (**a**) were subjected to immunofluorescence. The fluorescence originates from a specific antibody for the PI detected using FITC-conjugated anti-rabbit secondary antibody. The images show that the PI is mainly detectable as a large aggregate (which is known to localize in the vacuole), whereas in the presence of BFA, the PI is retained in the ER as shown by the labeling of the typical plant ER network, confirming that PI transport is blocked. Bars = 25 μm

support (e.g., two cut pipettes placed in parallel) to carry the slides, in order to avoid any contact with the water. The box must be closed with a lid.

3. Resuspend protoplasts at a concentration of 5×10^5 mL^{-1} in MaCa buffer (if they are in K3, add carefully 4 volumes of W5 medium, centrifuge for 10 min at $60 \times g$, and remove supernatant).

4. Drop (always use cut pipette tips) carefully 300 μL of protoplast suspension onto the center of a polylysine-coated slide and let it adhere for 30 min.

5. After 30 min, tilt the long edge of the slides to a 45° angle on a paper towel to eliminate excess MaCa buffer with unadhered protoplasts. Adopt this technique for every washing step in the experiment (*see* **Note 11**).

6. Fix for 30 min the adhered protoplasts adding carefully 400 μL fixative buffer with 4 % paraformaldehyde. Do not pour the solution directly over protoplasts but place the pipette tip alongside the cells and carefully release the buffer. Add solutions in this way in all the following steps of the immunofluorescence procedure.

7. Permeabilize the protoplasts by washing three times with 500 μL TSW buffer for 10 min each time (*see* **Note 12**).

8. Dilute the primary antibody at appropriate concentration in 400 μL TSW buffer and incubate protoplasts for 1 h.

9. Eliminate the primary antibody excess by washing three times with 500 μL W5 buffer for 10 min each time.

10. Dilute the fluorophore conjugate secondary antibody at appropriate concentration in 400 μL TSW buffer and incubate protoplasts for 1 h. As the fluorescent secondary antibody is light sensible, place the humidity chamber in the dark or cover it, for example by wrapping aluminum foil around the box and the lid.

11. Eliminate the secondary antibody excess by washing three times with 500 μL W5 buffer for 10 min each time.

12. Cover cells, first with a mounting medium optimized for fluorescent samples and then with a glass cover slip. Visualize the protoplasts under UV light using a microscope with the proper UV filters (Fig. 2b).

3.4 Evaluation of Golgi-Mediated Protein Traffic by Endoglycosidase H (Endo-H) Treatment

Some secretory proteins are glycosylated into the lumen of the ER and the most common glycosylation is the N-linked glycosylation. N-linked glycans undergo extensive remodeling in the Golgi apparatus modifying their cleavage sensitivity to the Endo-H. This enzyme cleaves the high-Man-type N-linked glycans, but it does not digest complex glycans derived from glycan modification in the Golgi apparatus. Therefore, the presence of N-glycosylated polypeptides sensitive to Endo-H action indicates that their traffic does not depend on Golgi-mediated delivery.

1. Radiolabel protoplasts and then immunoprecipitate the N-glycosylated protein of interest as described above (Subheadings 3.3.1 and 3.3.2), with the exception that, after **step 7** of Subheading 3.3.2, wash the PAS beads twice with water.

2. Release proteins bounded to the PAS denaturing at 95 °C for 4 min in 50 μL of release buffer.

3. Centrifuge samples for 5 min at $14,500 \times g$ and recover the supernatant.

4. Dilute the supernatant tenfold with 0.1 M sodium citrate, pH 5.5, supplemented with 1.5 mM PMSF, a broad-spectrum inhibitor mixture and 50 µg BSA.

5. Divide the sample into two aliquots, one of which should be treated with 20 mU Endo-H and the other with the same volume of water.

6. Incubate for 18 h at 37 °C.

7. Precipitate the proteins adding trichloroacetic acid (TCA).

8. Add 50 µL 1× loading buffer to the pellet and analyze by SDS-PAGE and fluorography as already described.

3.5 N-Linked Glycan's Involvement in Protein Transport to the Vacuole Can Be Verified Using Tunicamycin

Tunicamycin is an inhibitor of N-linked glycosylation in the ER and, in plants, treatment with this chemical is used to study drug-induced ER stress [11] or the secretory pathway [8]. To investigate if the deliver to the vacuole of a glycosylated protein involves its N-linked glycans, protoplasts expressing this protein can be incubated with tunicamycin and then analyzed by pulse-chase, SDS-PAGE, and fluorography. In the absence of the inhibitor, the recovery of the immunoselected protein of interest should decrease over time, as a result of its vacuolar delivery. In the presence of tunicamycin, the protein of interest traffic could be unchanged or blocked, strongly suggesting that protein N-linked glycans play no specific role in the targeting to the vacuole or that protein N-linked glycans play a specific role in this route, respectively.

1. Preincubate protoplasts 1×10^6 mL^{-1} in K3 supplemented with 50 µg/mL tunicamycin (from a 5 mg/mL of 10 mM NaOH stock solution) or equivalent quantities of solvent for the controls, for 45 min in the dark, and maintain the same conditions for the whole experiment.

2. Incubate protoplasts to perform the pulse-chase analysis and then immunoprecipitate the N-glycosylated protein of interest as described in Subheadings 3.3.1 and 3.3.2, respectively. Evaluate by SDS-PAGE and fluorography the intensity of the immunoselected protein over time comparing the tunicamycin-treated protoplasts with untreated ones.

4   Notes

1. The term "water" refers to ultrapure water that has been deionized such as that dispensed from a Millipore water purification system.

2. Check for precipitates when unlabeled Met/Cys 10× stock solution is melted because after several freezing and melting the solution may become cloudy. If this occurs, discard and prepare a new stock.

3. If the denaturation buffer is too viscous, heat it at 50 °C for several minutes. To prepare a loading buffer 1×, add to one volume of the loading buffer 2× an identical volume of water.

4. To dissolve paraformaldehyde it is necessary to heat the water to 60 °C, but take care to keep water temperature or the paraformaldehyde solution temperature under 70–80 °C; otherwise potential paraformaldehyde degradation may occur.

5. The powder will not immediately dissolve into solution; hence dissolve the paraformaldehyde (which is clear when dissolved but some fine particles will remain) with few drops of 2 N NaOH.

6. Due to their small size tobacco seeds can be sucked up by the Pasteur pipette. To avoid this, the Pasteur pipette internal diameter can be reduced by stretching glass Pasteur pipettes over a flame. Alternatively, use a sterile stainless steel to separate the seeds from the solution.

7. It is extremely important to use young fully expanded leaves for protoplast isolation because they guarantee a high number of healthy protoplasts, whereas less protoplasts can be obtained from old leaves.

8. If a microscope with a UV lamp is not available, viable protoplasts can be counted using bright light and they will appear as round, intact cells. This method, of course, can only offer a rough estimation of the viable protoplast number.

9. Protoplast incubation (pulse) can be reduced to 30 min or less if the secreted proteins to be analyzed have a short half-life.

10. Protein labeling mix containing both ^{35}S-methionine and ^{35}S-cysteine is commercially available. The intensity of the radioactive signals is correlated to the number of methionine and cysteine residue in the protein of interest. If the studied protein does not have cysteine residues, only ^{35}S-methionine can be used.

11. At the beginning, the number of protoplasts is more than needed for the experiment. So do not mind if many of them will remain unattached to the polylysine-coated slide, because the attached ones are still enough for the immunofluorescence analysis.

12. TSW buffer permeabilizes protoplasts and causes, consequently, chlorophyll loss. At the end of the three washes protoplasts will result almost transparent and difficult to be visualized on the slide.

References

1. Brandizzi F, Barlowe C (2013) Organization of the ER-Golgi interface for membrane traffic control. Nat Rev Mol Cell Biol 14:382–392

2. Hwang I (2008) Sorting and anterograde trafficking at the Golgi apparatus. Plant Physiol 148:673–683

3. Xiang L, Etxeberria E, Van den Ende W (2013) Vacuolar protein sorting mechanisms in plants. FEBS J 280:979–993

4. Robinson DG, Brandizzi F, Hawes C, Nakano A (2015) Vesicles versus tubes: Is endoplasmic reticulum-Golgi transport in plants fundamentally different from other Eukaryotes? Plant Physiol 168:393–406

5. Herman EM, Schmidt M (2004) Endoplasmic reticulum to vacuole trafficking of endoplasmic reticulum bodies provides an alternate pathway for protein transfer to the vacuole. Plant Physiol 136:3440–3446

6. Honig A, Avin-Wittenberg T, Ufaz S, Galili G (2012) A new type of compartment, defined by plant-specific Atg8-interacting proteins, is induced upon exposure of Arabidopsis plants to carbon starvation. Plant Cell 24:288–303

7. De Marchis F, Bellucci M, Pompa A (2013) Unconventional pathways of secretory plant proteins from the endoplasmic reticulum to the vacuole bypassing the Golgi complex. Plant Signal Behav 8, e25129

8. De Marchis F, Bellucci M, Pompa A (2013) Traffic of human α mannosidase in plant cells suggests the presence of a new endoplasmic reticulum-to-vacuole pathway without involving the Golgi complex. Plant Physiol 161:1769–1782

9. Murashige T, Skoog F (1962) A revised medium for rapid growth and bioassays with tobacco tissue cultures. Physiol Plant 15:473–497

10. Gamborg OL, Miller RA, Ojima K (1968) Nutrient requirements of suspension cultures of soybean root cells. Exp Cell Res 50:151–158

11. Watanabe N, Lam E (2007) BAX inhibitor-1 modulates endoplasmic reticulum stress-mediated programmed cell death in Arabidopsis. J Biol Chem 283:3200–3210

Chapter 5

From Cytosol to the Apoplast: The Hygromycin Phosphotransferase (HYG^R) Model in Arabidopsis

Haiyan Zhang and Jinjin Li

Abstract

The process by which proteins are secreted via endoplasmic reticulum (ER)/Golgi-independent mechanism is conveniently called unconventional protein secretion. Recent studies have revealed that unconventional protein secretion operates in plants, but little is known about its underlying mechanism and function. This chapter provides methods we have used to analyze unconventional character of hygromycin phosphotransferase (HYG^R) secretion in plant cells. Following isolation of protoplasts from *HYG^R-GFP*-transgenic plants and incubation with brefeldin A (BFA), an inhibitor of conventional secretory pathway, we easily obtain protein extracts from protoplasts and culture medium separately. These proteins are separated by sodium dodecyl sulfate-polyacrylamide gel electrophoresis (SDS-PAGE), followed by Western blot analysis with anti-GFP antibodies.

Key words Unconventional protein secretion, Hygromycin phosphotransferase, Protoplast isolation, Trichloroacetic acid precipitation, Western blot, GFP antibodies

1 Introduction

The classical or conventional secretory proteins often contain an N-terminus signal sequence that directs the nascent protein to co-translate and vectorially transfer across the membrane of the endoplasmic reticulum (ER). Such secretory proteins are then transported to the extracelluar space or the plasma membrane through the ER-Golgi secretory pathway [1, 2]. However, a large number of proteins have been identified to be secreted without any apparent signal sequence. In addition, the secretion of these proteins is not affected by the presence of brefeldin A (BFA), a drug that blocks conventional ER/Golgi-dependent secretory transport [3, 4]. This phenomenon, termed unconventional secretion, was found in eukaryotes approximately 25 years ago [5]. Plant secretomics studies have revealed that secretory proteins without signal sequence can account for more than half of the total identified proteins. Furthermore, the unconventional protein secretion in plants seems

Andrea Pompa and Francesca De Marchis (eds.), *Unconventional Protein Secretion: Methods and Protocols*, Methods in Molecular Biology, vol. 1459, DOI 10.1007/978-1-4939-3804-9_5, © Springer Science+Business Media New York 2016

to be involved in biotic and abiotic stresses [6]. The resistance gene coding for hygromycin B phosphotransferase (HYG^R, E.C. 2.7.1.119) has been mainly used as a positive selective marker for transgenic cells [7]. We found that HYG^R lacks a signal sequence and its secretion is not sensitive to BFA treatment, therefore HYG^R-is secreted via the unconventional secretory pathway [8]. Here we describe a step-wise protocol to detect unconventional secretory proteins in extracellular space with an example of HYG^R in plants. Arabidopsis plants stably expressing a fusion protein containing HYG^R and green fluorescent protein (GFP) are used to isolate mesophyll protoplasts from leaf tissues. After incubated with BFA, the protoplasts and the culture medium are collected respectively. Trichloroacetic acid (TCA) precipitation method is used to extract soluble proteins from the culture medium. The effectiveness of BFA on the conventional ER/Golgi pathway was verified by measuring the activity of acid phosphatase (AcPase) [9] in the medium and protoplast lysates. Finally, the protoplast lysates and medium proteins were separated by SDS-PAGE and immunoblotted with anti-GFP antibodies. The protocol is based on some previously published protocols or methods [10–16] with some modifications. It consists of protoplast isolation, BFA treatment, TCA protein precipitation, Western blot analysis, and AcPase assay.

2   Materials

2.1   Arabidopsis Mesophyll Protoplast Isolation

1. Plant materials: Wild-type, *Arabidopsis thaliana* ecotype Columbia; *HYGR-GFP* transgenic plants, Arabidopsis expressing HYG^R-GFP fusion protein under the control of constitutive cauliflower mosaic virus (CaMV) 35S promoter [8].

2. Ethanol (70%): mix 70 mL 100% ethanol in 30 mL ddH$_2$O.

3. Bleach containing 0.05% (v/v) Tween 20: dissolve 50 μL Tween 20 into 100 mL bleach (5.25–6.15% Sodium hypochlorite).

4. Sterile agar (0.1%): 0.1 g agar in 100 mL ddH$_2$O. Sterilize by autoclaving.

5. Murashige and Skoog (MS) agar medium (0.5×): 2.15 g MS salts to 900 mL ddH$_2$O stir to dissolve. Check and adjust the pH to 5.6 using 1 M KOH. Dilute to final volume of 1000 mL. Add 8 g agar and 10 g sucrose. Sterilize by autoclaving.

6. Mannitol solution (1 M): dissolve 182 g mannitol into a 1000 mL volumetric flask and add 600 mL ddH$_2$O. After mannitol is completely dissolved, dilute the solution with ddH$_2$O to 1000 mL and mix (*see* **Note 1**).

7. Enzyme solution: 1% cellulase R10, 0.25% macerozyme, 0.4 M mannitol, 20 mM KCl, 20 mM MES (pH 5.7), 10 mM

CaCl$_2$, 5 mM β-mercaptoethanol (optional), 0.1 % BSA. Resolve the components completely. Adjust the pH to 5.6 with 1 M KOH. Filter the enzyme solution into new 50 mL falcon tube through 0.45 μm filter (*see* **Note 2**).

8. Sucrose solution (21 %): Dissolve 21 g of sucrose in ddH$_2$O to make total volume of 100 mL.

9. W5 solution: 154 mM NaCl, 125 mM CaCl$_2$, 5 mM KCl, 2 mM MES. Dissolve the components completely and adjust the pH to 5.6 with 1 M KOH.

2.2 Brefeldin A (BFA) Treatment

1. BFA stock solution: Prepare stock solution (50 mM) in DMSO and stored at or below –20 °C (*see* **Note 3**).

2.3 Trichloroacetic Acid Protein Precipitation

1. Trichloroacetic acid (TCA) (100 %): dissolve 500 g of solid TCA into 350 mL ddH$_2$O. Maintain in dark bottle at room temperature.

2. Acetone (100 %).

2.4 Western Blot

1. Tris buffer (1.5 M, pH 8.8): Weigh 90.68 g Tris and transfer to a glass beaker. Add ddH$_2$O to a volume of 400 mL. Mix and adjust the pH to 8.8 with concentrated HCl. Make up to 500 mL with ddH$_2$O. Store at 4 °C.

2. Tris buffer (0.5 M, pH 6.8): Weigh 30.29 g Tris and prepare a 500 mL solution as in previous step. Store at 4 °C.

3. Sodium dodecyl sulfate (SDS, 10 %): 10 g SDS in 100 mL ddH$_2$O.

4. Ammonium persulphate (APS, 10 %): 100 mg APS in 1 mL ddH$_2$O (prepared before use).

5. Tris-buffered saline with Tween 20 (TBST) buffer (1×): 20 mM Tris, 150 mM NaCl, 0.1 % Tween 20, Check the pH and adjust to 7.5.

6. SDS-PAGE sample buffer (2×): 4 % SDS, 5 % β-mercaptoethanol, 10 % glycerol, 0.02 % bromophenol blue, 0.2 M Tris–HCl. Check the pH and adjust to 6.8.

7. Tris-Glycine running buffer (1×): 25 mM Tris, 190 mM glycine (pH 8.3), 0.1 % SDS.

8. Transfer Buffer (1×): 25 mM Tris, 190 mM glycine, 20 % methanol.

9. Separation gel buffer (12 %) (For 10 mL): 3.35 mL ddH$_2$O, 2.5 mL 1.5 M Tris buffer (pH 8.8), 100 μL 10 % SDS, 5 μL *N*,*N*,*N'*,*N'*-Tetramethylethylenediamine (TEMED), 4 mL 30 % acrylamide-bisacrylamide mix (Acr-Bis, Arc : Bis = 29:1), 50 μL 10 % APS.

10. Stacking gel buffer (5%) (For 5 mL): 3.05 mL ddH$_2$O, 1.25 mL 0.5 M Tris buffer (pH 6.8), 50 µL 10% SDS, 5 µL TEMED, 0.65 mL 30% Acr-Bis, 50 µL 10% APS.

11. Blocking Buffer: 5% (w/v) nonfat dry milk in 1× TBST.

12. Antibody dilution buffer: Both the primary and the secondary antibodies are diluted in 1× TBST supplemented with 0.5% (w/v) nonfat dry milk.

13. Primary antibodies: Rabbit monoclonal anti-GFP antibody (1:4000); anti-tubulin (1:1000) (*see* **Note 4**).

14. Secondary antibody: Horseradish-peroxidase-conjugated goat anti rabbit IgG (1:4000).

15. Enhanced chemiluminescence (ECL) reagents.

2.5 Acid Phosphatase Assay

1. MES-Tris (40 mM, pH 5.5): Dissolve 7.8 g MES in 800 mL ddH$_2$O. After adjustment of the pH to 5.5 with 1.7 M Tris, adjust the volume to 1000 mL with ddH$_2$O.

2. Extraction buffer: 40 mM MES-Tris (pH 5.5), 1 mM EDTA, 1 mM DTT, 1% (w/v) polyvinyl polypyrrolidone (PVPP), store at 4 °C.

3. Reaction buffer: 5 mM *p*-nitrophenyl phosphate (*p*-NPP), 40 mM MES-Tris (pH 5.5), 10 mM MgCl$_2$ (*see* **Note 5**).

4. Stop solution: Dissolving 1.6 g of NaOH in ddH$_2$O to make total volume of 100 mL.

3 Methods

3.1 Arabidopsis Plant Growth

1. Incubate Arabidopsis seeds expressing *HYG-GFP* and wild type in 70% ethanol for 5 min by inverting tubes several times.

2. Remove ethanol and add bleach. Incubate for 15 min, invert tubes as above.

3. Remove as much bleach as possible.

4. Wash 5 times with a large volume of sterile ddH$_2$O.

5. Take seeds up in 0.1% sterile Agar and plate on 0.5× MS agar medium.

6. Grow in a growth chamber for 15–20 days at 22 °C under 16-h photoperiod and120 µmol/m^2/s.

3.2 Arabidopsis Protoplast Isolation

1. Use leaves from agar-plate-cultured Arabidopsis plants (*see* **Note 6**). Cut leaves in half and then into four pieces by cross-sectioning with a sharp blade and put them into 50 mL falcon tube containing 15 mL 1 M mannitol. Incubate 30 min at room temperature (*see* **Note 7**).

2. Remove 1 M mannitol and add 15 mL prepared enzyme solution. Incubate on a platform shaker (40 rpm) for 3–6 h at room temperature (*see* **Note 8**).

3. Check for the release of protoplasts in the solution under the microscope (*see* **Note 9**).

4. Filter the protoplasts suspension through a 70-μm nylon mesh to remove undigested leaf tissues (The size of Arabidopsis mesophyll protoplasts is approximately 30–50 μm).

5. Load protoplasts onto 10 mL 21% sucrose in 50 mL falcon tube. Centrifuge the tube at $100 \times g$ for 3 min.

6. Take the intact floating protoplasts concentrated in gradient interface and at the top of the tube with a broad-mouthed pipet into 50 mL falcon tube containing 15 mL W5 solution.

7. Collect protoplasts by centrifugation at $100 \times g$ for 3 min.

8. Remove as much supernatant as possible and wash the protoplasts five times with 15 mL of W5 solution each. Resuspend the protoplasts in 15 mL W5 solution.

9. Resuspend the protoplast pellet obtained after the final wash in 2 mL of W5 solution. Take a small aliquot of protoplast suspension and count the protoplast using a hemocytometer under a light microscope. Adjust the protoplast density to 5×10^5 cells/mL by adding more W5 solution (*see* **Note 10**).

3.3 BFA Treatment

1. Add 50 mM BFA stock solution into protoplast suspension to the final concentration of 25 μM.

2. Incubate the protoplasts with or without BFA at 23 °C for 5 h. During the incubation, sample aliquots (500 μL) of protoplast suspension at hourly intervals for measuring the activity of AcPase (*see* Subheading 3.6) (*see* **Notes 11** and **12**).

3. After incubation, take a small aliquot of protoplast suspension and observe if the protoplasts are intact under a light microscope (*see* **Note 13**).

4. Collect protoplasts by centrifugation at $100 \times g$ for 3 min. Transfer carefully the supernatant into a fresh tube without disturbing the protoplast pellet for protein precipitation from the culture medium (*see* Subheading 3.4).

5. Wash protoplasts twice with W5 solution. Harvest protoplasts by centrifugation at $100 \times g$ for 2 min and discard the supernatant.

6. Lyse protoplasts by adding equal volume of 2× SDS-PAGE sample buffer and vortex vigorously.

7. Centrifuge the samples at $15,000 \times g$ for 15 min at 4 °C. Collect the supernatants into a fresh tube for Western blot analysis (*see* Subheading 3.5).

3.4 Protein Precipitation from Protoplast Culture Medium

1. Chill trichloroacetic acid (TCA) (100%) and acetone at –20 °C.

2. Precipitate proteins in protoplast culture medium by adding 100% TCA to the protoplast culture supernatant. The final concentration of TCA should be about 12%.

3. Mix well and keep the solution overnight at 4 °C (*see* **Note 14**).

4. Spin at 15,000×*g* for 15 min at 4 °C and immediately decant or siphon off the supernatant. The pellet will appear as a slight white residue along with the outer side of the microcentrifuge tube.

5. Wash pellet by adding 5 mL prechilled acetone slowly along the sides of the tube. Incubate at –20 °C for 5 min.

6. Spin again for 15 min at 4 °C.

7. Discard the supernatant very carefully at this stage since the pellet is not adhered tightly to the tube.

8. Leave the tubes open in a fume cupboard for 10–20 min or apply vacuum in a SpeedVac briefly to remove traces of solvent (*see* **Note 15**).

9. Dissolve the pellet in 100 µL 2× SDS-PAGE sample buffer by repeatedly pipetting up and down to break up the pellet. The buffer solution should be blue in color (*see* **Note 16**).

10. Heat samples at 90 °C for 10 min before analyzing by Western blot.

3.5 Western Blot

3.5.1 Sample Loading and Electrophoresis

1. Prepare an appropriate percentage (generally 10–15%) polyacrylamide gel depending on the estimated molecular weight of the protein of interest and assemble electrophoresis cell (*see* **Note 17**).

2. Fill the upper (inner) and lower (outer) buffer chamber with 1× running buffer.

3. Load 10–20 µL of the protein samples prepared from protoplast lysates and the culture medium, and molecular mass markers (prestained protein molecular weight standard) (*see* **Note 18**).

4. Run the gel in 1× Tris-Glycine buffer at 60–120 V for 1–3 h at room temperature.

5. Rinse the gel three times for 5 min each with 100 mL of ddH$_2$O.

3.5.2 Protein Transfer

1. Pre-wet materials such as gel, filter paper, and sponge in 1× transfer buffer.

2. Incubate PVDF membrane in methanol for 10 s to 1 min and then moved to 1× transfer buffer (*see* **Note 19**).

3. Prepare the transfer cassette as following: case (black side), sponge, filter paper, gel, PVDF membrane, sponge, case (clear side).

4. Place in the transfer apparatus with black side facing black.

5. Transfer in cooling environment with cold 1× transfer buffer at a constant 200 mA current transfer for 40 min (*see* **Note 20**).

6. Incubate the membrane in 1× TBST for 15 min before blocking the membrane (facing up) with 5 % nonfat milk in 1× TBST, either for 2 h at room temperature or overnight at 4 °C on a shaker.

3.5.3 Immunoblotting

1. Prepare a working dilution of the primary antibodies (anti-GFP or anti-tubulin) in 10 mL antibody dilution buffer. Dilution ratio for each antibody should be optimized according to the results.

2. Incubate the membrane in the diluted primary antibody for 2 h at room temperature or overnight at 4 °C.

3. Wash the membrane in 1× TBST three times for 10 min on a shaker at room temperature.

4. Incubate the membrane in an appropriately diluted secondary antibody solution prepared in 10 mL antibody dilution buffer. Incubate the membrane for 2 h at room temperature.

5. Wash the membrane in 1× TBST three times for 10 min on a shaker at room temperature.

6. Prepare the chemiluminescent reagent (0.125 mL of chemiluminescence reagent per cm^2 of membrane) by mixing equal volumes of the Enhance Luminol Reagent and the Oxidizing Reagent (*see* **Note 21**).

7. Incubate the membrane in the chemiluminescence reagent for 1 min.

8. Remove excess chemiluminescence reagent by draining and place the membrane in a plastic sheet protector.

9. Expose to X-ray film for 30 s. Develop the film and, if necessary, use the result to determine an optimum exposure.

10. Use the bands of the molecular weight marker as a reference to determine the mass of the protein in each lane.

3.6 Acid Phosphatase Assay

1. Centrifuge the aliquots (*see* Subheading 3.3, **step 2**) of protoplast suspension at $100 \times g$ for 3 min. Transfer the supernatants into fresh tubes without disturbing the protoplast pellet and keep them on ice.

2. Wash the protoplasts twice with W5 solution. Harvest protoplasts by centrifugation at $100 \times g$ for 2 min and discard the supernatant.

3. Freeze the protoplasts in liquid N$_2$ for 1 min and thaw on ice. Lyse protoplasts by adding extraction buffer (EB) and vortex vigorously.

4. Centrifuge the samples at $15,000 \times g$ for 10 min at 4 °C. Remove the supernatants into fresh tubes and keep them on ice.

5. Mix 100 µL of enzyme samples (including the protoplast medium and extracts from the protoplasts) with 900 µL of reaction buffer and incubate at 30 °C for 30 min in dark. A blank reaction (reaction buffer without enzyme samples) is run in parallel (*see* **Note 22**).

6. Stop the reactions with 2 mL of stop solution (*see* **Note 23**).

7. Transfer the reaction mixture to a cuvette and measure the absorption at 405 nm.

8. Calculate the AcPase activity in the sample according to the following equations: Units/mL $= [(A_{405}$ [sample] $- A_{405}$ [blank]$) \times 3]/[18.3 \times \text{Time} \times \text{Venz}]$ (for 3 mL cuvette assay). A_{405} [sample]: Absorbance of the sample; A_{405} [blank]: Absorbance of the blank; Time $=$ Time of incubation at 30 °C in minutes; Venz $=$ Volume of enzyme sample added to the assay in mL; 18.3: Millimolar extinction coefficient (ε^{mM}) of p-nitrophenol at 405 nm; 3: 3 mL, the total assay volume in the cuvette, including the stop solution. (Unit definition: one unit of acid phosphatase will hydrolyze 1 mmol of p-nitrophenyl phosphate per minute at pH 5.5 at 30 °C).

4   Notes

1. Keep mannitol solution sterile as there is a high risk of bacterial contamination. All working solutions containing mannitol should be prepared fresh.

2. The enzyme solution should be dissolved in protoplast culture medium at 4 °C overnight or alternatively it can be dissolved at 30 °C with shaking at 130 rpm for 1 h. Finally, the enzyme solution should be clear light brown (*see* ref. 11). The enzyme solution should be prepared fresh.

3. It is advisable to store the stock solution as aliquots in tightly sealed vials at −20 °C.

4. In our research, tubulin was used as an intracellular marker to detect contamination of the medium with intracellular proteins due to breakage of protoplasts.

5. The Reaction buffer should be freshly prepared.

6. Plant health is vitally important for producing robust and reliable protoplasts. We generally use plants grown on 0.5× MS agar medium to avoid any possible stresses from soil culture. Do not use pale, discolored, curled, or vitrified leaves.

7. Use clean and sharp cut to avoid crushing of the leaf and to decrease the amount of debris in the protoplast preparation. Change razor blade when sharpness diminishes (*see* ref. 11).

8. The optimal time of incubation may vary because the activity of enzyme may differ from different suppliers. It is advisable to determine the requirements for optimal protoplasting if other manufacturers' enzymes are used.

9. Check protoplasts by inspecting an aliquot under a light microscope. At this stage of enzyme digestion, the cell suspension should be green in color and healthy protoplasts are spherical and float freely.

10. Protoplasts must be handled very carefully throughout the whole protocol. When adding a solution to the protoplasts, pipette it slowly along the tube wall. When pipetting protoplast suspension, the end of the pipet tip can be cut to slightly increase the size of the tip hole.

11. It has been demonstrated that AcPase is secreted via conventional secretory pathway and its secretion is sensitive to BFA treatment (*see* ref. 9). Other reported conventional secretory proteins can also be used to verify the effectiveness of BFA.

12. The incubation time is dependent on the protein of interest. If the expression level is low or the target protein is not secreted efficiently, the incubation time can be prolonged but not longer over 16 h, because protoplasts start dying after 16-h incubation and will be less effective to export proteins.

13. It is very important to confirm that the protoplasts maintain intact during the whole experimental process for secretory protein detection.

14. The incubation time is dependent on the target protein. Incubate the solution at –20 °C for 5–30 min if the protein concentration is high.

15. Do not dry excessively to avoid difficulty in solubilizing the pellet. At this stage, samples can be stored at –20 °C until you are ready to run electrophoresis.

16. A greenish-yellow sample indicates presence of residual TCA. You can correct it by adding 1 µL aliquots of 1 M Tris solution.

17. High percentage gels (15–20%) are required for best resolution of low molecular weight proteins, whereas 4–6% gels is applicable to resolving large proteins (>200 kDa). If your protein of interest has multiple isoforms of low to high molecular sizes, gradient gels would be your best option for achieving efficient separation of proteins.

18. Generally, load a volume of protein samples containing between 2.5 and 25 µg of total protein, depending on anticipated relative purity of the target protein(s).

19. For proteins with a molecular weight less than 30 kDa, use 0.2 µm PVDF membrane, otherwise 0.45 µm PVDF is recommended.

20. Transfer current and time should be optimized according to blotting system manufacturer's recommendations. Low molecular weight proteins (<30 kDa) require a short transfer time to avoid pulling the protein through the membrane.

21. ECL reagents and X-ray films optimized for ECL-detection can be purchased from several vendors.

22. If acid phosphatase activity of the test sample is low, the incubation time can be extended up to 60 min.

23. NaOH is used to stop the reaction (denatures AcPase) while simultaneously converting *p*-nitrophenyl (*p*-NP) product into the yellow-colored *p*-nitrophenol. The colored solution is stable for several hours (*see* ref. 15).

References

1. Burgess TL, Kelly RB (1987) Constitutive and regulated secretion of proteins. Annu Rev Cell Biol 3:243–293

2. Ding Y, Robinson DG, Jiang LW (2014) Unconventional protein secretion (UPS) pathways in plants. Curr Opin Cell Biol 29:107–115

3. Nickel W, Rabouille C (2009) Mechanisms of regulated unconventional protein secretion. Nat Rev Mol Cell Biol 10:148–155

4. Drakakaki G, Dandekar A (2013) Protein secretion: how many secretory routes does a plant cell have? Plant Sci 203:74–78

5. Rubartelli A, Cozzolino F, Talio M, Sitia R (1990) A novel secretory pathway for interleukin-1-beta, a protein lacking a signal sequence. EMBO J 9:1503–1510

6. Agrawal GK, Jwa NS, Lebrun MH, Job D, Rakwal R (2010) Plant secretome: unlocking secrets of the secreted proteins. Proteomics 10:799–827

7. McGaha SM, Champney WS (2007) Hygromycin B inhibition of protein synthesis and ribosome biogenesis in *Escherichia coli*. Antimicrob Agents Chemother 51:591–596

8. Zhang H, Zhang L, Gao B, Fan H, Jin JB, Botella MA et al (2011) Golgi apparatus-localized synaptotagmin 2 is required for unconventional secretion in Arabidopsis. PLoS One 6, e26477

9. Kaneko TS, Sato M, Osumi M, Muroi M, Takatsuki A (1996) Two isoforms of acid phosphatase secreted by tobacco protoplasts:

differential effect of brefeldin A on their secretion. Plant Cell Rep 15:409–413

10. Zhang Y, Su J, Duan S, Ao Y, Dai J, Liu J et al (2011) A highly efficient rice green tissue protoplast system for transient gene expression and studying light/chloroplast-related processes. Plant Methods 7:30

11. Yoo SD, Cho YH, Sheen J (2007) Arabidopsis mesophyll protoplasts: a versatile cell system for transient gene expression analysis. Nat Protoc 2:1565–1572

12. Miao YS, Jiang LW (2007) Transient expression of fluorescent fusion proteins in protoplasts of suspension cultured cells. Nat Protoc 2:2348–2353

13. Jin JB, Kim YA, Kim SJ, Lee SH, Kim DH, Cheong GW et al (2001) A new dynamin-like protein, ADL6, is involved in trafficking from the trans-Golgi network to the central vacuole in Arabidopsis. Plant Cell 13:1511–1525

14. Selevsek N, Matondo M, Carbayo MS, Aebersold R, Domon B (2011) Systematic quantification of peptides/proteins in urine using selected reaction monitoring. Proteomics 11:1135–1147

15. Pfeiffer W (1996) Auxin induces exocytosis of acid phosphatase in coleoptiles from *Zea mays*. Physiol Plant 98:773–779

16. Ibrahim H, Pertl H, Pittertschatscher K, Fadl-Allah E, el-Shahed A, Bentrup FW et al (2002) Release of an acid phosphatase activity during lily pollen tube growth involves components of the secretory pathway. Protoplasma 219:176–183

Chapter 6

Following the Time-Course of Post-pollination Events by Transmission Electron Microscopy (TEM): Buildup of Exosome-Like Structures with Compatible Pollinations

Darya Safavian, Jennifer Doucet, and Daphne R. Goring

Abstract

In the Brassicaceae, the dry stigma is an initial barrier to pollen acceptance as the stigmatic papillae lack surface secretions, and consequently rapid cellular responses are required to accept compatible pollen. Regulated secretion with secretory vesicles or multivesicular bodies is initiated in the stigmatic papillae towards the compatible pollen grain. In self-incompatible species, this basal compatible pollen response is superseded by the self-incompatibility signaling pathway where the secretory organelles are found in autophagosomes and vacuole for destruction. In this chapter, we describe a detailed protocol using the Transmission Electron Microscope to document the rapid cellular changes that occur in the stigmatic papillae in response to compatible versus self-incompatible pollen, at the pollen–stigma interface.

Key words Pollen–pistil interactions, Self-incompatibility, Vesicles, Multivesicular bodies, Exosomes, Autophagy, Transmission electron microscopy

1 Introduction

In flowering plants, successful fertilization is guided by a series of interactions between the pollen and the pistil. These interactions ultimately guide the growing pollen tube to an ovule within the pistil where sperm cells are released for double fertilization with the egg cell and the central cell [1, 2]. In the Brassicaceae, the dry stigma is covered with unicellular papillae that provide a first point of contact for pollen grains [3]. When a compatible pollen grain contacts a stigmatic papilla, this leads to pollen adhesion and the activation of a recognition system to deliver resources for pollen germination [4–9]. In some Brassicaceae species, the self-incompatibility system exists to reject self-pollen and prevent inbreeding. Both the basal compatible pollen pathway and the self-incompatibility pathway function in the stigmatic papillae following pollen contact. The components and dynamics of these cellular pathways have been

Andrea Pompa and Francesca De Marchis (eds.), *Unconventional Protein Secretion: Methods and Protocols*, Methods in Molecular Biology, vol. 1459, DOI 10.1007/978-1-4939-3804-9_6, © Springer Science+Business Media New York 2016

previously reviewed [1, 10, 11], and here we focus on one aspect, the secretory activity in the stigmatic papillae following compatible pollinations versus self-incompatible pollinations.

Earlier studies using transmission electron microscopy (TEM) on *Brassica oleracea* stigmatic papillae treated with either compatible pollen grains or the surface pollen coat extracted from compatible pollen grains had uncovered vesicle-like structures near or in the cell wall of the treated stigmatic papillae. The papillar cell wall was also expanded in response to this compatible pollen coat treatment. This suggested that polar vesicle secretion was part of the stigmatic papilla response towards the compatible pollen, presumably to facilitate pollen hydration, pollen germination, and pollen tube penetration of the stigmatic surface. However, the localization of vesicle-like structures within the cell wall was puzzling [6, 7, 12]. Subsequently, we identified Exo70A1 as required in the stigmatic papillae for the basal compatible pollen pathway, and a strong candidate to be inhibited by the self-incompatibility pathway to reject self-pollen [13–15]. Recently, we have shown that the remaining 7 exocyst subunit genes are also required in the stigmatic papillae to accept compatible pollen [16]. Thus, with the implication of the exocyst complex and polarized secretion [17] during compatible pollen–stigma interactions, we undertook a detailed time-course study of post-pollinations events using the TEM for three Brassicaceae species: *Brassica napus*, *Arabidopsis lyrata*, and *Arabidopsis thaliana* [13, 18]. This highly informative work uncovered cellular responses that had not been previously known to occur during pollen–stigma interactions in the Brassicaceae. These included the discovery of multivesicular bodies (MVBs) fusing to the *Brassica napus* stigmatic papillar plasma membrane for the secretion of exosomes for compatible pollinations, and the relocation of these MVBs to the vacuole for degradation with self-incompatible pollinations [13]. Secretory vesicle-like structures were observed fusing to the plasma membrane for the Arabidopsis species during compatible pollinations, but unexpectedly self-incompatible pollinations were associated with signs of autophagy in the stigmatic papillae [13, 18, 19]. Thus, this study demonstrated the utility of TEM studies as a starting point to document cellular changes during pollen–pistil interactions, and will provide an important context to interpret future studies using fluorescent protein tags to study stigmatic papillar endomembrane dynamics following compatible and self-incompatible pollinations. In this chapter, we describe a detailed protocol using the TEM to document subcellular changes in the stigmatic papillae during the early stages of pollen–stigma interactions.

2 Materials

1. The plant materials used in this study were Brassicaceae species: *A. thaliana, A. lyrata,* and *B. napus.*

2. Glutaraldehyde 8 % aqueous solution; 10 ml ampoules.

3. Paraformaldehyde 16 % aqueous solution; 10 ml ampoules.

4. Osmium tetroxide 4 % $OsO4$ aqueous solution; 2 ml ampoules.

5. Low Viscosity Embedding Media Kit (Spurr's resin) consist of: Vinylcyclohexene Dioxide—225 ml (10 g), Diglycidyl Ether of Polypropylene Glycol—225 ml (8 g), Nonenyl Succinic Anhydride—450 ml (26 g), Dimethylaminoethanol—25 ml (0.4 g) Mix according kit instructions. It is advised to mix fresh (can store overnight at –20 °C).

6. Uranyl Acetate solution: Add 10 g Uranyl Acetate into 100 ml methanol. Mix until the powder is fully dissolved. Store at room temperature.

7. Lead Citrate solution: Dissolve 0.1 g Lead Citrate into 100 ml freshly distilled water. Add 1 ml of 10 N NaOH after lead has dissolved in water. Store at room temperature.

8. Toluidine blue solution: Add 0.1 g toluidine blue in 100 ml distilled water. Store at room temperature.

9. 1 % Formvar in a volumetric flask: Dissolve 1 g formvar in 100 ml chloroform. Store at room temperature.

10. Glass knife for thick sectioning.

11. Microscope slides and cover slips to observe the thick sections under the light microscope.

12. Diamond knife for ultrathin sectioning.

13. Microtome to perform thick and thin sectioning.

14. TEM grids: single slotted copper Grids (2×1 mm, EMS2010-Cu) to collect ultrathin sections.

15. Grid Tweezers and TEM Grid Storage Box.

16. Separatory funnel.

17. Slides.

18. lint-free cloths/kimwipes.

19. Grids and Forceps.

20. Small beaker.

21. Deep staining dish.

22. Prepare all solutions using ultrapure water (prepared by purifying deionized water).

23. 0.2 M phosphate buffer pH 7.2: Prepare stock solutions of 0.2 Na_2HPO_4 (dibasic) and 0.2 M NaH_2PO_4 (monobasic). For a

0.2 M working solution, mix together 72 ml of the 0.2 M Na$_2$HPO$_4$ (dibasic) solution with 28 ml of the 0.2 M NaH$_2$PO$_4$ (monobasic) solution. Measure the pH to ensure that your 0.2 M phosphate buffer is at pH 7.2 at room temperature (to lower the pH, add 0.2 M NaH$_2$PO$_4$; to increase the pH, add 0.2 M Na$_2$HPO$_4$). Store at room temperature.

24. 0.1 M phosphate buffer pH 7.2: Dilute 0.2 M NaPO$_4$ pH 7.2 solution with an equal volume of sterile water.

25. Primary Fixative: 2.7% glutaraldehyde, 2.7% paraformaldehyde, 0.1 M NaPO4 pH 7.2. Mix together 10 ml of 8% glutaraldehyde, 5 ml of 16% paraformaldehyde, and 15 ml of 0.2 M NaPO4 pH 7.2. Make fresh fixative before each use.

26. Secondary Fixative: 1% Osmium tetroxide, 0.1 M NaPO4 pH 7.2. Working in a fume hood, mix together 2 ml of 4% Osmium tetroxide, 4 ml of 0.2 M NaPO4 pH 7.2, and 2 ml of ddH$_2$O. Can be stored at 4 °C in a dark bottle.

27. Graded Ethanol Series: (30, 50, 70, 80, 90, and 100%). Use deionized water to make the ethanol solutions at each concentration. Store at room temperature.

3   Methods

3.1 Sample Preparation

1. On day 1, emasculate late stage 12 flower buds by removing petals and anthers and protect with a small amount of plastic wrap (*see* **Note 1**).

2. On the second day, the stage 13 pistils (Fig. 1) are hand-pollinated in the morning: anthers from other freshly opened flowers are removed and 1–4 anthers are gently brushed across the stigmatic surface (*see* **Note 2**).

3. Remove the pollinated pistils and immediately place in the primary fixative (*see* **Note 3**).

4. Fix pistils in 1 ml primary fixative for 1 h through vacuum infiltration to allow thorough penetration of the fixative into the samples (*see* **Note 4**) and then remove the primary fixative.

5. Rinse the pistils three times by adding 1 ml of 0.1 M phosphate buffer and let the samples sit for 10 min each time.

6. Remove the final rinse.

7. Working in a fume hood, add 1 ml of the secondary fixative and fix for 1 h (*see* **Note 5**) and then remove the secondary fixative.

8. Rinse the pistils once with 1 ml distilled water for 5 min and then remove the water.

9. Dehydrate the pistils by using an ethanol gradient series of 30, 50, 70, 80, and 90%. That is, start by adding 1 ml of 30% ethanol and let the samples sit for 15 min.

Fig. 1 Stage 13 *Arabidopsis thaliana* flower. (**a**) A flower showing the position of the pistil and long anthers. At stage 13, the flower buds have opened (anthesis) and anther dehiscence has occurred. Some parts of the flower have been removed to give a better view of the pistil and anthers. (**b**) Close-up of the top half of the pistil and the long anthers from the flower in (**a**). Note that the stigmatic papillae are elongated, anther dehiscence has occurred, but the long anthers have not extended above the stigma yet (*see* [20] for flower staging)

10. Remove the 30 % ethanol, add 1 ml of 50 % ethanol and let the samples sit for 15 min. Continue through the ethanol gradient series (for 15 min each) and end with the dehydration step in 100 % ethanol for 15 min.

11. Remove the 100 % ethanol, and embed the samples in Spurr's Resin as follows:

    (a) Add 1 ml of 3:1 100 % ethanol/Spurr's Resin, and infiltrate for 30 min. Remove this solution.

    (b) Add 1 ml of 1:1 100 % ethanol/Spurr's and infiltrate for 30 min. Remove this solution.

    (c) Add 1 ml of 100 % Spurr's resin (samples should be in a closed microcentrifuge tube or capped vial) and leave overnight on a slow rotator.

12. The next day, replace the 100 % Spurr's resin with fresh 100 % Spurr's resin and infiltrate the pistils 5 h. Then, embed the samples in 100 % Spurr's resin by placing the pistils in molds and placing the mold with 100 % Spurr's resin in a 65 °C oven to polymerize overnight (Fig. 2) (*see* **Note 6**).

3.2 Formvar Coating Grids

1. For previously used TEM grids, sonicate grids in 95 % ethanol in a small beaker to clean.

2. Rinse TEM grids with distilled water.

3. Transfer the TEM grids onto filter paper and let dry.

4. **Steps 1–3** can be skipped if new TEM grids are being used.

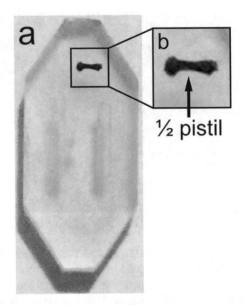

Fig. 2 An *Arabidopsis lyrata* pistil embedded in the resin mold. (**a**) The top half of the pistil was collected and placed horizontally in the resin mold for sectioning of the pollen–stigma interface. (**b**) Close-up of the region in the resin mold containing the top half of the pistil

5. Assemble a separatory funnel and fill with the 1 % formvar solution (Fig.3a).

6. Polish a microscope slide with a lint-free cloth or kimwipes until shinny/slippery, and wipe clean (*see* **Note 7**).

7. Submerge the polished slide into the 1 % formvar solution (Fig 3a).

8. Drain off the 1 % formvar solution in the separatory funnel and remove the slide after 30 s. Let the slide stand to dry (*see* **Note 8**).

9. Overfill a staining dish with distilled water and make sure the surface is clean.

10. Once the coated slide is dry, score the slide edges with a sharp razor blade on both sides and the bottom to release the film from the slide.

11. Exhale on the slide to get a layer of water vapor on and under the film. Immediately float the film onto the water surface by touching the bottom of the slide. Slowly lower the slide into the bath, teasing the film off as necessary (Fig. 3b) (*see* **Note 9**).

12. With a pair of tweezers pick up a TEM grid and gently lay the grid, dull side down, on the floating formvar film in the regions that are of the proper thickness (silver to pale gold color; Fig. 3c). Continue this step until all the TEM grids are laid on the formvar film.

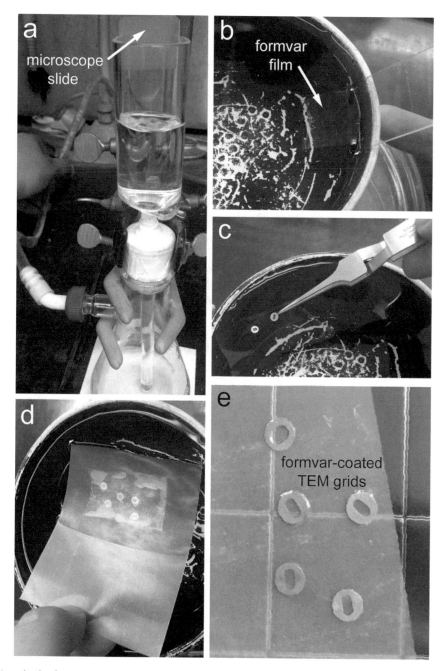

Fig. 3 Steps in the formvar coating of TEM grids. (**a**) Separatory funnel with a microscope slide submerged in the 1 % formvar solution. (**b**) Floating the fomvar film from the microscope slide. (**c**) Laying the TEM grids on the floating fomvar film. (**d**) Using parafilm to pick up the formvar film with the grids lying top. (**e**) Formvar-coated TEM grids left to dry

13. Use a piece of parafilm to cover and pick up the formvar film with the TEM grids on top (Fig. 3d).

14. Place the parafilm in a petri dish with the TEM grids facing up, and let the formvar-coated TEM grids dry several hours to overnight before using them (Fig. 3e).

3.3 Sectioning

1. Using an Ultramicrotome with a glass knife, take serial thick sections of 60 μm thickness from the embedded pistil in the area of interest (the stigma).

2. As you take the thick sections, stain the sections with the Toluidine blue solution to determine whether you have reached the area of interest for the TEM (the pollen–stigma contact point). To do this, use a loop to collect and place the thick sections on a glass slide. Place a drop of toluidine blue on the sections on the slide, wait for 1 min, and then place a drop of water on the slide to wash the sections. Examine the stained sections in a light microscope. Continue the steps of sectioning and staining the sections until the stained sections start to show the pollen–stigmatic papillar interface. The entire pollen grain is stained dark blue and in contact with a papilla which has a light stain with a dark blue outline. Once this is observed in the stained section, move onto the next step.

3. From the area of interest identified from the previous step in the embedded pistil, take ultrathin sections of 80–90 nm thickness using an Ultramicrotome with a diamond knife.

4. Collect the ultrathin serial sections on the formvar-coated TEM grids. This is done by bringing the formvar-coated grid to the vicinity of thin sections on the diamond knife to collect the sections on the grid.

3.4 Staining and TEM Observations

1. Stain the ultrathin sections collected on formvar-coated TEM grids with uranyl acetate by submerging the grids in 70–100 μl of uranyl acetate for 40 min.

2. Rinse the grids in three changes of distilled water for 10 min each time.

3. Stain the sections with lead citrate by submerging the grids in 100 μl of lead citrate for 3 min.

4. Rinse the grids in three changes of distilled water for 10 min each time. Let the grids with the stained ultrathin sections dry overnight before examining them in the TEM.

5. For our work, sections were examined and photographed using either a Hitachi H-7000 TEM or a Hitachi HT7700 TEM at 75 kV (an example shown in Fig. 4).

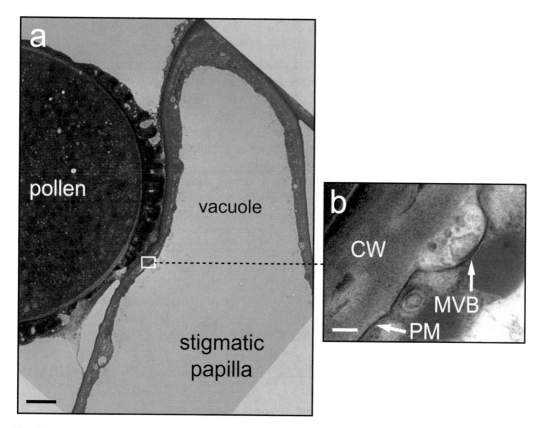

Fig. 4 TEM image of a *Brassica napus* stigmatic papilla in contact with a compatible pollen grain. (**a**) A stigmatic papilla at 10 min post-pollination with compatible pollen. The *white boxed area* is shown in (**b**). Scale bar = 2 μm. (**b**) A multivesicular vesicle (MVB) is fusing to the stigmatic papillar plasma membrane (PM), in the region adjacent to the compatible pollen grain, to release exosomes into the stigmatic papillar cell wall (CW). Scale bar = 100 nm

4   Notes

1. These emasculated flower buds will reach maturity overnight (i.e. Stage 13 is where flower buds open and the stigma is receptive to pollen (Fig. 1). Prior to pollination, make sure that the flower staging is correct by looking for elongated stigmatic papillae (shorter papillae indicate an immature flower bud) [20]. Then, pollinate lightly so that the stigmatic papillae are not damaged and not overly pollinated. If you have not done manual pollinations before, it may be useful to practice and monitor the quality by aniline blue staining *see* [21] for a detailed protocol. With a compatible pollination, the aniline blue-stained pistils should show elongated stigmatic papillae with abundant pollen tubes. Overly bright callose deposits in the stigmatic papillae indicate wounding and possibly over-pollination.

2. The pollen can be left for different time points (we examined 5, 10, and 20 min post-pollination.

3. Unpollinated pistils should also be prepared to compare the ultrastructural features of stigmatic papillae before and after pollination. In comparison to the *Arabidopsis* species, the *B. napus* flowers have larger pistils, and so to facilitate the penetration of fixative, *B. napus* pistils were sectioned longitudinally (cut in half) with a razor blade after pollination and prior to fixation. The procedures that follow from this point should be performed at room temperature and in the fume hood unless otherwise specified. As well, to avoid damaging the pistils, do not place samples on a shaker or rotator during fixing and buffer wash steps unless otherwise specified.

4. Following vacuum infiltration, fix tissue for another hour under normal pressure. Pistils generally sink to the bottom of the tubes, indicating the fixative has fully penetrated the samples. Sometimes, bubbles in the tissue will cause the pistils to float, but as long as the stigmas are facing downwards into solution, they should be properly fixed.

5. The pistils will turn dark brown to black following this step (Toxic chemicals: All fixing steps should be performed in the fume hood.)

6. The pistils are oriented horizontally in the resin mold to allow for sectioning of the pollen–stigma interface for the TEM (Fig. 2).

7. If there is some dust on the slide, the dust can be blown off using compressed air.

8. The thickness of the formvar films produced is regulated by the speed with which the funnel drains and the concentration of the formvar. Use films that are a silver to very pale gold color.

9. Be sure your forceps are cleaned with ethanol.

Acknowledgements

D.S. and J.D. were supported by Ontario Graduate Scholarships (OGS). This research was supported by grants from Natural Sciences and Engineering Research Council of Canada to D.R.G.

References

1. Qu LJ, Li L, Lan Z, Dresselhaus T (2015) Peptide signalling during the pollen tube journey and double fertilization. J Exp Bot 66(17):5139–5150

2. Higashiyama T, Takeuchi H (2015) The mechanism and key molecules involved in pollen tube guidance. Annu Rev Plant Biol 66:393–413

3. Heslop-Harrison Y, Shivanna K (1977) The receptive surface of the angiosperm stigma. Ann Bot 41(6):1233–1258

4. Roberts IN, Stead AD, Ockendon DJ, Dickinson HG (1980) Pollen stigma interactions in Brassica oleracea. Theor Appl Genet 58(6):241–246

5. Chapman LA, Goring DR (2010) Pollen-pistil interactions regulating successful fertilization in the Brassicaceae. J Exp Bot 61(7):1987–1999

6. Dickinson H (1995) Dry stigmas, water and self-incompatibility in *Brassica*. Sex Plant Reprod 8(1):1–10

7. Elleman CJ, Dickinson HG (1996) Identification of pollen components regulating pollination-specific responses in the stigmatic papillae of *Brassica oleracea*. New Phytol 133(2):197–205

8. Edlund AF, Swanson R, Preuss D (2004) Pollen and stigma structure and function: the role of diversity in pollination. Plant Cell 16(Suppl):S84–S97

9. Swanson R, Edlund AF, Preuss D (2004) Species specificity in pollen-pistil interactions. Annu Rev Genet 38:793–818

10. Sawada H, Morita M, Iwano M (2014) Self/non-self recognition mechanisms in sexual reproduction: new insight into the self-incompatibility system shared by flowering plants and hermaphroditic animals. Biochem Biophys Res Commun 450(3):1142–1148

11. Indriolo E, Safavian D, Goring DR (2014) Signaling events in pollen acceptance or rejection in the *Arabidopsis* species. In: Sawada H, Inoue N, Iwano M (eds) Sexual reproduction in animals and plants. SpringerOpen, Tokyo, pp 255–271

12. Elleman CJ, Dickinson HG (1990) The role of the exine coating in pollen-stigma interactions in *Brassica oleracea* L. New Phytol 114(3):511–518

13. Safavian D, Goring DR (2013) Secretory activity is rapidly induced in stigmatic papillae by compatible pollen, but inhibited for self-incompatible pollen in the Brassicaceae. PLoS One 8(12), e84286

14. Safavian D, Jamshed M, Sankaranarayanan S, Indriolo E, Samuel MA, Goring DR (2014) High humidity partially rescues the *Arabidopsis thaliana exo70A1* stigmatic defect for accepting compatible pollen. Plant Reprod 27(3):121–127

15. Samuel MA, Chong YT, Haasen KE, Aldea-Brydges MG, Stone SL, Goring DR (2009) Cellular pathways regulating responses to compatible and self-incompatible pollen in *Brassica* and *Arabidopsis* stigmas intersect at Exo70A1, a putative component of the exocyst complex. Plant Cell 21(9):2655–2671

16. Safavian D, Zayed Y, Indriolo E, Chapman L, Ahmed A, Goring D (2015) RNA silencing of exocyst genes in the stigma impairs the acceptance of compatible pollen in *Arabidopsis*. Plant Physiol 169(4):2526–2538

17. Synek L, Sekeres J, Zarsky V (2014) The exocyst at the interface between cytoskeleton and membranes in eukaryotic cells. Front Plant Sci 4:543

18. Indriolo E, Safavian D, Goring DR (2014) The ARC1 E3 ligase promotes two different self-pollen avoidance traits in *Arabidopsis*. Plant Cell 26(4):1525–1543

19. Safavian D, Goring D (2014) Autophagy in the rejection of self-pollen in the mustard family. Autophagy 10(12):2379–2380

20. Smyth DR, Bowman JL, Meyerowitz EM (1990) Early flower development in *Arabidopsis*. Plant Cell 2:755–767

21. Balanza V, Ballester P, Colombo M, Fourquin C, Martinez-Fernandez I, Ferrandiz C (2014) Genetic and phenotypic analyses of carpel development in *Arabidopsis*. Methods Mol Biol 1110:231–249

Part III

UPS Contemplates Multidisciplinary Approaches for Animals and Fungi

Chapter 7

Investigating Alternative Transport of Integral Plasma Membrane Proteins from the ER to the Golgi: Lessons from the Cystic Fibrosis Transmembrane Conductance Regulator (CFTR)

Margarida D. Amaral, Carlos M. Farinha, Paulo Matos, and Hugo M. Botelho

Abstract

Secretory traffic became a topical field because many important cell regulators are plasma membrane proteins (transporters, channels, receptors), being thus key targets in biomedicine and drug discovery. Cystic fibrosis (CF), caused by defects in a single gene encoding the CF transmembrane conductance regulator (CFTR), constitutes the most common of rare diseases and certainly a paradigmatic one.

Here we focus on five different approaches that allow biochemical and cellular characterization of CFTR from its co-translational insertion into the ER membrane to its delivery to the plasma membrane.

Key words Secretory traffic, Biochemistry, CFTR, Cystic fibrosis, Endoplasmic reticulum, Golgi, High-throughput microscopy, Plasma membrane, N-glycosylation, ABC transporters

1 Introduction

Secretory traffic became a topical field because many key cell regulators are plasma membrane (PM) proteins (transporters, channels, receptors), thus being key targets in biomedicine and drug discovery. Cystic fibrosis (CF), which affects well over 30,000 sufferers in Europe with a further 50 thousand worldwide, constitutes the most common of rare diseases and certainly a paradigmatic one. CF is caused by defects in a single gene encoding the CF transmembrane conductance regulator (CFTR), a chloride/bicarbonate channel that is also a member of the ATP-binding cassette ABC transporter superfamily.

Like most membrane proteins, CFTR biogenesis and trafficking follow the secretory pathway, from the endoplasmic reticulum (ER) to the plasma membrane through the Golgi apparatus. After

Andrea Pompa and Francesca De Marchis (eds.), *Unconventional Protein Secretion: Methods and Protocols*, Methods in Molecular Biology, vol. 1459, DOI 10.1007/978-1-4939-3804-9_7, © Springer Science+Business Media New York 2016

insertion into the ER membrane, newly synthesized CFTR—both wild type (wt) or bearing a mutation (e.g., F508del, the most common disease-causing variant)—is N-glycosylated through the addition of a 14-unit oligosaccharide on two asparagine residues located in its fourth extracellular loop. This addition generates the immature, core-glycosylated form of CFTR (known as band B). After undergoing correct folding, which is assessed by the ER quality control (ERQC), the mature core-glycosylated form of wt-CFTR is exported in COPII vesicles. When passing through the Golgi, its glycan moieties undergo processing producing the mature form (known as band C). In contrast, CFTR bearing F508del is retained in the ER due to misfolding and prematurely targeted for degradation by the ubiquitin-proteasome pathway (UPP). Thus, this variant never acquires the fully glycosylated pattern [1]. The model that we have proposed to explain the early stages of CFTR trafficking through the secretory pathway [2] involves several checkpoints to assess CFTR folding status, namely (a) two checkpoints involving the cytosolic Hsp70 and ER calnexin chaperones that recognize CFTR structural/glycan cues, respectively [3]; (b) negative selection at the ER exit sites mediated by arginine-framed tripeptides (AFTs) [4, 5]; and (c) positive selection upon exposure of a DAD motif [6, 7].

This movement of CFTR from the ER to its final destination, the PM, involves transport through a series of distinct vesicular compartments. The early CFTR traffic pathways so far described include (a) conventional anterograde traffic from the ER exit sites into COPII vesicles, (b) retrograde recycling from the *cis* Golgi to the ER, and (c) non-conventional trafficking via tubular structures migrating peripherally to the Golgi cisternae. This "unconventional" traffic route for CFTR transport to the PM was described to be insensitive to blocking of conventional ER-to-Golgi traffic and appears to involve the SNARE protein syntaxin 13 [8]. In this pathway however, the protein still travels back to the *cis* Golgi where it undergoes oligosaccharide processing to the complex form.

More recently, another unconventional pathway was described to occur during ER stress involving GRASP (Golgi reassembly stacking proteins). GRASP55/65 were shown to be tethering factors that are involved in the ER stress-induced non-conventional secretion. In this situation, CFTR reaches the PM in its core-glycosylated form [9]—in fact, non-glycosylated (either chemically or genetically) CFTR was also shown to reach the cell surface and to be functional [3].

The complex regulatory processes affecting CFTR traffic cannot be easily rationalized owing to the large number of potential mediators: at least 200 different proteins have been shown to interact with CFTR using a proteomics approach [10]. This figure does not reflect the potentially larger number of indirect players, as we need to determine the functional relationship of these interactors. Loss-of-function assays employing RNA interference (RNAi) are instrumental for dissecting such complex networks [11]. These assays are typically based on measuring a control CFTR "cellular phenotype" and performing systematic RNAi experiments to identify genes whose knockdown/knockout affects that cellular phenotype. Such genes are then postulated as candidate regulators. In this regard, high-content microscopy (HCM) is a powerful tool as it allows thorough characterization of a biological process [12], overcoming many of the drawbacks of plate reader-based assays. In the context of CF, an HCM protocol enabled identifying diacylglycerol kinase isoform iota (DGKɪ) and ciliary neurotrophic factor receptor (CNTFR) as novel regulators of the epithelial sodium channel (ENaC), a major contributor to CF morbidity [13]. Because HCM approaches are phenotype based, they can be readily applied to the drug discovery pipeline to identify novel CFTR correctors, as recently demonstrated [14].

In this chapter, we describe a set of biochemical and cell biology methods that allow evaluation of CFTR trafficking to the cell surface—assessing its steady-state levels, turnover and stability, localization at the plasma membrane, glycan processing, and trafficking efficiency.

2 Materials

2.1 Western Blot Detection of CFTR

1. Buffer: 31.25 mM Tris–HCl pH 6.8, 1.5% (w/v) SDS, 5% glycerol, 0.01% (w/v) bromophenol blue, 0.5 mM DTT (*see* **Note 1**).

2. Bradford reagent.

3. Spectrophotometer.

4. Mini sodium dodecyl sulfate-polyacrylamide-gel (SDS-PAGE) gel system.

5. Benzonase: 25 U/ml in sample buffer.

6. 7–9% Laemmli SDS-PAGE gel (either a pre-cast gel, or a in-house prepared with 4% (w/v) stacking and 7–9% (w/v) separating gel, prepared with acrylamide:bisacrylamide mixture 37.5:1).

7. Running buffer: 0.025 M Tris, 0.192 M glycine, 0.1% (w/v) SDS.

8. Transfer buffer: 0.025 M Tris–HCl pH 8.3, 0.192 M glycine, 20% (v/v) methanol.

9. SDS-PAGE molecular weight standards.

10. PVDF membrane (pore size: 0.45 μm).

11. 0.1% (v/v) Tween 20 in phosphate-buffered saline, PBS (PBST).

12. 5% (w/v) skimmed milk in PBST.

13. Anti-CFTR monoclonal antibody (CFF reference 596): Working solution is 1: 3000 in 5% milk-PBST.

14. Anti-mouse IgG horseradish-peroxidase-conjugated secondary antibody (Bio-Rad): Working solution is 1:3000 in 5% milk-PBST.

15. Western chemiluminescent substrate system.

16. Anti-α-tubulin monoclonal antibody produced in mouse (clone clone B-5-1-2, T5168—Sigma Aldrich).

17. X-ray film.

2.2 Pulse-Chase of CFTR

1. Hanks' balanced salt solution (HBSS).

2. Medium without methionine.

3. Protein G agarose beads.

4. Protease inhibitor cocktail.

5. PBS.

6. L-[^{35}S]/Methionine L-[^{35}S] cysteine protein labeling mix.

7. Methionine: Prepare a 100 mM solution and filter sterilize.

8. Radioimmunoprecipitation assay (RIPA) buffer: 1.0% (w/v) sodium deoxycholate, 1.0% (v/v) Triton X-100, 0.1% (w/v) SDS, 150 mM NaCl, 50 mM Tris pH 7.4 (*see* **Note 2**).

9. Sodium salicylate 1 M.

2.3 Cell Surface Biotinylation and Endocytosis of CFTR

1. EZ-Link Sulfo-NHS-SS-Biotin.

2. Streptavidin-agarose beads.

3. Protease inhibitor cocktail.

4. L-Glutathione reduced.

5. PBS.

6. PBS-CM: PBS, 0.1 mM $CaCl_2$, 1 mM $MgCl_2$.

7. Quenching buffer: 100 mM Tris pH 8.0, 150 mM NaCl, 0.1 mM $CaCl_2$, 1 mM $MgCl_2$, 10 mM glycine, 1% BSA (w/v).

8. Pull-down buffer (PD buffer): 50 mM Tris pH 7.5, 100 mM NaCl, 10% glycerol (v/v), 1% NP-40 (v/v).

9. Wash buffer: 100 mM Tris pH 7.5, 300 mM NaCl, 1% TX-100 (v/v).

10. Stripping buffer: 60 mM L-Glutathione, 90 mM NaCl, 1 mM $MgCl_2$, 0.1 mM $CaCl_2$, 90 mM NaOH, 10% FBS (v/v).

2.4 Assessment of CFTR Glycosylation

1. PNGase F (500,000 U/ml).

2. PNGase F buffer (G7)—supplied—0.5 M sodium phosphate pH 7.5.

3. NP-40 10% (v/v).

4. Endoglycosidase H (500,000 U/ml).

5. Endoglycosidase H buffer (G5)—supplied—0.5 M sodium citrate pH 5.5.

6. Neuraminidase.

7. Neuraminidase buffer—50 mM sodium phosphate buffer pH 6.0.

8. Fucosidase.

9. Fucosidase buffer—50 mM sodium phosphate buffer pH 5.0.

10. Labeling and lysis reagents as in Subheading 2.2.

2.5 Microscopy-Based Assays to Analyze CFTR Traffic

1. CFBE or A549 cell lines expressing a Tet-ON mCherry-Flag-CFTR construct (wt or F508del variants, described in [12]).

2. DMEM high glucose with L-glutamine supplemented with 10% fetal bovine serum, 10 µg/ml blasticidin, and 2 µg/ml puromycin.

3. Doxycycline.

4. VX-809 (*see* **Note 3**).

5. Dulbecco's PBS, supplemented with 0.7 mM $CaCl_2$ and 1.1 mM $MgCl_2$ (DPBS++) (*see* **Note 4**).

6. 3% Paraformaldehyde (PFA), freshly diluted in DPBS++ (*see* **Note 5**).

7. Bovine serum albumin.

8. Mouse monoclonal anti-FLAG antibody (Sigma F1804): 1:500 (2 µg/ml), prepared in DPBS++ supplemented with 1% BSA.

9. Donkey anti-mouse antibody, Alexa Fluor® 647 conjugate (Life Technologies A-31571): 1:500 (2 µg/ml), prepared in DPBS++ supplemented with 1% BSA.

10. 0.2 µg/ml Hoechst 33342, in DPBS++.

11. 10 cm Cell culture Petri dishes.

12. Chambered cover slips or multiwell plates: Plates can be used as is or after coated with siRNA/lipofectamine for reverse transfection [12, 15] (*see* **Note 6**).

13. Non-targeting scrambled siRNA.

14. COPB1 siRNA.

15. Multidrop™ Combi dispenser.

16. Liquidator™ 96 Manual pipette: Only required for 96- or 384-well plates.

17. Automated wide-field epifluorescence microscope equipped with a 10× objective.

18. Personal computer running CellProfiler (http://www.cellprofiler.org/ [16]).

3   Methods

3.1   Western Blot to Determine Steady-State Levels of Immature and Mature Forms of CFTR

The use of biochemical techniques to study CFTR has been previously described as a workflow [17]. The Western blot technique allows the assessment of steady-state levels of the different forms of CFTR (in general, the immature—with a molecular mass of about 140–150 kDa—and the mature form—with a molecular mass of 170–180 kDa) and, from quantification of these forms, an evaluation of the efficiency of processing.

1. Grow CFTR-expressing cells on 60 mm Petri dishes until confluence.

2. Wash the cells three times in PBS.

3. Make a cell lysate: Solubilize cells in 200 μl sample buffer per dish supplemented with 25 U/ml benzonase.

4. Quantify total protein by Bradford's assay [18] or another appropriate protein quantification method.

5. Load the cell lysate (~30 μg/well) on a mini SDS-PAGE (stacking 4% (w/v), separating 7% (w/v) acrylamide:bisacrylamide 37.5:1) at constant voltage (100–120 V), for 3 h. Use 8–10 μl of an appropriate molecular weight standard (*see* **Note 7**).

6. Transfer proteins to a PVDF membrane using wet blotting (Bio-Rad Mini-PROTEAN) for 1 h 30 min at 400 mA, constant current. The transfer should be done on ice (mixed with water) or with refrigeration (4 °C).

7. Wash the PVDF membrane with PBST for 15 min.

8. Block the PVDF membrane by incubating in 5% (w/v) skimmed milk in PBST for 2 h.

9. Cut the membrane around the 75 kDa molecular weight marker. Use the higher molecular weight part to probe with the anti-CFTR antibody (below). The lower molecular weight segment can be probed with anti-tubulin antibody as an internal loading control.

10. Incubate anti-CFTR 596 monoclonal antibody overnight (diluted 1:3000 in 5% (w/v) skimmed milk in PBST) at 4 °C with gentle mixing.

11. Wash three times for 5 min with PBST.

12. Incubate with anti-mouse horseradish peroxidase (HRP)-conjugated secondary antibody (diluted 1:3000 5% (w/v) skimmed milk in PBST) for 1 h.

13. Wash three times for 5 min with PBST.

14. Detect protein bands using the chemiluminescent substrate as per kit instructions by exposing X-ray films for the appropriate time (1–5 min) (Fig. 1a).

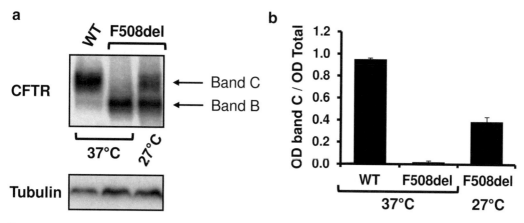

Fig. 1 Western blot detection of CFTR and calculation of processing efficiency. (**a**) BHK cells stably expressing wt- or F508del-CFTR were cultured at 37 °C and either directly lysed or placed for 48 h at 27 °C before lysis, to allow partial rescue of F508del-CFTR maturation. Following electrophoretic separation, proteins were transferred to PVDF membranes and probed with antibodies anti-CFTR and also anti-α-tubulin, as a loading control. Note the partial conversion of immature F508del-CFTR (band B) to mature protein (band C) upon incubation at 27 °C. (**b**) WB films were scanned and the intensity of the bands corresponding to tubulin and CFTR immature and mature forms was quantified using an appropriate software (ImageJ in this example). The graph expresses the efficiency of CFTR processing, calculated as the ratio between the OD values for band C and for the total amount of CFTR (band C + band B). Data are represented as means ± SEM, $n = 7$–9

15. Scan the gel and quantify the bands corresponding to CFTR immature and mature forms using an appropriate software, for example ImageJ, GE ImageQuant 1D, Bio-Rad Quantity One, and Bio-Rad Image Lab.

16. Calculate the efficiency of processing as the ratio between the amount of band C and the total amount of CFTR (Fig. 1b).

3.2 Pulse-Chase Experiments to Determine Turnover Rate of CFTR Immature Form and Efficiency of Maturation

The pulse-chase technique is a well-described and common procedure that, with the use of a labeled amino acid (or amino acids), allows the determination of the turnover of CFTR immature form (band B) and also of the conversion of the immature into the mature form (band C).

Cells expressing the CFTR variant of interest should be grown to loose confluency. After a period of starvation, cells are then incubated, in general, with a mixture of [^{35}S]-methionine/cysteine. This period is called the pulse. After this period, during which the labeled amino acids are incorporated into nascent proteins, the medium containing the radioactive amino acids is removed and replaced by the chase medium that contains an excess of unlabeled methionine. At different times after medium replacement, the cells are lysed and CFTR is immunoprecipitated. Samples are run in an SDS-PAGE gel which after fluorography and drying is exposed to an X-ray film.

Results are then obtained by quantifying CFTR mature and immature forms. The ratio between the amount of labeled immature CFTR at a specific time (P) and the amount of labeled immature

CFTR at the end of pulse (P_0) plotted against the duration of chase corresponds to the turnover of CFTR. The ratio between the amount of labeled mature CFTR at a specific time and the amount of labeled immature CFTR at the end of pulse plotted against the duration of chase corresponds to the efficiency of processing (Fig. 2).

1. Seed BHK cells expressing CFTR on 60 mm dishes 24 h before experiment. The seeding density should be such that cells are at sub-confluency at the time of starting the experiment (*see* **Notes 8** and **9**).

2. Remove media and wash twice with 2 ml of HBSS.

3. Incubate in methionine-free medium for 30 min.

4. Remove media and pulse with 1 ml methionine-free medium containing 150 µCi/ml [^{35}S]-methionine/cysteine for 30 min.

5. After the pulse period, remove the medium containing the labeled amino acids and wash twice with 1 ml HBSS.

6. Feed the cells with their regular medium supplemented with the appropriate amount of fetal bovine serum, 1 mM methionine, and, if needed, 25 µg/ml of cycloheximide (*see* **Note 10**).

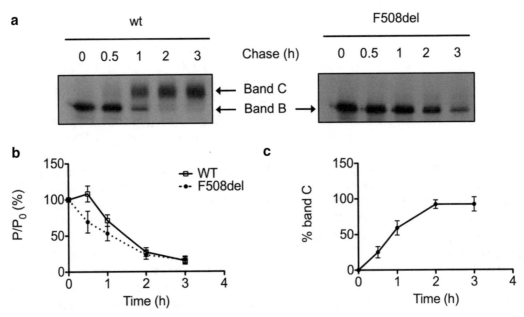

Fig. 2 Turnover and processing of wt- and F508del CFTR. (**a**) BHK cells stably expressing wt- or F508del-CFTR were pulse-labeled with [^{35}S]-methionine and chased for 0, 0.5, 1, 2, and 3 h. Cells were then lysed and immunoprecipitated with an anti-CFTR Ab. Following electrophoretic separation and fluorography, immature (band B) and mature (band C) forms of CFTR were quantified. (**b**) Turnover of the core-glycosylated form (band B) of wt- and F508del-CFTR is shown as the ratio between P, the amount of band B at time t, and P_0, the amount of band B at the start of the chase (i.e., at the end of pulse). (**c**) The efficiency of conversion of the core-glycosylated form (band B) into the fully glycosylated form of wt-CFTR (band C) is determined as the ratio between the amount of band C at time t and the amount of band B at the start of the chase (P_0)

7. At appropriate intervals of incubation with the chase medium, harvest cells. For this, wash them twice with 1 ml ice-cold PBS, add 1 ml RIPA buffer supplemented with protease inhibitors, and incubate for 30 min at 4 °C.

8. Scrape cells off the plate and pass ten times through a 1 ml pipet tip and then ten additional times through a 200 μl pipet tip.

9. Transfer the cell lysates to a 1.5 ml microcentrifuge tube with O-ring.

10. Centrifuge in a microcentrifuge at $14,000 \times g$ for 30 min, 4 °C.

11. Transfer the supernatant to a new microcentrifuge and discard the pellet (*see* **Note 11**).

12. Add the appropriate amount of antibody (1:1000 if using anti-CFTR antibody 596 or 570, provided by CFFT) and 40 μl of Protein G beads (*see* **Note 12**).

13. Incubate overnight in cold room (4 °C) with shaking.

14. Wash the beads three times with 1 ml ice-cold RIPA (*see* **Note 13**).

15. Add 70 μl of sample buffer (*see* Subheading 2.1) and incubate for 30 min with continuous mixing at room temperature.

16. Spin at $14,000 \times g$ for 2 min.

17. Collect supernatant (*see* **Note 11**).

18. Load samples onto a 20 cm gel and run overnight at approximately 75 V (*see* **Note 14**).

19. Fix the gel in 30 % (v/v) methanol/10 % (v/v) acetic acid for 30 min.

20. Wash four times for 15 min in bidistilled H_2O.

21. For fluorography, treat the gel for 1 h with 1 M sodium salicylate.

22. Dry the gel and expose to an X-ray film.

23. Develop in 24/48 h.

24. Scan the film and quantify the bands corresponding to CFTR immature and mature forms using an appropriate software: ImageJ, GE ImageQuant 1D, Bio-Rad Quantity One, or Bio-Rad Image Lab.

3.3 Cell Surface Protein Biotinylation to Assess CFTR Plasma Membrane Abundance and Endocytosis

The end point of the successful processing and trafficking of CFTR is its delivery to and function at the plasma membrane. A number of methods have been proposed to enrich, purify, and quantify the amount of CFTR at the cell surface. Among them, usage of biotinylating reagents and exploitation of the strong interaction between biotin and streptavidin for the purification of biotinylated surface proteins have rapidly gained in popularity and allowed some of the most significant progresses in evaluating and quantifying the efficacy of CFTR trafficking to the plasma membrane and the rate of its turnover from the cell surface.

The most common targets for modifying protein molecules are primary amine groups that are present as lysine side-chain epsilon-amines and N-terminal alpha-amines. Hence, N-hydroxysuccinimide (NHS) esters are among the most widely used amine-reactive biotinylation reagents. Their poor solubility in aqueous solutions has been overcome by the addition of a sulfonate group on the N-hydroxysuccinimide ring, which also made them ideal as surface biotinylation reagents, because sulfo-NHS-esters do not penetrate the cell membrane.

The protocol described herein makes use of another NHS-biotin derivate—the sulfo-NHS-SS-biotin, a reagent that includes a disulfide bond in the spacer arm separating the sulfo-NHS and biotin groups. The S-S bond can be cleaved using reducing agents, enabling the biotin group to be disconnected ("stripped") from the labeled proteins at the plasma membrane. This reagent can thus be used to analyze both the steady-state amount of CFTR at the cell surface and to follow the rate of CFTR internalization from the plasma membrane (Fig. 3). In

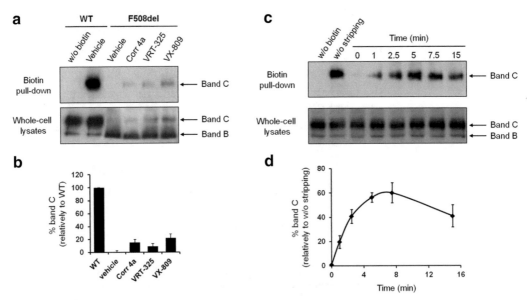

Fig. 3 Abundance and endocytosis of CFTR at the cell surface. (**a**) CFBE cells stably expressing wt- or F508del-CFTR were incubated for 24 h with 5 μM VRT-325, 10 μM Corr-4a, or 3 μM VX-809. CFTR abundance at plasma membrane was analyzed by surface protein biotinylation followed by western blot with an anti-CFTR Ab. Shown are representative images of immunoblots for the biotinylated fraction of mature (band C) CFTR at the plasma membrane (biotin pull-down) and of whole-cell lysates showing the steady-state abundance of immature (band B) and mature (band C) forms of CFTR upon the different treatments. (**b**) Quantification of the abundance of mature F508del-CFTR at the cell surface after treatment with the indicated corrector compounds, relatively to wt-CFTR. Data are mean values ± SEM of $n = 4$ independent experiments. (**c**) Proteins at the surface of BHK cells stably expressing wt-CFTR were labeled with biotin at 0–4 °C and allowed to internalize by placing the cells at 37 °C for 0, 1, 2.5, 5, 7.5, and 15 min. Following striping of the remaining biotin labels at the cell surface, cells were lysed and the internalized, biotin-labeled CFTR proteins were isolated by streptavidin-mediated capture and analyzed by Western blot as in (a). (**d**) Quantification of the amount of CFTR internalized at the different time points, expressed as a percentage of the total amount of CFTR biotinylated at the cell surface (w/o stripping). Data are mean values ± SEM of $n = 6$ independent experiments

the first case cells are placed on ice to stop all exocytic and endocytic trafficking and labeled with a solution of sulfo-NHS-SS-biotin in PBS. After quenching of the reaction with an amine-rich buffer, the cells are thoroughly washed and lysed and biotin-labeled proteins at the cell surface isolated by pull-down with streptavidin-coated agarose beads. To assess the rate of CFTR endocytosis the first protein labeling step is identical but after quenching the reaction the labeled proteins are allowed to internalize by replacing the cells at 37 °C for increasing time intervals, followed by striping the biotin labels from the remaining labeled proteins at the surface with a reducing agent (such as glutathione). In this way, only internalized CFTR proteins, protected inside the cell from the reducing reagent action, will be isolated by streptavidin-mediated capture. Captured proteins are then eluted with dithiothreitol (DDT)-containing Lacmmli buffer, allowing the efficient cleavage of S-S bond and their release for analysis by Western blot as described in Subheading 3.1.

The following protocol was successfully used in our lab to analyze CFTR plasma membrane levels and endocytosis in BHK, CFBE, HeLa, and HEK-293 cells [19–22].

1. Seed cells expressing CFTR on 60 mm dishes 24 h before the experiment. The seeding density should be such that cells are at near confluence at the time of starting the experriment.

2. Before starting prepare all required solutions (see Subheading 2.3) and pre-block the streptavidin-agarose beads as follows:

   (a) Wash 50 μl/dish of streptavidin bead slurry three times with PBS.

   (b) Remove the supernatant and add at least twice the dry bead volume of PD-buffer (see Subheading 2.3) supplemented with 2 % of skimmed milk.

   (c) Rotate for at least 1 h at 4 °C.

   (d) Wash three times with pull-down buffer using $1000 \times g$ spins and remove the supernatant.

   (e) Add one volume of PD-buffer supplemented with protease inhibitors to the dry beads.

3.3.1 From Here on Procedures Should Be Carried Out on Ice, Preferably in a Cold Room (4 °C)

1. Wash cells 3–5 times with 2 ml of ice-cold PBS-CM to remove all medium contaminants (washing procedure depends on the cell type used (see **Note 15**)).

2. Keep one dish with PBS-CM to function as the "without-biotin" control (to assess the amount of contaminant proteins precipitating with the beads alone) and incubate the remaining dishes with 1.5 ml of ice-cold PBS-CM + 0.5 mg/ml EZ-Link sulfo-NHS-SS-biotin for 30 min (see **Notes 16–20**).

3. Aspirate all the labeling buffer and discard. Rinse cells twice with 2 ml of ice-cold quenching buffer.

4. Quench the reaction for 10 min with 2 ml of fresh quenching buffer (ice cold).

5. Wash 3× with 2 ml of ice-cold PBS-CM.

6. Stop here and proceed to subheading 3.3 (**Step 15**) for the endocytosis assay procedure.

7. Lyse cells on ice with 250 μl of ice-cold PD-buffer supplemented with protease inhibitors, scrape cells, and collect whole lysates to 1.5 ml microcentrifuge tubes. Centrifuge for 5 min at $10,000 \times g$ at 4 °C.

8. Save 40 μl of cleared lysates to new microcentrifuge tubes containing 40 μl of 2× Laemmli buffer (to assess total CFTR levels in the samples).

9. Pass 200 μl of the cleared lysates to new 1.5 ml microcentrifuge tubes.

10. Add 50 μl of pre-blocked streptavidin bead slurry to each tube and rotate for 1 h at 4 °C.

11. Centrifuge for 1 min at $5000 \times g$, discard supernatant, and wash 3–5 times with wash buffer.

12. Remove the supernatant, dry the beads, and elute captured protein with 25 μl of DTT-containing 2× Laemmli buffer.

13. Analyze samples by Western blot as described in Subheading 3.1.

3.3.2 For Assessing the Rate of CFTR Endocytosis Proceed as Follows

1. Before starting, warm a flask containing the necessary volume of culture medium (~2 ml/dish) to 37 °C.

2. After biotinylation of cell surface proteins (Subheading 3.3.1, **Steps 1–5**, *see* above) place all the 60 mm dishes on ice in a styrofoam box with lid and take them near an incubator at 37 °C.

3. Keep the dishes corresponding to the time, "with out biotin," "without striping," (which will return the total amount of CFTR initially labeled at the cell surface), and "0 min" (where no surface-labeled CFTR is yet internalized) on ice.

4. Replace the buffer in all remaining dishes with 2 ml of warm medium and incubate at 37 °C for the required periods, e.g., 1, 2.5, 5, 7.5, and 15 min.

5. At each time point, quickly discard the buffer, add 2 ml of ice-cold PBS-CM to the dish, and place it back on the ice box. Keep the lid closed and return the dishes to the cold room as soon as possible.

6. Back in the cold room, discard the buffer of all dishes, and wash with 2 ml of fresh, ice-cold PBS-CM.

7. Add 2 ml of stripping buffer to the 0, 1, 2.5, 5, 7.5, and 15 min dishes and incubate on ice for 15 min at 4 °C. Repeat this step three times, discarding the buffer and adding 2 ml of fresh stripping buffer each time to each dish (total stripping time: 45 min divided in three independent washes).

8. Aspirate all the stripping buffer and discard. Rinse cells 3–5 times with 2 ml of ice-cold PBS-CM and proceed with cell lysis and pull-down as above (Subheading 3.3.1, **Steps 7–13**).

3.4 Glycosylation Assessment with Specific Glycosidases to Assess Trafficking Through or Out of the Golgi

CFTR is N-glycosylated at two asparagine residues located in the fourth extracellular loop (N894 and N900). As occurs with membrane and secreted proteins, glycosylation occurs co-translationally at the ER and the 14-unit glycan is then processed through its trafficking in the endoplasmic reticulum and Golgi complex.

Treatment of CFTR with different glycosidases can then be used to assess if the protein passed through different compartments along its secretory trafficking, as shown for wt- and F508del-CFTR (Fig. 4).

Treatment with endoglycosidase H (endoH) assesses protein exit from the ER and reaching the *cis* Golgi. The ER immature forms are sensitive to endoH whereas later/fully processed forms are endoH resistant.

Treatment with PNGase F (N-glycanase) is used to assess the presence of all sorts of N-linked glycans, as the enzyme hydrolyzes the N-glycosidic bond connecting the glycan to the protein's asparagine residues.

Treatment with either neuraminidase or fucosidase will identify if the protein reached the latter cisternae of the Golgi complex where sialic (N-acetylneuraminic) acid or fucose residues are attached to protein-linked glycans. Thus, resistance to treatment with either enzyme will indicate that the protein does not possess such residues, whereas a decrease in molecular weight upon treatment will indicate that they are present in the glycan unit.

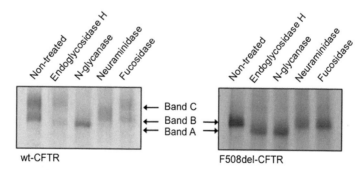

Fig. 4 Analysis of the glycan moieties of wt- and F508del-CFTR with different glycosidases (endoglycosidase H, PNGase F/N-glycanase, neuraminidase, and fucosidase). Following endoH-treatment, a shift is observed in CFTR immature but not in CFTR mature form. Following N-glycanase treatment, all forms are digested, this corresponding to removal of all N-linked glycans, resulting in the de-glycosylated primary CFTR amino acidic chain, also called band A. Following treatment with neuraminidase or fucosidase, there is a shift only in wt-CFTR mature form

1. Seed BHK cells expressing CFTR on 60 mm dishes 24 h before experiment. The seeding density should be such that cells are at sub-confluency at the time of starting the experiment.

2. Label cells for 3 h with 150 µCi of [^{35}S]-methionine/cysteine (for details *see* above Subheading 3.2).

3. After labeling, wash cells twice with ice-cold PBS, lyse them in 1 ml of complete RIPA buffer as above, and add anti-CFTR antibody and protein G-sepharose beads as above (*see* Subheading 3.2).

4. Incubate overnight at 4 °C with rotation.

5. Wash three times with 1 ml of RIPA and elute in 40 µl of sample buffer (*see* Subheading 2.1).

6. Prepare the glycosidase reactions as follows:

Endoglycosidase H	Neuraminidase
30 µl of protein sample	30 µl of protein sample
5 µl of buffer G5	5 µl Sodium phosphate buffer 0.5 M pH 6
1 µl of endoglycosidase H	1 µl of neuraminidase
Bidistilled H$_2$O to 50 µl	Bidistilled H$_2$O to 50 µl
PNGase F	*Fucosidase*
30 µl of protein sample	30 µl of protein sample
5 µl of buffer G7	5 µl of sodium phosphate buffer 0.5 M pH 5
5 µl of NP-40 10%	1 µl of fucosidase
1 µl of PNGase F	Bidistilled H$_2$O to 50 µl
Bidistilled H$_2$O to 50 µl	

7. Incubate the reactions overnight at 37 °C.

8. Run the samples in a 20 cm gel (*see* above Subheading 3.2) (*see* **Note 21**).

9. Treat, dry, and expose gel as above (Subheading 3.2) (*see* **Note 22**).

3.5 Microscopy-Based Assays to Determine Traffic Efficiency and Endocytosis in Inducible Systems

The CFTR traffic assay is based on the quantification of CFTR traffic efficiency in CFBE or A549 cell lines which express a double-tagged CFTR construct: an mCherry-CFTR fusion molecule (wt- or F508del-) harboring a Flag tag insertion at the fourth extracellular loop (Fig. 5), as described in [12]. By performing immunofluorescence labeling in unpermeabilized fixed cells, Flag tags (i.e., CFTR molecules) located at the PM—but not elsewhere—can be detected (Alexa 647) (*see* **Note 23**). Then, the fraction of CFTR molecules delivered to the PM—i.e., the traffic efficiency—is determined in each cell as follows:

$$\text{CFTR traffic efficiency} = \frac{\text{PM CFTR}}{\text{Total CFTR}} = \frac{\text{Alexa Fluor}^{\circledR}\ 647\ \text{Integrated Fluorescence}}{\text{mCherry Integrated Fluorescence}}$$

This assay can be coupled to RNAi or chemical compound treatments. In these cases, the inducible Tet-ON promoter allows initiating CFTR expression only after the onset of knockdown or compound effect (Fig. 5).

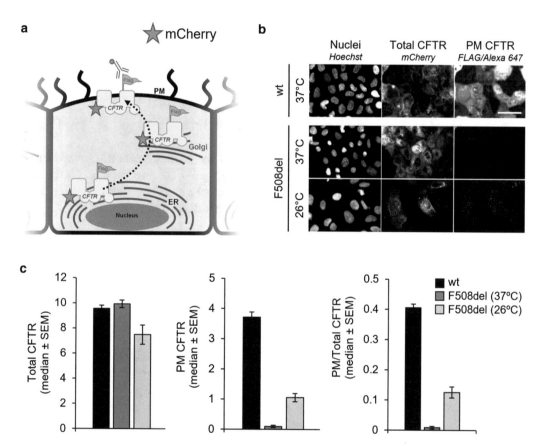

Fig. 5 Overview of the microscopy-based CFTR traffic assay. (**a**) The basis for the traffic assay is a CFTR traffic reporter which is a double-tagged (mCherry and Flag) CFTR molecule (wt- or F508del-variants), whose gene has been stably integrated into A549 or CFBE cell lines. The Flag tag resides at CFTR's fourth extracellular loop and only becomes extracellular if the protein successfully traffics to the PM. By immunostaining the Flag tag in unpermeabilized cells, the fraction of CFTR molecules at the PM can be detected. (**b**) Fluorescence microscopy imaging of the mCherry-Flag-CFTR constructs. wt-CFTR was expressed at 37 °C and F508del-CFTR was expressed at either 37 or 26 °C. No siRNAs or correctors were used. Images were acquired using a metal halide light source (Leica EL6000) at maximum brightness, a 10× N.A. 0.4 objective, a Leica DFC365 FX CCD camera, and the following filter cubes: Leica N.21 (mCherry) and a custom excitation BP 645/30; emission LP 670 filter cube (Alexa 647). Under these conditions, the exposure times for Cherry and Alexa 647 were 1.3 and 8 s, respectively. Scale bar = 50 μm. (**c**) Quantification of the images in panel **b**, showing the rescue of mCherry-Flag-F508del-CFTR at low temperature

3.5.1 CFTR Traffic Assay

1. Culture CFBE/A549 mCherry-Flag-CFTR cells (wt- or F508del-) to confluency in 10 cm dishes using DMEM supplemented with FBS, blasticidin, and puromycin (*see* **Note 24**).

2. Split cells to ~50% confluency.

3. 24 h later, trypsinize and seed cells: 20,000 cells/well (8-well chambered cover slips or 96-well plates), or 3000 cells/well (384-well plates) (*see* **Note 25**). This defines time $t=0$ h (*see* **Note 26**). For the negative controls add scrambled siRNA or DMSO. For the positive control, add COPB1 siRNA (wt-CFTR) or 3 μM VX-809 (F508del-CFTR). Seeding of 96- and 384-well plates must be performed with an automated dispenser (*see* **Note 27**).

4. Grow cells at 37 °C and 5% CO_2.

5. Induce CFTR expression with doxycycline (1 μg/ml) at $t=48$ h (wt-CFTR) or $t=24$ h (F508del-CFTR).

6. Incubate cells until $t=72$ h and immunostain extracellular Flag tags.

3.5.2 Fluorescence Staining of Extracellular Flag Tags

When 96- or 384 well plates are used, all pipetting steps can be significantly expedited by using a 96-channel pipette. In this case, all solutions must be previously dispensed in 96-well source plates. DPBS++ can be dispensed from a deep-well reservoir.

1. Wash the plate once with ice-cold DPBS++ (*see* **Note 28**).

2. Add mouse anti-Flag antibody: 45 μl/well (8 well), 30 μl/well (96 well), or 15 μl/well (384 well) (*see* **Note 29**).

3. Spin down plates (maximal centrifugal field: ~50×*g*) (*see* **Note 30**).

4. Incubate plate for 1 h at 4 °C.

5. Rinse three times with ice-cold DPBS++.

6. Fix the cells with 3% (w/v) paraformaldehyde for 20 min at 4 °C (*see* **Notes 31** and **32**). Typical volumes are 100 μl/well (8 well), 80 μl/well (96 well), and 40 μl/well (384 well). All further steps are performed at room temperature.

7. Wash three times with DPBS++.

8. Add goat anti-mouse Alexa 647-conjugated antibody. Use the same volumes as in **step 2**.

9. Spin down plates (maximal centrifugal field: ~50×*g*).

10. Incubate plate for 1 h at room temperature.

11. Wash three times with DPBS++.

12. Add Hoechst 33342 for nuclei staining. Use the same volumes as in **step 6**.

13. Incubate plate for 1 h at room temperature.

14. Wash three times with DPBS++.

15. Submerge cells in DPBS++. Use the same volumes as in **step 6**.

16. Incubate DPBS^{++} overnight at 4 °C in the dark (*see* **Note 33**).

17. Store plates at 4 °C in the dark until imaging (*see* **Note 34**).

3.5.3 Wide-Field Fluorescence Image Acquisition

Imaging of multiwell plates can be performed with any wide-field fluorescence screening microscope (*see* **Note 35**).

1. Input the exact dimensions of the plate's well matrix into the microscope software.

2. Set up multiposition imaging: For statistical reasons, each well should be imaged in at least four sub-positions.

3. Set up autofocus based on the nuclei (Hoechst) staining.

4. Choose exposure time and filter sets for Hoechst, mCherry, and Alexa 647 (visualization of nuclei, total CFTR, and PM CFTR) (*see* **Note 36**).

5. Start automated image acquisition.

3.5.4 Image Analysis

1. Quantify mCherry and Alexa 647 fluorescence as well as the mCherry/Alexa 647 ratio on a cell-by-cell basis using CellProfiler. A previously published analysis pipeline [12] is schematically depicted in Fig. 6.

2. It is convenient to express the overall quantification of a given image by the mean or median value of all cells in an image.

3. The quantification result can be expressed as the absolute value or the deviation to negative controls. HTM Explorer, an excellent software tool for interactive data visualization, quality control, and statistical analysis of HCM data, has been created by Dr. Christian Tischer (EMBL, Heidelberg) and is freely available at https://github.com/tischi/HTM_Explorer. This tool is based on and requires the installation of the R software (https://cran.r-project.org/).

4   Notes

1. Store at −20 °C as single-use vials.

2. When preparing RIPA buffer, be aware that sodium deoxycholate is an irritant, so wear a mask.

3. The compound is stable up to 6 months (DMSO solubilized, −80 °C) or 3 years (powder, −20 °C).

4. Add $CaCl_2$ and $MgCl_2$ just before use.

5. Alternatively, 3% solutions can be stored at −20 °C.

6. This is a lyophilized gelatine/sucrose/lipofectamine/siRNA coating which is quickly rehydrated and uptaken by cells at the moment of seeding. It is stable at room temperature if stored under desiccation conditions [15].

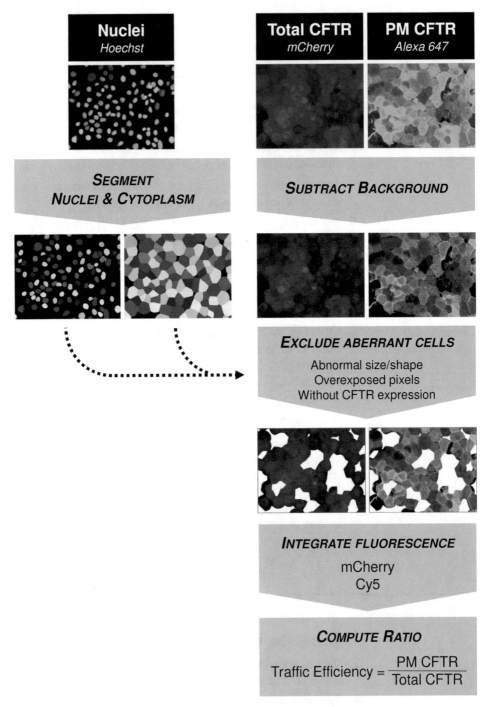

Fig. 6 Image analysis workflow for the quantification of CFTR traffic efficiency. The experimental outputs of the CFTR traffic assay are wide-field fluorescence microscopy images for three fluorophores: Hoechst 33342 (nuclei), mCherry (Total CFTR), Alexa 647 (CFTR molecules located at the PM). The Hoechst image is used for nuclei segmentation. The cytoplasm is segmented through dilation of the nuclei objects. Both CFTR channels are background corrected (via flat-field correction or subtracting illumination functions generated in silico). Using the segmentation data, cells presenting apoptotic- or mitotic-like morphology, without CFTR expression, or containing saturated pixels are excluded. Finally, a set of "approved cells" is established, all of them being characterized by the integrated mCherry and Alexa 647 fluorescence intensity as well as by the corresponding traffic efficiency (PM/Total CFTR). The analysis pipeline can be readily implemented in CellProfiler to enable the simultaneous analysis of large image datasets

7. In some cell types increasing acrylamide concentration up to 9 % can decrease the smearing of band C allowing for better densitometric analysis of band intensity. The gel can be run until the 50 kDa marker is 0.5 cm above the bottom of the gel.

8. The pulse-chase procedure can be applied to different cell lines expressing CFTR. However, the duration of the pulse may need to be adjusted to obtain enough initial labeling.

9. In the pulse-chase experiment, each time point (durations of chase) corresponds to one P60 cell culture dish.

10. Addition of cycloheximide during the chase period is done to inhibit incorporation of the radioactive amino acids that were taken into the cell but not used for protein synthesis during the pulse period. As it is toxic, its use is optional and may depend on the cell line. So, perform first a test to assess if labeled CFTR levels go up in the first 30 min of chase. If so, use then cycloheximide.

11. Samples can be frozen (at –70 °C) either after clearing the lysates or before the immunoprecipitated proteins are loaded into the gel.

12. Other antibodies can be used in this procedure as long as they are efficient in CFTR immunoprecipitation.

13. To wash the beads, use a 2 or 5 ml syringe with a 21 G needle to aspirate RIPA. Do not vacuum aspiration as this tends to lead to sample loss.

14. In the first times you run a set of samples in a gel, you may consider using a radiolabeled protein molecular weight standard ([methyl-^{14}C] methylated, protein molecular weight markers, Perkin-Elmer NEC811001UC). The CFTR pattern is usually very clear, so in general there is no need for continued use of the molecular weight standard.

15. The initial washing steps should be carried out gently with easily detachable cells (such as BHK or HEK-293 cells) but vigorously with highly adherent cells (such as HeLa or specially CFBE cells) to ensure the removal of all unviable, biotin-permeable cells, so as to minimize contamination of the assays with biotin-labeled intercellular proteins.

16. CAUTION: Solubilizing the sulfo-NHS-SS-biotin in DMSO may increase its cell permeability leading to contamination of the assay with intracellular proteins. The water solubility of sulfo-NHS-esters eliminates the need to dissolve the reagent in an organic solvent.

17. Sulfo-NHS esters should be dissolved in water just before use because they hydrolyze easily.

18. The purity of the reagent is of the utmost importance to avoid the isolation of intracellular protein. Cheaper reagents may

contain high proportion of desulfonated NHS-biotin that is cell membrane permeable.

19. There is some flexibility in the conditions for conjugating NHS-esters to primary amines. Incubation temperatures can range from 4 to 37 °C, pH values in the reaction range from 7 to 9, and incubation times range from a few minutes to 12 h.

20. Buffers containing amines (such as Tris or glycine) must be avoided because they compete with the protein biotinylation reaction.

21. In some protocols, after the hydrolysis with the glycosidases, there is an additional protein precipitation step with either ethanol or trichloroacetic acid. The precipitate is then dissolved and loaded in the SDS-PAGE gel. This step is dispensable as the solubilization is usually difficult, and most of the times this reflects negatively on the band pattern obtained. The hydrolysis reaction can be loaded directly.

22. Alternatively to the use of extracts from metabolically labeled cells, the treatment with glycosidases can be done using total extracts (prepared as in Subheading 3.1). After cell lysis and protein quantification, the reactions can be prepared as in Subheading 3.4, **step 6**, using 50 μg of total protein. After overnight incubation at 37 °C, the samples can be analyzed by Western Blot with an anti-CFTR antibody (*see* above Subheading 3.1). Autofluorescence is minimized by choosing blue and red/infrared-emitting fluorophores.

23. Autofluorescence is minimized by choosing blue or red/infrared-emitting fluorophores.

24. Blasticidin and puromycin are the selection agents for the mCherry-Flag-CFTR construct.

25. Cell amounts are chosen such that a near-confluent cell monolayer is formed by the end of the assay.

26. From this point onwards, blasticidin and puromycin can be withdrawn from the culture media.

27. This ensures seeding homogeneity throughout all wells in a plate.

28. A 4 °C temperature is required to inhibit CFTR endocytosis.

29. These volumes minimize antibody usage and are provided as a suggestion

30. A low centrifugal field ensures that the cells are not disrupted.

31. Fixation masks the Flag epitope and therefore can only be performed after adding the primary antibody.

32. Paraformaldehyde solutions are irritating and carcinogenic. Such solutions must be handled in a chemical fume hood and disposed of in dedicated containers.

33. Overnight incubation is required for the Hoechst dye to fully diffuse into the nuclei, since cells are not permeabilized during this protocol.

34. Multiple plates can be stained in parallel. If stained plates are stored at 4 °C in the dark, imaging can occur within 1 week of staining without appreciable fluorescence loss.

35. A confocal microscope is not desirable because the assay readout is whole-cell fluorescence.

36. mCherry and Alexa 647 fluorescence cannot be spectrally separated with most conventional filter sets. To avoid mCherry fluorescence bleed-through, we use a custom filter cube for the Alexa 647 channel: excitation BP 645/30; emission LP 670.

Acknowledgements

Work supported by UID/MULTI/04046/2013 center grant (to BioISI) from FCT/MCTES/PIDDAC, Portugal. H.M.B. is recipient of SFRH/BPD/93017/2013 postdoctoral fellowship (FCT, Portugal) and P.M. is supported by grant IF/2012 (FCT, Portugal).

References

1. Farinha CM, Matos P, Amaral MD (2013) Control of cystic fibrosis transmembrane conductance regulator membrane trafficking: not just from the endoplasmic reticulum to the Golgi. FEBS J 280:4396–4406

2. Bannykh SI, Bannykh GI, Fish KN, Moyer BD, Riordan JR, Balch WE (2000) Traffic pattern of cystic fibrosis transmembrane regulator through the early exocytic pathway. Traffic 1:852–870

3. Farinha CM, Amaral MD (2005) Most F508del-CFTR is targeted to degradation at an early folding checkpoint and independently of calnexin. Mol Cell Biol 25:5242–5252

4. Chang XB, Cui L, Hou YX, Jensen TJ, Aleksandrov AA, Mengos A, Riordan JR (1999) Removal of multiple arginine-framed trafficking signals overcomes misprocessing of delta F508 CFTR present in most patients with cystic fibrosis. Mol Cell 4:137–142

5. Roxo-Rosa M, Xu Z, Schmidt A, Neto M, Cai Z, Soares CM, Sheppard DN, Amaral MD (2006) Revertant mutants G550E and 4RK rescue cystic fibrosis mutants in the first nucleotide-binding domain of CFTR by different mechanisms. Proc Natl Acad Sci U S A 103:17891–17896

6. Farinha CM, King-Underwood J, Sousa M, Correia AR, Henriques BJ, Roxo-Rosa M, Da Paula AC, Williams J, Hirst S, Gomes CM, Amaral MD (2013) Revertants, low temperature, and correctors reveal the mechanism of F508del-CFTR rescue by VX-809 and suggest multiple agents for full correction. Chem Biol 20:943–955

7. Nishimura N, Balch WE (1997) A di-acidic signal required for selective export from the endoplasmic reticulum. Science 277:556–558

8. Yoo JS, Moyer BD, Bannykh S, Yoo HM, Riordan JR, Balch WE (2002) Non-conventional trafficking of the cystic fibrosis transmembrane conductance regulator through the early secretory pathway. J Biol Chem 277:11401–11409

9. Gee HY, Noh SH, Tang BL, Kim KH, Lee MG (2011) Rescue of DeltaF508-CFTR trafficking via a GRASP-dependent unconventional secretion pathway. Cell 146:746–760

10. Wang X, Venable J, LaPointe P, Hutt DM, Koulov AV, Coppinger J, Gurkan C, Kellner W, Matteson J, Plutner H, Riordan JR, Kelly JW, Yates JR 3rd, Balch WE (2006) Hsp90 cochaperone Aha1 downregulation rescues misfolding of CFTR in cystic fibrosis. Cell 127:803–815

11. Simpson JC, Joggerst B, Laketa V, Verissimo F, Cetin C, Erfle H, Bexiga MG, Singan VR, Heriche JK, Neumann B, Mateos A, Blake J, Bechtel S, Benes V, Wiemann S, Ellenberg J, Pepperkok R (2012) Genome-wide RNAi screening identifies human proteins with a regulatory function in the early secretory pathway. Nat Cell Biol 14:764–774

12. Botelho HM, Uliyakina I, Awatade NT, Proença MC, Tischer C, Sirianant L,

Kunzelmann K, Pepperkok R, Amaral MD (2015) Protein traffic disorders: an effective high-throughput fluorescence microscopy pipeline for drug discovery. Sci Rep 5:9038

13. Almaça J, Faria D, Sousa M, Uliyakina I, Conrad C, Sirianant L, Clarke LA, Martins JP, Santos M, Heriché JK, Huber W, Schreiber R, Pepperkok R, Kunzelmann K, Amaral MD (2013) High-content siRNA screen reveals global ENaC regulators and potential cystic fibrosis therapy targets. Cell 154:1390–1400

14. Van Goor F, Hadida S, Grootenhuis PD, Burton B, Cao D, Neuberger T, Turnbull A, Singh A, Joubran J, Hazlewood A, Zhou J, McCartney J, Arumugam V, Decker C, Yang J, Young C, Olson ER, Wine JJ, Frizzell RA, Ashlock M, Negulescu P (2009) Rescue of CF airway epithelial cell function in vitro by a CFTR potentiator, VX-770. Proc Natl Acad Sci U S A 106:18825–18830

15. Erfle H, Neumann B, Liebel U, Rogers P, Held M, Walter T, Ellenberg J, Pepperkok R (2007) Reverse transfection on cell arrays for high content screening microscopy. Nat Protoc 2:392–399

16. Kamentsky L, Jones TR, Fraser A, Bray MA, Logan DJ, Madden KL, Ljosa V, Rueden C, Eliceiri KW, Carpenter AE (2011) Improved structure, function and compatibility for cell profiler: modular high-throughput image analysis software. Bioinformatics 27:1179–1180

17. Amaral MD, Lukacs GL (2011) Introduction to section III: biochemical methods to study CFTR protein. Methods Mol Biol 741:213–218

18. Bradford MM (1976) A rapid and sensitive method for the quantitation of microgram quantities of protein utilizing the principle of protein-dye binding. Anal Biochem 72:248–254

19. Mendes AI, Matos P, Moniz S, Jordan P (2010) Protein kinase WNK1 promotes cell surface expression of glucose transporter GLUT1 by regulating a Tre-2/USP6-BUB2-Cdc16 domain family member 4 (TBC1D4)-Rab8A complex. J Biol Chem 285:39117–39126

20. Mendes AI, Matos P, Moniz S, Luz S, Amaral MD, Farinha CM, Jordan P (2011) Antagonistic regulation of cystic fibrosis transmembrane conductance regulator cell surface expression by protein kinases WNK4 and spleen tyrosine kinase. Mol Cell Biol 31:4076–4086

21. Moniz S, Sousa M, Moraes BJ, Mendes AI, Palma M, Barreto C, Fragata JI, Amaral MD, Matos P (2013) HGF stimulation of Rac1 signaling enhances pharmacological correction of the most prevalent cystic fibrosis mutant F508del-CFTR. ACS Chem Biol 8:432–442

22. Loureiro CA, Matos AM, Dias-Alves A, Pereira JF, Uliyakina I, Barros P, Amaral MD, Matos P (2015) A molecular switch in the scaffold NHERF1 enables misfolded CFTR to evade the peripheral quality control checkpoint. Sci Signal 8:48

Chapter 8

Quantification of a Non-conventional Protein Secretion: The Low-Molecular-Weight FGF-2 Example

Tania Arcondéguy, Christian Touriol, and Eric Lacazette

Abstract

Quantification of secreted factors is most often measured with enzyme-linked immunosorbent assay (ELISA), Western Blot, or more recently with antibody arrays. However, some of these, like low-molecular-weight fibroblast growth factor-2 (LMW FGF-2; the 18 kDa form), exemplify a set of secreted but almost non-diffusible molecular actors. It has been proposed that phosphorylated FGF-2 is secreted via a non-vesicular mechanism and that heparan sulfate proteoglycans function as extracellular reservoir but also as actors for its secretion. Heparan sulfate is a linear sulfated polysaccharide present on proteoglycans found in the extracellular matrix or anchored in the plasma membrane (syndecan). Moreover the LMW FGF-2 secretion appears to be activated upon FGF-1 treatment. In order to estimate quantification of such factor export across the plasma membrane, technical approaches are presented (evaluation of LMW FGF-2: (1) secretion, (2) extracellular matrix reservoir, and (3) secretion modulation by surrounding factors) and the importance of such procedures in the comprehension of the biology of these growth factors is underlined.

Key words LMW FGF-2 secretion, $NaClO_3$, Heparinase II, FGF-1 stimulation

1 Introduction

FGF-2 is a prototype of the heparin-binding growth factor family and includes five isoforms resulting from alternative translation initiation at four noncanonic initiation CUG codons and one AUG codon [1]. The smallest isoform (LMW FGF-2, 18 kDa) is mainly cytoplasmic and released from the cell despite the absence of any signal peptide [2]. LMW FGF-2 is exported via a non-vesicular mechanism based on direct translocation of cytoplasmic proteins across the plasma membrane [3]. This unconventional secretion involves cell surface heparan sulfate proteoglycans components and phosphorylation of LMW FGF-2 by the Tec kinase complex (Fig. 1) [4, 5]. In addition, it has been proposed that FGF-1 stimulation could induce the LMW FGF-2 secretion pathway [6]. LMW FGF-2 amount is under the limit of detection (using the Western blot method) in conditioned media unless they are significantly overexpressed, and then only in

Andrea Pompa and Francesca De Marchis (eds.), *Unconventional Protein Secretion: Methods and Protocols*, Methods in Molecular Biology, vol. 1459, DOI 10.1007/978-1-4939-3804-9_8, © Springer Science+Business Media New York 2016

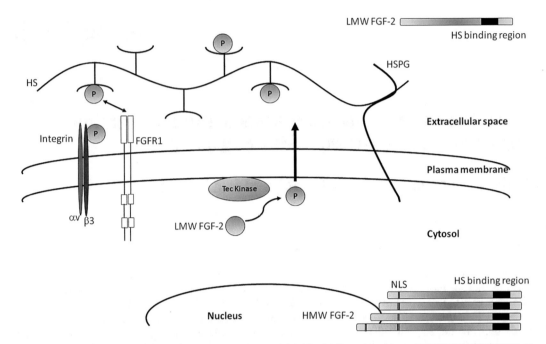

Fig. 1 Schematic representation. The 22, 22.5, 24, and 34 kDa high-molecular-weight FGF-2 (HMW FGF-2) exhibit one or two nuclear localization sequence (NLS) in their extended amino-terminus and are addressed to the nucleus (intracrine response) whereas the 18 kDa form or low-molecular-weight FGF-2 (LMW FGF-2) is phosphorylated by the Tec Kinase and secreted by an unconventional mechanism, namely direct translocation. The latter form can interact with heparan sulfate (HS) and constitute a reservoir, but can also interact with the FGR-1 receptor (autocrine and paracrine response); they only express FGFR receptor at detectable levels among FGR receptors in Hela cells. Specific binding of LMW FGF-2 to its cognate receptor parallels binding to heparan sulfate. Heparan sulfate proteoglycan (HSPG) anchored in the plasma membrane such as Syndecan 4 binds the LMW FGF-2 and possesses an FGF-2 noncanonic signalization pathway (endothelial cells)

specific cells [2, 7, 8]. Interestingly, a higher plasma level of FGF-2 has been found in patients treated with anti-vascular endothelial growth factor therapies [9–13]. These findings highlight the role of LMW FGF-2 in potentially mediating resistance to these therapies, the interplay between these angiogenic factors, and the importance of the surrounding molecular actors in vivo. LMW FGF-2 differs from the other major angiogenic growth factor namely vascular endothelial growth factor A (VEGF-A) where alternative splicing modulation can yield a highly diffusible isoform lacking the heparin-binding domain [14]. On the other side, LMW FGF-2 exhibits a non-spliced heparan sulfate-binding region leading to the formation of a reservoir in the extracellular matrix. These molecules have also been viewed as accessory co-receptors serving to facilitate FGF-2 tyrosine kinase receptor binding. LMW FGF-2 binds to heparan sulfate from proteoglycans in the extracellular matrix or anchored in the plasma membrane (syndecan 4) with a dissociation constant of 10^{-8}–10^{-9} but also to integrins [15].

Consequently, estimation of LMW FGF-2 secretion needs to take account of the reservoir but also the outcome of surrounding factors such as FGF-1 proposed to increase release of LMW FGF-2 into conditioned media [6]. The methods presented in this chapter summarize current protocols to efficiently quantify LMW FGF-2 reservoir and LMW FGF-2 secretion in conditioned media but also an approach to estimate the role of surrounding factors. Sodium chlorate pretreatment (24 h) of the cell culture ensures the release into the conditioned media during the following 24 h and consequently enables to evaluate secretion of LMW FGF-2. Heparinase II treatment allows estimating LMW FGF-2 trapped in the extracellular matrix. Finally, induction with FGF-1 enables to evaluate the effect of a surrounding factor on LMW FGF-2 secretion in conditioned media taking into account that the extracellular matrix is already saturated. To quantify LMW FGF-2 the ELISA method has been chosen and allows detection in a range of 12.5–800 pg/ml with a minimum measurable dose at 7.5 pg/ml.

2 Materials

2.1 Cell Type

1. Hela cells were obtained from ATCC (ref® CCL-2™). They arc adenocarcinoma cells deriving from cervix epithelial cells expressing FGF-2 [16] and they only express FGFR1 receptor at detectable levels among FGFRs [17].

2. Cells were propagated in 60.1 cm^2 culture dishes.

3. LMW FGF-2 secretion was estimated in 6-well (8.96 cm^2/well) tissue culture test plates.

2.2 Media and Solutions

Solutions are prepared with MilliQ water and analytical grade reagents. Prepare and store all reagents at room temperature (unless indicated otherwise). Follow all waste disposal regulations. $NaClO_3$ is a corrosive agent, so wear gloves and a mask to weigh powder.

1. Propagation medium: Dulbecco's modified Eagle's medium (DMEM) 1 g/L glucose supplemented with l-glutamine 100×, 10% fetal bovine serum (FBS), and 100× penicillin/streptomycin.

2. Cell passaging and splitting were performed with the use of trypsin–EDTA solution and Dulbecco's phosphate-buffered saline.

3. 3 M $NaClO_3$ (100×) solution: Weigh 6.39 g $NaClO_3$ in a glass beaker. Add about 15 ml water in the glass beaker and dissolve with a magnetic stirrer. Transfer to a 50 ml graduated cylinder and make up to 20 ml with water. Filter sterilize with a syringe and a 0.22 μm filter. Store at room temperature.

4. DMEM 10% FBS with and without sodium chlorate ($NaClO_3$): Add 5 ml FBS, 500 µl penicillin/streptomycin, and 500 µl glutamine to 50 ml DMEM 1 g/l glucose and 500 µl or not of 3 M $NaClO_3$ (30 mM final concentration) in 50 ml Falcon tubes.

5. DMEM 0,1% FBS with and without sodium chlorate ($NaClO_3$): Add 50 µl FBS, 500 µl penicillin/streptomycin, and 500 µl glutamine to 50 ml DMEM 1 g/l glucose and 500 µl or not of 3 M $NaClO_3$ (30 mM final concentration) in 50 ml Falcon tubes.

6. FGF-1 1 µg/ml (100×) solution: Add 10 ml sterile water for 10 µg of lyophilized FGF-1 (ref: GTX65081; Euromedex), shake the container to dissolve powder, and make aliquots of 250 µl. Store at –20 °C.

7. Heparinase II solution (100×): Reconstitute lyophilized powder of a 25 units vial with 250 µl sterile water. Store at –20 °C.

8. 50 ml Syringe (TERUMO).

9. 0.22 µm Filters.

10. PBS (Dulbecco's phosphate-buffered saline).

2.3 Incubator and Growth Condition

1. Cells were incubated in a humid CO_2 incubator maintained at 37 °C and 5% CO_2. A subcultivation ratio between 1:2 and 1:6 was maintained and medium was renewed 2–3 times per week. All media or PBS 1× is kept at 4 °C and preheated at 37 °C before use.

2.4 ELISA Test

1. LMW FGF-2 concentration was measured with the human bFGF/FGF2 (basic fibroblast growth factor) commercial ELISA kit (ref: MBS2505069; BioSource).

2. 100 µl of sample per well is used; consequently to work in the detection range (12.5–800 pg/ml), under some conditions culture media can be concentrated by Centricon Plus-70 centrifugal Filter Units-10 kDa (ref: UFC701008; Millipore) according to the manufacturer's instructions (*see* **Note 1**).

3. Absorbance measurements for ELISA were performed at 450 nm with background subtraction at 620 nm using an Asys UVM340 microplate reader (BIOCHROM®). Data were collected and analyzed with the MikroWin™ 2000 software.

3 Methods

3.1 Sodium Chlorate Treatment: Evaluation of LMW FGF-2 Secretion During 24 h

Chlorate ion is a competitive and a reversible inhibitor of proteoglycan sulfation because it competes for sulfate recognition by ATP sulfurylase [18]. LMW FGF-2 interacts with a specific HS sequence that consists of a hexasaccharide containing 2-O-sulfated iduronic acid and N-sulfated glucosamine [19]. Twenty-four-hour pretreatment of cell cultures with 30 mM sodium chlorate will suppress sulfation and consequently impair LMW FGF-2 binding to heparan sulfate (*see* **Note 2**).

1. Wash the desired wells with 2 ml PBS 1× of the subconfluent cell culture in 6-well tissue culture test plates.

2. Pretreat cells for 24 h in the incubator with 30 mM sodium chlorate in DMEM 10% FBS: Add 2 ml of supplemented DMEM 10% FBS plus sodium chlorate. Untreated cells are incubated in the same media without sodium chlorate.

3. Remove the media and wash each well twice with 2 ml PBS 1×.

4. Add 2 ml of supplemented DMEM 0.1% FBS and 30 mM of sodium chlorate and incubate for 24 additional hours in the incubator. Untreated cells are incubated in the same media without sodium chlorate (*see* **Note 3**).

5. Recover the media in a 1.5 ml test tube.

6. Centrifuge at $500 \times g$ during 10 min in order to clear the sample.

7. Use 100 μl of the cleared media to perform the ELISA test according to the manufacturer's recommendations (*see* **Notes 4** and **5**).

3.2 Heparinase II Treatment: Evaluation of the Matrix Reservoir

1. Wash the desired wells twice with 2 ml PBS 1× of the subconfluent cell culture in 6-well tissue culture test plates.

2. Add 2 ml of DMEM 0.1% FBS or DMEM 10% FBS without NaClO$_3$ in the desired wells and incubate for 24 h in the incubator (*see* **Notes 6** and **7**).

3. Wash the wells twice with 2 ml PBS 1×.

4. Add in the desired wells 1 ml of heparinase II solution (1 ml of PBS with 10 μl of the heparinase II stock solution). In control wells (for each condition), add 1 ml of PBS. Incubate for 1 h at 37 °C.

5. Recover the supernatant in a 1.5 ml test tube.

6. Centrifuge at $500 \times g$ during 10 min in order to clear the sample.

7. Use 100 μl of the cleared supernatant to perform the ELISA test according to the manufacturer's recommendations.

3.3 FGF-1 Stimulation: Evaluation of the Outcome of Surrounding Factors in Hela Cells

1. Wash the desired wells twice with 2 ml PBS 1× of the subconfluent cell culture in 6-well tissue culture test plates.

2. Add 2 ml of DMEM 0.1% FBS in each well and incubate for 24 h in the incubator.

3. Wash the wells twice with 2 ml PBS 1×.

4. Add in the desired wells 2 ml DMEM 0.1% FBS with 20 μl FGF-1 100× solution (stimulation). In control wells, add 2 ml DMEM 0.1% FBS (unstimulated conditions). Incubate for 24 h the 6-well tissue culture test plates in the incubator.

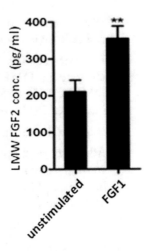

Fig. 2 LMW FGF-2 secretion after FGF-1 stimulation. Hela cells were stimulated or not by FGF-1 and LMW FGF-2 was measured by ELISA. **$P < 0.005$ (t-test) Bars ± SEM

5. Recover the media in a 1.5 ml test tube.

6. Centrifuge at $500 \times g$ during 10 min in order to clear the sample.

7. Use 100 µl of the cleared media to perform the ELISA test according to the manufacturer's recommendations.

8. The induction is represented in a graph as pg/ml under unstimulated or stimulated conditions (Fig. 2).

4 Notes

1. In the case of Centricon use, concentration calculation is necessary; results must take into account the final volume obtained after concentration.

2. Using the sodium chlorate method, binding experiments have shown that this treatment reduced the specific binding of FGF-2 to heparan sulfate by more than 80 % [15].

3. This experiment is presented when cells are not stimulated by surrounding molecular actors but can also be performed in DMEM 10 % FBS when cells are continuously stimulated by surrounding factors.

4. For better accuracy, perform each condition in triplicate using 100 µl aliquots for ELISA measurement.

5. For ELISA analysis, the standard curve should be examined for linearity; and experimental measurements should be comprised in the range of the FGF-2 standard, i.e., between 12.5 and 800 pg/ml.

6. Incubation in DMEM 0.1 % FBS or DMEM 10 % FBS is useful to estimate the saturation of the reservoir as LMW FGF-2 is stable and protected from inactivation when it is bound to heparan sulfate [20]. In DMEM 10 % FBS, stimulation by surrounding factors is continuous.

7. Always prepare desired conditions with control wells as they will be treated or not by heparinase II.

Acknowledgements

This work was supported by "La ligue contre le cancer" (2FI10869NEBD).

References

1. Touriol C, Bornes S, Bonnal S, Audigier S, Prats H, Prats AC, Vagner S (2003) Generation of protein isoform diversity by alternative initiation of translation at non-AUG codons. Biol Cell 95(3-4):169–178

2. Bikfalvi A, Klein S, Pintucci G, Rifkin DB (1997) Biological roles of fibroblast growth factor-2. Endocr Rev 18(1):26–45. doi:10.1210/edrv.18.1.0292

3. Schafer T, Zentgraf H, Zehe C, Brugger B, Bernhagen J, Nickel W (2004) Unconventional secretion of fibroblast growth factor 2 is mediated by direct translocation across the plasma membrane of mammalian cells. J Biol Chem 279(8):6244–6251. doi:10.1074/jbc.M310500200

4. Zehe C, Engling A, Wegehingel S, Schafer T, Nickel W (2006) Cell-surface heparan sulfate proteoglycans are essential components of the unconventional export machinery of FGF-2. Proc Natl Acad Sci U S A 103(42):15479–15484. doi:10.1073/pnas.0605997103

5. Ebert AD, Laussmann M, Wegehingel S, Kaderali L, Erfle H, Reichert J, Lechner J, Beer HD, Pepperkok R, Nickel W (2010) Teckinase-mediated phosphorylation of fibroblast growth factor 2 is essential for unconventional secretion. Traffic 11(6):813–826. doi:10.1111/j.1600-0854.2010.01059.x

6. Meunier S, Navarro MG, Bossard C, Laurell H, Touriol C, Lacazette E, Prats H (2009) Pivotal role of translokin/CEP57 in the unconventional secretion versus nuclear translocation of FGF2. Traffic 10(12):1765–1772. doi:10.1111/j.1600-0854.2009.00985.x

7. Szebenyi G, Fallon JF (1999) Fibroblast growth factors as multifunctional signaling factors. Int Rev Cytol 185:45–106

8. Trudel C, Faure-Desire V, Florkiewicz RZ, Baird A (2000) Translocation of FGF2 to the cell surface without release into conditioned media. J Cell Physiol 185(2):260–268. doi:10.1002/1097-652(200011)185:2<260::AID-JCP11>3.0.CO;2-X

9. Batchelor TT, Sorensen AG, di Tomaso E, Zhang WT, Duda DG, Cohen KS, Kozak KR, Cahill DP, Chen PJ, Zhu M, Ancukiewicz M, Mrugala MM, Plotkin S, Drappatz J, Louis DN, Ivy P, Scadden DT, Benner T, Loeffler JS, Wen PY, Jain RK (2007) AZD2171, a pan-VEGF receptor tyrosine kinase inhibitor, normalizes tumor vasculature and alleviates edema in glioblastoma patients. Cancer Cell 11(1):83–95. doi:10.1016/j.ccr.2006.11.021

10. Cenni E, Perut F, Granchi D, Avnet S, Amato I, Brandi ML, Giunti A, Baldini N (2007) Inhibition of angiogenesis via FGF-2 blockage in primitive and bone metastatic renal cell carcinoma. Anticancer Res 27(1A):315–319

11. Kopetz S, Hoff PM, Morris JS, Wolff RA, Eng C, Glover KY, Adinin R, Overman MJ, Valero V, Wen S, Lieu C, Yan S, Tran HT, Ellis LM, Abbruzzese JL, Heymach JV (2010) Phase II trial of infusional fluorouracil, irinotecan, and bevacizumab for metastatic colorectal cancer: efficacy and circulating angiogenic biomarkers associated with therapeutic resistance. J Clin Oncol 28(3):453–459. doi:10.1200/JCO.2009.24.8252

12. Porta C, Paglino C, Imarisio I, Ganini C, Sacchi L, Quaglini S, Giunta V, De Amici M (2013) Changes in circulating pro-angiogenic cytokines, other than VEGF, before progression to sunitinib therapy in advanced renal cell carcinoma patients. Oncology 84(2):115–122. doi:10.1159/000342099

13. Sharpe K, Stewart GD, Mackay A, Van Neste C, Rofe C, Berney D, Kayani I, Bex A, Wan E, O'Mahony FC, O'Donnell M, Chowdhury S, Doshi R, Ho-Yen C, Gerlinger M, Baker D, Smith N, Davies B, Sahdev A, Boleti E, De Meyer T, Van Criekinge W, Beltran L, Lu YJ, Harrison DJ, Reynolds AR, Powles T (2013) The effect of VEGF-targeted therapy on biomarker expression in sequential tissue from patients with metastatic clear cell renal cancer. Clin Cancer Res 19(24):6924–6934. doi:10.1158/1078-0432.CCR-13-1631

14. Arcondeguy T, Lacazette E, Millevoi S, Prats H, Touriol C (2013) VEGF-A mRNA processing, stability and translation: a paradigm for intricate regulation of gene expression at the post-transcriptional level. Nucleic Acids Res 41(17):7997–8010. doi:10.1093/nar/gkt539

15. Rapraeger AC, Krufka A, Olwin BB (1991) Requirement of heparan sulfate for bFGF-mediated fibroblast growth and myoblast differentiation. Science 252(5013):1705–1708

16. Rifkin DB, Quarto N, Mignatti P, Bizik J, Moscatelli D (1991) New observations on the intracellular localization and release of bFGF. Ann N Y Acad Sci 638:204–206

17. Haugsten EM, Sorensen V, Brech A, Olsnes S, Wesche J (2005) Different intracellular trafficking of FGF1 endocytosed by the four homologous FGF receptors. J Cell Sci 118(Pt 17):3869–3881. doi:10.1242/jcs.02509

18. Farley JR, Nakayama G, Cryns D, Segel IH (1978) Adenosine triphosphate sulfurylase from Penicillium chrysogenum equilibrium binding, substrate hydrolysis, and isotope exchange studies. Arch Biochem Biophys 185(2):376–390

19. Guimond S, Maccarana M, Olwin BB, Lindahl U, Rapraeger AC (1993) Activating and inhibitory heparin sequences for FGF-2 (basic FGF). Distinct requirements for FGF-1, FGF-2, and FGF-4. J Biol Chem 268(32):23906–23914

20. Vlodavsky I, Miao HQ, Medalion B, Danagher P, Ron D (1996) Involvement of heparan sulfate and related molecules in sequestration and growth promoting activity of fibroblast growth factor. Cancer Metastasis Rev 15(2):177–186

Chapter 9

Human Primary Keratinocytes as a Tool for the Analysis of Caspase-1-Dependent Unconventional Protein Secretion

Gerhard E. Strittmatter, Martha Garstkiewicz, Jennifer Sand, Serena Grossi, and Hans-Dietmar Beer

Abstract

Inflammasomes comprise a group of protein complexes, which activate the protease caspase-1 upon sensing a variety of stress factors. Active caspase-1 in turn cleaves and thereby activates the pro-inflammatory cytokines prointerleukin (IL)-1β and -18, and induces unconventional protein secretion (UPS) of mature IL-1β, IL-18, as well as of many other proteins involved in and required for induction of inflammation. Human primary keratinocytes (HPKs) represent epithelial cells able to activate caspase-1 in an inflammasome-dependent manner upon irradiation with a physiological dose of ultraviolet B (UVB) light. Here, we describe the isolation of keratinocytes from human skin, their cultivation, and induction of caspase-1-dependent UPS upon UVB irradiation as well as its siRNA- and chemical-mediated inhibition. In contrast to inflammasome activation of professional immune cells, UVB-irradiated HPKs represent a robust and physiological cell culture system for the analysis of UPS induced by active caspase-1.

Key words Unconventional protein secretion, Keratinocytes, Inflammasomes, Caspase-1, UVB irradiation, Interleukin-1

1 Introduction

Unconventional protein secretion (UPS), although known for several decades, is still a poorly characterized and understood pathway (Nickel, 2003). Several proteins, such as interleukin (IL)-1α or fibroblast growth factor (FGF) 2, which are believed to be secreted independently of a signal peptide, play important roles in inflammation and repair, suggesting that UPS is linked to these processes (Monteleone et al., 2015; Nickel & Rabouille, 2009). Inflammation and subsequent repair are induced by different types of tissue stress, which can also cause cell death and lysis associated with the passive release of intracellular proteins. This complicates examination of stress-induced UPS, since it is difficult to discriminate between regulated secretion of leaderless proteins and their passive release upon cell lysis.

Andrea Pompa and Francesca De Marchis (eds.), *Unconventional Protein Secretion: Methods and Protocols*, Methods in Molecular Biology, vol. 1459, DOI 10.1007/978-1-4939-3804-9_9, © Springer Science+Business Media New York 2016

Inflammasomes represent a group of innate immune complexes, which are able to sense a variety of different stress factors (Strowig et al., 2012). This sensing induces complex assembly and activation of the protease caspase-1. Once activated caspase-1 cleaves and thereby activates the pro-inflammatory cytokines proIL-1β and -18, which both lack a signal peptide and whose secretion induces an inflammatory response. Caspase-1 activity can also induce a lytic form of cell death termed pyroptosis. Pyroptosis proceeds independently of classic apoptotic caspases but requires caspase-1 or caspase-11 only(Bergsbaken et al., 2009; Jorgensen & Miao, 2015). Inflammasomes have been mainly characterized in immune cells such as macrophages and dendritic cells. However, also human primary keratinocytes (HPKs), epithelial cells forming the outermost layer of the skin, express inflammasome proteins. Irradiation of HPKs with a physiological dose of UVB light results in inflammasome-dependent caspase-1 activation and in turn in secretion of active IL-1β and IL-18 (Feldmeyer et al., 2007). This is most likely of physiological significance, since skin inflammation (sunburn) is reduced in UVB-irradiated mice lacking caspase-1 expression in comparison to wild-type mice (Feldmeyer et al., 2007). Experiments with UVB-irradiated HPKs revealed caspase-1 activity as an inducer of UPS (Keller et al., 2008). In addition, a proteomics approach allowed the identification of more than 50 proteins without a signal peptide released by the unconventional pathway induced by caspase-1 activity. These proteins included already known unconventionally secreted polypeptides as well as others, whose secretion was not yet described at this time. Several of these novel members of the UPS family play important roles in inflammation, repair, cell death, and cytoprotection (Keller et al., 2008). Caspase-1-dependent UPS of several proteins could be confirmed in the human monocytic cell line THP-1 as well as in human and murine fibroblasts and macrophages, demonstrating that the identified pathway is of broad relevance (Keller et al., 2008). However, an open question is whether certain (stress) conditions (perhaps only in certain cell types) induce UPS of a limited number of leaderless proteins. IL-1α secretion for example is induced upon caspase-1 activation (Keller et al., 2008; Kuida et al., 1995; Li et al., 1995) but is also released independently of the protease (Freigang et al., 2013; Gross et al., 2012).

Here, we provide detailed protocols for the isolation of HPKs from human skin and the cultivation of these cells (Fig. 1). In addition, we describe the induction of UPS upon inflammasome activation by UVB irradiation and the inhibition of caspase-1 activity mediated by siRNA-mediated knockdown of caspase-1 expression or treatment with caspase-1 inhibitors.

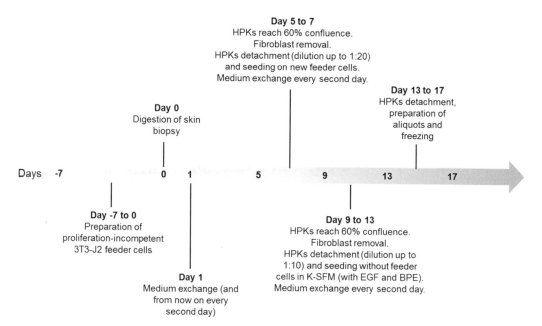

Fig. 1 Timescale for the isolation of HPKs

2   Materials

2.1   Complete Keratinocyte Medium (CKM)

Stock solutions (storage temperature and order information):

1. Dulbecco's modified Eagle medium (DMEM), store at 4 °C. Gibco, Life Technologies, Carlsbad, CA (21765)

2. F-12 Nutrient Mix (Ham's), store at 4 °C. Gibco, Life Technologies, Carlsbad, CA (21765)

3. Heat-inactivated fetal bovine serum (FBS), long-term storage −80 °C, short-term 4 °C. Gibco, Life Technologies, Carlsbad, CA (10500)

4. Antibiotic-antimycotic (Anti-Anti, 100×), long-term storage −20 °C, short-term 4 °C. Gibco, Life Technologies, Carlsbad, CA (15240)

5. 1.2 mg/ml Adenin: Dissolve in ddH$_2$O (sterile filtered), store at −20 °C. Sigma, Munich, Germany (A2786)

6. 5 mg/ml Apo-Transferrin: Dissolve in ddH$_2$O (sterile filtered), store at −20 °C. Sigma, Munich, Germany (T1147)

7. 2 μM 3,3′,5-Triiodo-L-thyronine: Dissolve in ddH$_2$O with 1/50 volume of 1 M NaOH (sterile filtered), store at −20 °C. Sigma, Munich, Germany (T6397)

8. 50 μg/ml Hydrocortisone: Dissolve 0.1 mg in 0.1 ml ethanol, then add 4.9 ml of sterile ddH$_2$O, and store at −20 °C. Sigma, Munich, Germany (H0888)

9. 1 µM Cholera toxin: Dissolve in ddH$_2$O (sterile filtered), and store at 4 °C. Sigma, Munich, Germany (C8052)

10. 2 mg/ml Insulin: Dissolve in hydrochloric acid, pH = 2.5 (sterile filtered), and store at –20 °C. Sigma, Munich, Germany (I6634)

11. 10 µg/ml Epidermal growth factor (EGF): Dissolve in DMEM with 10 % FBS (sterile filtered), and store at –20 °C. Sigma, Munich, Germany (E4127)

12. Working solution: 375 ml DMEM, 125 ml F-12 Nutrient Mix, 55 ml FBS, 5.5 ml Anti-Anti, 10 ml adenine, 0.6 ml apo-transferrin, 0.6 ml 3,3′,5-triiodothyronine, 4.8 ml hydrocortisone, 60 µl cholera toxin, 1.5 ml insulin, 0.6 ml EGF (stored at 4 °C).

2.2 Transfection with siRNA

1. 40 µM solutions of siRNA (e.g., from Sigma, Munich, Germany) in RNase-free ddH$_2$O (sterile filtered) are stored at –20 °C in aliquots (Fig. 2).

2. INTERFERin, store at 4 °C. Polyplus, Illkirch, France (409-01)

3. Keratinocyte-SFM (K-SFM), store at 4°C. Gibco, Life Technologies, Carlsbad, CA (17005-042)

4. Scrambled (scr): UUCUCCGAACGUGUCACGU[dTdT].

5. Caspase-5: GUGGCUGGACAAACAUCUA[dTdT].

6. VEGF-A: CUGAUGAGAUCGAGUACAU[dTdT].

7. Caspase-1_S1: GGCAGAGAUUUAUCCAAUA[dTdT].

8. Caspase-1_S2: AAGAGAUCCUUCUGUAAAGGU[dTdT].

2.3 Other Materials

1. UVB irradiation: Waldmann (Villingen-Schwenningen, Germany) UV Therapy System UV 802 L with 80–90 mJ/cm^2 broadband UVB and a distance of 6 cm from the lamp.

2. Trypsin/EDTA solution, store at 4 °C.

3. K-SFM with EGF and BPE. After addition of EGF and BPE the medium is stored at 4 °C. (see **Note 2**). Gibco, Life Technologies, Carlsbad, CA (17005-42)

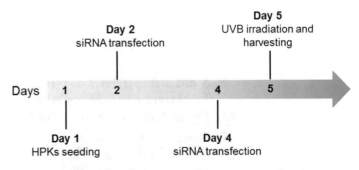

Fig. 2 Timescale for knock-down of gene expression by siRNA transfection

4. Sterile PBS.

5. LDH assay: CytoTox 96 Cytotoxicity Assay (Promega, Madison, WI).

6. IL-1β ELISA: DuoSet (R&D Systems, Minneapolis, MN, DY201).

7. 2× SDS loading buffer: 100 mM TRIS/HCl pH 8.0, 20 % v/v glycerol, 10 % w/v SDS, 0.01 % w/v bromophenol blue, 0.2 M DTT.

8. Antibodies (with dilutions): proIL-1β (R&D Systems, Minneapolis, MN, MAB201, 1:250); proIL-18 (MBL, Woburn, MA, PM014, 1:1000); caspase-1 (Santa Cruz, Santa Cruz, CA, sc-622, 1:1000); β-actin (Sigma, Munich, Germany, A5441, 1:5000); Gelsolin (Sigma, Munich Germany, G4896, 1:1000); Bid (Santa Cruz, Santa Cruz, CA, sc-11423, 1:500), Bcl-$_{XL}$ (Cell Signaling, Danvers, MA, 2764S, 1:1000); ASC (Enzo Life Sciences, Lausanne, Switzerland, ALX-210-905, 1:1000); cytochrome C (Cell Signaling, Danvers, MA, 4280, 1:1000). 3T3J2 mouse embryonic fibroblast cell line (Rhewinwald and Green, 1975). (*see* **Note1**).

10 μg/ml mitomycin C in 3T3-J2 medium, sterile filtered, −20°C: Sigman, Munich, Germany (M4287).
100 mM Ac-YVAD-CMK (Enzo Life Sciences, Lausanne, Switzerland, ALX-260-028), a caspase-1 inhibitor, in DMSO, aliquots stored at −20°C.
20 mM z-VAD-FMK (ALX-260-20), a pan-caspase inhibitor, in DMSO, aliquots are stored at −20°C.
3T3-J2 medium: DMEM with 10% FBS and 1% Anti-Anti

---

3   Methods

3.1 Isolation of Human Primary Keratinocytes (HPKs)

Isolation of keratinocytes from human skin was first described in 1975 (Rheinwald & Green, 1975) (*see* **Note 3**).

1. For the preparation of proliferation-incompetent 3T3-J2 cells incubate a confluent culture of cells for 2 h in prewarmed 3T3-J2 medium with 10 μg/ml mitomycin C.

2. Remove the medium, wash the cells three times with PBS, and detach them by incubation in trypsin/EDTA solution for about 2 min.

3. Seed the cells at two- to threefold lower density in 6 cm dishes in fresh 3T3-J2 medium (*see* **Notes 4** and **5**).

4. Start with a piece of (fore)skin of approximately 0.5–1 cm². However, successful isolation of HPKs is also well possible with a smaller piece of skin and an accordingly adjusted protocol (*see* **Notes 6** and **7**).

5. If necessary, disinfect biopsy by a short incubation with 70% ethanol and remove traces of ethanol by triple washing in sterile PBS (*see* **Note 8**).

6. Incubate biopsy for 10 min in PBS with 10% Anti-Anti (*see* **Note 9**).

7. Wash biopsy three times for 10 min in PBS with 1% Anti-Anti.

8. Wash for 10 min in PBS.

9. If necessary, remove fat using a scalpel.

10. Cut the biopsy in a 10 cm dish containing 8 ml trypsin/EDTA solution with tweezers, scalpel, and scissors in squares of 1–4 mm².

11. Incubate for 30 min at 37 °C in an incubator.

12. Mix the skin pieces with a 10 or 25 ml pipette by pipetting up and down avoiding air bubbles.

13. Pour the medium with cells and skin pieces through a 70 μm cell strainer. The skin pieces are kept for further processing (*see* **step 16**).

14. Add 12 ml of prewarmed CKM to the trypsin/EDTA solution containing HPKs and centrifuge for 3 min at $150 \times g$ and RT.

15. Remove the supernatant and resuspend the cells in 5 ml prewarmed CKM. Remove the medium from a 6 cm dish containing mitomycin C-treated 3T3-J2 feeder cells and add the cell suspension. Place the dish in an incubator and change the medium every second day.

16. Incubate the remaining skin pieces (from **step 13**) in 8 ml trypsin/EDTA solution and repeat the procedure (from **steps 11** to **15**) a second, third, and fourth time (**Note 10**).

17. After about 3 days the first small colonies of HPKs become visible. After 6–7 days about 60% of the surface is covered with keratinocytes.

18. When reaching a confluency of about 60%, wash the cells three times with PBS and incubate with 1 ml of trypsin/EDTA solution for about 5–10 min or longer (**Notes 11** and **12**).

19. Add the cell suspension directly onto new feeder cells in CKM. HPKs of one 6 cm dish can be diluted in up to three feeder-containing 15 cm dishes. Change the medium every second day.

20. When the culture has reached a cell density of about 60% detach HPKs as described (**step 18**). Try to remove fibroblasts as completely as possible.

21. After inactivation of trypsin by the addition of the equivalent volume of FBS, centrifuge the cells for 3 min at $150 \times g$ and RT and resuspend in prewarmed K-SFM (with EGF and BPE). HPKs can be diluted at this step up to 1:10. Change the medium every second day.

22. When the culture reaches a confluency of about 75 % wash the cells three times with PBS and remove HPKs with trypsin/EDTA solution (2.5 ml per T175 flask). Add the same volume of FBS and centrifuge the suspension for 3 min at $150 \times g$ and RT. Resuspend HPKs in K-SFM (with EGF and BPE) with 10 % DMSO (3×10^6 cells/ml), aliquot, freeze, and store in liquid nitrogen.

3.2 Culture of HPKs

Cryopreserved cells isolated according to Subheading 3.1 proliferate with a doubling time of about 1.5 days and gradually undergo senescence and terminal differentiation, which prevents proliferation completely after about 2 weeks. Since individual cells differentiate "completely" rather than all cells "a bit," the HPK culture becomes with time more and more heterogenous containing a smaller fraction of cells able to proliferate. Therefore, experiments should be performed with HPKs of the same, ideally low passage (second after thawing).

1. Thaw cryopreserved HPKs (from **step 22**, Subheading 3.1) at 37 °C, add the cell suspension (without centrifugation) to 10 ml prewarmed K-SFM (with EGF and BPE) in a T75 flask, and incubate at 37 °C in an incubator. After about 10 h wash the cells carefully with 10 ml prewarmed K-SFM (with EGF and BPE) and incubate in 10 ml of fresh medium.

2. Change the medium at the second day and if necessary every second day.

3. After one or up to three days HPKs reach a confluency of about 75–90 %. Wash cells three times with PBS and detach with 1 ml trypsin/EDTA solution. Add 1 ml FBS, centrifuge the suspension for 3 min at $150 \times g$ and RT, and resuspend in prewarmed K-SFM (with EGF and BPE). Dilute HPKs up to 1:10 at this step.

4. Change the medium every second day.

5. After 3–5 days (depending on the dilution in **step 3**) detach HPKs as described (**step 3**), count, and seed in new dishes at a density of 0.3–0.4×10^5 cells/ml (e.g., 1 ml volume in a 12-well dish).

6. Change the medium every second day.

3.3 Knockdown of Gene Expression by siRNA Transfection

1. Seed cells (as described in Subheading 3.2, **step 5**) in the evening.

2. Next morning pipet to 0.2 ml prewarmed K-SFM (without EGF and BPE) 0.3 µl siRNA (stock 40 µM) and mix the solution by vortexing (*see* **Note 13**). Add 1 µl INTERFERin and vortex for 15 s. Incubate at RT for 10 min. Since siRNA has off-target effects use at least two different siRNAs (separately) for the knock down of expression of a certain gene.

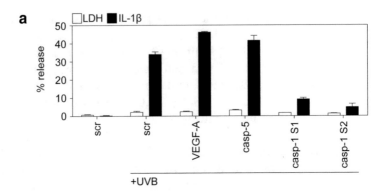

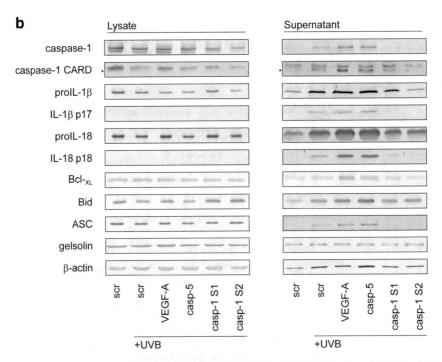

Fig. 3 Knockdown of caspase-1 expression inhibits UPS by UVB-irradiated HPKs. (**a** and **b**) At the first and third days after seeding, HPKs were transfected with siRNA as indicated, scrambled (scr) and siRNA targeting caspase-5 and VEGF-A expression served as non-related controls. At day four HPKs were irradiated with 86.4 mJ/cm^2 UVB and harvested 5 h later. (**a**) Lysates and supernatants were analyzed for IL-1β production and expression by ELISA, cell integrity was determined by an LDH assay. (**b**) Lysates and supernatants were analyzed for secretion, processing, and expression of the indicated proteins by Western blot. Caspase-1, proIL-1β, proIL-18, Bcl-$_{XL}$, Bid, and ASC are secreted dependent on caspase-1. ProIL-1β and proIL-18 are substrates of caspase-1. The secretion of the isolated CARD domain reflects activation and processing of caspase-1. Gelsolin, β-actin, and cytochrome C served as controls (Keller et al., [2008]; Sollberger et al., [2015])

3. Change the medium (1 ml K-SFM with EGF and BPE) of HPKs.

4. Carefully pipet the transfection mix onto the cells (final siRNA concentration 10 nM). Mix medium by carefully swinging the plate. Incubate HPKs at 37 °C in an incubator (5 % CO$_2$).

5. Repeat transfection 2 days after first transfection.

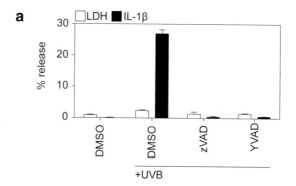

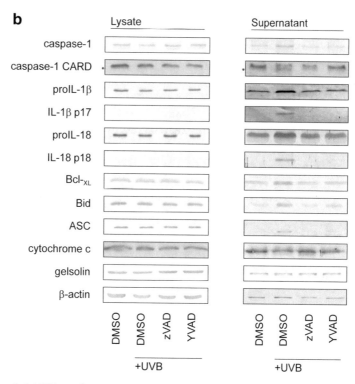

Fig. 4 Inhibition of caspase-1 activity prevents UVB-induced UPS by UVB-irradiated HPKs. (**a** and **b**) Four days after seeding, HPKs were treated with the caspase-1 inhibitor YVAD (50 μM), with the pan-caspase inhibitor zVAD (10 μM), or with the solvent DMSO. Cells were mock treated or irradiated with 86.4 mJ/cm² UVB and harvested 5 h later. (**a**) Lysates and supernatants were analyzed for IL-1β production and expression by ELISA, cell integrity was determined by LDH assay. (**b**) Lysates and supernatants were analyzed for secretion, processing, and expression of the indicated proteins by Western blot

6. Perform experiment the day after (*see* **Notes 14** and **15**).

3.3.1 Caspase-1 Inhibitor Treatment of HPKs

1. If using cells as described in Subheading 3.2, **step 5**, treat HPKs after 4 days (*see* **Note 13**). Remove the medium and add 1 ml of fresh prewarmed K-SFM (with EGF and BPE).

2. Treat cells either with 0.5 μl YVAD (final concentration 50 μM) or zVAD (final concentration 10 μM). Perform experiment after 1 h (*see* **Note 16**).

3. Use DMSO-treated HPKs as control.

3.4 UVB Irradiation of HPKs

For UVB-induced inflammasome activation and UPS, HPKs should not be more confluent than 80% (after **step 5** in Subheading 3.3 or **step 2** in Subheading 3.3.1, *see* **Note 17**). The medium can be changed 1 h before irradiation (*see* **step 1** in Subheading 3.3.1), but this step is not necessary. HPKs are irradiated in K-SFM (with EGF and BPE) with a distance of 6 cm from the UVB lamp. Only broadband UVB but not narrowband induces inflammasome activation. Control cells (mock-treatment) are also removed from the incubator but irradiated with a cover of aluminum foil. Note that DMSO inhibits inflammasome activation (Ahn et al., 2014). Therefore, DMSO-treated HPKs represent an essential control, if caspase-1 inhibitors are used.

1. Irradiate HPKs in K-SFM (with EGF and BPE) with 80–90 mJ/cm^2.

2. Harvest conditioned medium and cells after 5 h (*see* **Note 18**).

3. Collect the medium and centrifuge for 3 min at low speed ($300 \times g$) to pellet cells and debris. The supernatant is frozen and stored at –20 °C (e.g., for ELISA measurement and LDH assay, *see* **Note 19**), or precipitated with 2.5 volumes of acetone and stored overnight or longer at –20 °C.

4. Centrifuge the acetone precipitation for 1 h at 4 °C and $13,000 \times g$ to pellet the proteins. Remove the liquid phase and air-dry the pellet for 15 min. Resupend in 1× SDS loading buffer and incubate for 5 min at 95 °C.

5. Harvest lysates either directly in 1× SDS loading buffer for Western blot analysis. Harvest the cells by scraping them off, then heat to 95 °C for 5 min, and sonicate. Alternatively, lyse HPKs by incubation for 10 min in prewarmed K-SFM with 10% Triton-X100, then pipet up and down, and store lysates at –20 °C (e.g., for ELISA and LDH assay).

4 Notes

1. Cells can be obtained from Kerafast, Inc. (EF3003, Boston, MA).

2. Aliquots of medium are prewarmed to 37 °C just before use.

3. This was achieved by plating trypsinized human skin on lethally irradiated 3T3 fibroblasts. Later, particularly the clone 3T3-J2 has been used for primary keratinocyte isolation and propagation (Rasmussen et al., 2013) [14]. The isolation of keratinocytes

with feeder cells is efficient and, depending on the age of the donor, a lifetime of the culture of 50 (Rheiwald & Green, 1975) to 160 passages (Rasmussen et al., 2013) can be achieved (with feeder cells). Contaminating fibroblast and feeder cells can be easily removed by a short incubation with trypsin/EDTA solution, as HPKs adhere to surfaces much stronger than connective tissue cells. In addition, cultivation in K-SFM suppresses growth of fibroblasts allowing experiments with an almost homogenous population of keratinocytes. However, in the absence of feeder cells HPKs undergo senescence and differentiation after some weeks. Due to the young age of the donor, foreskin is ideal for the isolation of HPKs, but keratinocytes can also successfully be isolated from skin of other donors and parts of the human body. Since the outer root sheath of the hair follicle contains keratinocytes, plucked hair can also be used for the isolation of HPKs (Aasen & Izpisua Belmonte, 2012; Raab et al., 2014). This is of particular importance, as plucked hair allows the isolation of HPKs by a noninvasive method. Keratinocytes (and fibroblasts) grow out on tissue culture plastic directly from plucked hair or after enzymatic digestion. However, even on feeder cells the number of keratinocytes that can be isolated with this method is significantly lower compared to HPKs from larger skin samples such as foreskin.

4. It should be taken into account that mitomycin C-treated feeder cells survive only few days in culture. For the isolation of keratinocytes from biopsies (starting from 2), when cells are going to be incubated for a week or longer in the same dish, a higher amount of 3T3-J2 feeder cells should be used (up to a 1:1 dilution from a confluent culture).

5. Mitomycin C-treated 3T3 feeder cells can be frozen in medium with 10 % DMSO and stored in liquid nitrogen. When required, cells can be thawed and directly used. However, efficacy of adherence of these cells must be determined experimentally.

6. The skin biopsies should be stored after surgery in DMEM with 1 % Anti-Anti at 4 °C. Isolation of HPKs is well possible with biopsies up to 2 days post-surgery.

7. Handle the biopsy carefully with sterile forceps.

8. Isolation of HPKs should be performed under sterile conditions in a cell culture hood at RT.

9. For the isolation of HPKs from plucked hair use kanamycin instead of Anti-Anti.

10. Usually, most HPKs are obtained in the second and third trypsin fraction. Note that later fractions contain more fibroblasts.

11. Contaminating fibroblasts can be removed by incubation with trypsin/EDTA solution for 2 min and washing with PBS.

12. It is difficult to remove HPKs completely at this step. Pipet the trypsin/EDTA solution up and down directly onto the cells, but avoid excessive shear stress. Try to increase the yield of HPKs by a second incubation with trypsin/EDTA solution.

13. The details are given for transfection of HPKs in a single well of a 12-well dish. Adjust amounts of reagents according to the surface.

14. At this day cell density should be 60–90 %. If this is not the case adjust the protocol accordingly.

15. Use scrambled siRNA and/or the knockdown of unrelated genes as controls.

16. YVAD is less efficient in inhibiting caspase-1 than zVAD. For experiments that require a longer (more than 8 h) inhibition of caspase-1 activity/activation, add YVAD a second time either directly to the medium or with new medium.

17. In confluent HPKs inflammasome activation and UPS are significantly less efficient than in exponentially growing cells.

18. Inflammasome activation starts about 2 h after UVB irradiation and is maximal after 4–5 h. Although HPKs produce some mature IL-1β after 5 h, cell lysis strongly increases at later time points.

19. Perform experiments, e.g., for ELISA and LDH measurement, at least in triplicates.

Acknowledgements

Our work was supported by the Swiss National Science Foundation (SNF 31003A_132450), the Swiss Cancer League, the European Science Foundation (EuroMEMBRANE), the OPO Stiftung, the Center for Clinical Research, University Hospital Zurich and University of Zurich, the Helmut Horten-Stiftung, the Wilhelm Sander-Stiftung and the Novartis Foundation. M.G., J.S., and S.G. are members of the Zurich graduate program in Molecular Life Sciences. We would like to thank Prof. D. Hohl (CHUV, Lausanne) for his help with the establishment of HPK isolation and propagation, and Prof. S. Werner (ETH Zurich) and Prof. L. E. French (University Hospital Zurich) for continuous support.

References

1. Aasen T, Izpisua Belmonte JC (2010) Isolation and cultivation of human keratinocytes from skin or plucked hair for the generation of induced pluripotent stem cells. Nat Protoc 5:371–382

2. Ahn H, Kim J, Jeung EB, Lee GS (2014) Dimethyl sulfoxide inhibits NLRP3 inflammasome activation. Immunobiology 219:315–322

3. Bergsbaken T, Fink SL, Cookson BT (2009) Pyroptosis: host cell death and inflammation. Nat Rev Microbiol 7:99–109

4. Feldmeyer L, Keller M, Niklaus G, Hohl D, Werner S, Beer HD (2007) The inflammasome mediates UVB-induced activation and secretion of interleukin-1beta by keratinocytes. Curr Biol 17:1140–1145

5. Freigang S, Ampenberger F, Weiss A, Kanneganti TD, Iwakura Y, Hersberger M, Kopf M (2013) Fatty acid-induced mitochondrial uncoupling elicits inflammasome-independent IL-1alpha and sterile vascular inflammation in atherosclerosis. Nat Immunol 14:1045–1053

6. Gross O, Yazdi AS, Thomas CJ, Masin M, Heinz LX, Guarda G, Quadroni M, Drexler SK, Tschopp J (2012) Inflammasome activators induce interleukin-1alpha secretion via distinct pathways with differential requirement for the protease function of caspase-1. Immunity 36:388–400

7. Jorgensen I, Miao EA (2015) Pyroptotic cell death defends against intracellular pathogens. Immunol Rev 265:130–142

8. Keller M, Ruegg A, Werner S, Beer HD (2008) Active caspase-1 is a regulator of unconventional protein secretion. Cell 132:818–831

9. Kuida K, Lippke JA, Ku G, Harding MW, Livingston DJ, Su MS, Flavell RA (1995) Altered cytokine export and apoptosis in mice deficient in interleukin-1 beta converting enzyme. Science 267:2000–2003

10. Li P, Allen H, Banerjee S, Franklin S, Herzog L, Johnston C, McDowell J, Paskind M, Rodman L, Salfeld J et al (1995) Mice deficient in IL-1 beta-converting enzyme are defective in production of mature IL-1 beta and resistant to endotoxic shock. Cell 80:401–411

11. Monteleone M, Stow JL, Schroder K (2015) Mechanisms of unconventional secretion of IL-1 family cytokines. Cytokine 74:213–218

12. Nickel W (2003) The mystery of nonclassical protein secretion. A current view on cargo proteins and potential export routes. Eur J Biochem 270:2109–2119

13. Nickel W, Rabouille C (2009) Mechanisms of regulated unconventional protein secretion. Nat Rev Mol Cell Biol 10:148–155

14. Raab S, Klingenstein M, Liebau S, Linta I, (2014) A comparative view on human somatic cell sources for iPSC generation. Stem Cells Int 2014:768391

15. Rasmussen C, Thomas-Virnig C, Allen-Hoffmann BL (2013) Classical human epidermal keratinocyte cell culture. Methods Mol Biol 945:161–175

16. Rheinwald JG, Green H (1975) Serial cultivation of strains of human epidermal keratinocytes: the formation of keratinizing colonies from single cells. Cell 6:331–343

17. Sollberger G, Strittmatter GE, Grossi S, Garstkiewicz M, Keller UAD, French LE, Beer HD (2015) Caspase-1 activity is required for UVB-induced apoptosis of human keratinocytes. J Investig Dermatol 135:1395–1404

18. Strowig T, Henao Mejia J, Elinav E, Flavell R (2012) Inflammasomes in health and disease. Nature 481:278–286

Chapter 10

A Reporter System to Study Unconventional Secretion of Proteins Avoiding *N*-Glycosylation in *Ustilago maydis*

Janpeter Stock, Marius Terfrüchte, and Kerstin Schipper

Abstract

Unconventional secretion of proteins in eukaryotes is characterized by the circumvention of the Endoplasmic Reticulum (ER). As a consequence proteins exported by unconventional pathways lack *N*-glycosylation, a post-transcriptional modification that is initiated in the ER during classical secretion. We are exploiting the well-established enzyme β-glucuronidase (GUS) to assay unconventional protein secretion (UPS). This bacterial protein is perfectly suited for this purpose because it carries a eukaryotic *N*-glycosylation motif. Modification of this residue by attachment of sugar moieties during the passage of the ER apparently causes a very strong reduction in GUS activity. Hence, this enzyme can only be secreted in an active state, if the export mechanism does not involve ER passage. Here, we describe a reporter system applied in the corn smut fungus *Ustilago maydis* that is based on this observation and can be used to test if candidate proteins are secreted to the culture supernatant via alternative pathways avoiding *N*-glycosylation. Importantly, this system is the basis for the establishment of genetic screens providing mechanistic insights into unknown UPS pathways in the future.

Key words β-Glucuronidase (GUS), Unconventional protein secretion (UPS), 4-Methylumbelliferyl-β-D-glucuronide (MUG), *N*-Glycosylation, *Ustilago maydis*

1 Introduction

Protein secretion by classical mechanisms is mediated by a signal peptide at the N-terminus of proteins. In eukaryotic cells, the presence of such signal directs proteins to the ER lumen where they are folded into the native state and where post-translational modifications occur. The latter includes, i.e., the attachment of sugar moieties in the process of *N*-glycosylation. Proteins then further traverse through the endomembrane system passing the Golgi apparatus and are finally secreted into the extracellular space via secretory vesicles [1, 2]. By contrast, multiple reports exist that describe the unconventional export of proteins in eukaryotic cells [3–5]. These proteins often are of fundamental physiological importance and famous examples include human interleukin 1β

Andrea Pompa and Francesca De Marchis (eds.), *Unconventional Protein Secretion: Methods and Protocols*, Methods in Molecular Biology, vol. 1459, DOI 10.1007/978-1-4939-3804-9_10, © Springer Science+Business Media New York 2016

and fibroblast growth factor-2 [6, 7]. UPS has also been observed in fungi. For example, the lipopeptide a-factor in *Saccharomyces cerevisiae* is exported by direct transfer across the plasma membrane using the ABC transporter Ste6p [8]. In the same organism, the acyl-binding protein Acb1 is secreted via alternative, autophagy-related vesicular processes and similar observations have been made for its orthologs in the yeast *Pichia pastoris* and the protozoa *Dictyostelium discoideum* [3]. However, due to the fact that proteins exported by UPS mechanisms often have dual functions (cytoplasmic and extracellular) or are secreted only in minor amounts, unambiguous experimental validation of their noncanonical secretion is not easy to achieve. We are studying unconventional secretion of chitinase Cts1 in the dimorphic fungus *Ustilago maydis* [9–11]. This eukaryotic model is the pathogenic agent of corn smut disease [12, 13]. *U. maydis* grows yeast-like during the saprotrophic stage of its life cycle. For infection, cells switch to filamentous growth with hyphae invading the plant tissue [14]. This switch is regulated by a central dimeric transcription factor and can be induced artificially in the laboratory. To this end, the active transcription factor is expressed in the presence of nitrate (strain variant AB33) [15]. Cts1 is part of the chitinolytic system of *U. maydis* [16, 17]. The enzyme functions during cytokinesis. During cell division in the yeast stage it acts in concert with a second, conventionally secreted chitinase Cts2 and mediates the physical separation of mother and daughter cell [16]. In the filamentous stage Cst1 accumulates in hyphae with disturbed long-distance mRNA transport [10, 18, 19]. Hyphae lacking the protein are aggregating, indicating that the cell wall composition is altered [16]. Remarkably, Cts1 is active at the cell exterior but lacks a predictable N-terminal signal peptide [9, 10], suggesting a noncanonical protein export. This led us to develop a reporter system to assay UPS in *U. maydis*. This system is based on the well-established bacterial reporter enzyme β-glucuronidase (GUS) [20]. This enzyme is known to show a strong activity reduction upon passage of the classical secretory pathway involving the ER, because it coincidentally carries a eukaryotic *N*-glycosylation site (Fig. 1) [21]. Further advantages of this enzyme are the availability of diverse substrates that can be used for activity assays with chromogenic or fluorometric readouts [20]. In our GUS reporter system, homologous recombination is used to generate strains carrying the transgene stably inserted at a defined locus in the cell. Strains express translational fusions of the candidate protein to be tested for UPS (Fig. 1). A strain in which GUS is expressed cytoplasmically serves as controls for cell lysis, while a strain producing GUS fused to a conventional signal peptide derived from the enzyme invertase serves as control for conventional secretion [9]. In all strains, GUS activity is determined using the fluorescent substrate 4-methylumbelliferyl-β-d-glucuronide (MUG) using cell extracts

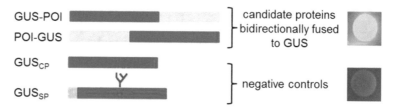

Fig. 1 GUS reporter system to study unconventional secretion of candidate proteins in *Ustilago maydis*. The bacterial enzyme GUS (*blue*) carries the eukaryotic *N*-glycosylation site (NLS) at amino acid position N_{358} that is modified by the addition of sugar moieties (*red*) during conventional secretion mediated by a signal peptide (GUS$_{SP}$, signal peptide in *orange*). Proteins of interest (POI, displayed in *grey*) are fused bidirectionally to GUS to investigate unconventional secretion with circumvention of *N*-glycosylation (GUS-POI and POI-GUS for N- and C-terminal GUS fusion proteins, respectively). If the protein is secreted with circumvention of the *N*-glycosylation machinery GUS activity can be determined in the culture supernatant. Cytoplasmic GUS (GUS$_{CP}$) does not meet the machinery either and is thus active and deals as a control for cell lysis. Gus activity of the depicted reporter proteins growing in the yeast form can be assayed on MUG-containing plates (modified from [12]). Cells convert the substrate supplied in an agar plate to a fluorescent product, if the enzyme is actively secreted

and cell-free culture supernatants. Active GUS converts this substrate to MU which upon excitation with 365 nm light emits fluorescence at 465 nm. Assaying UPS in yeast-like cells of *U. maydis* by the reporter system comprises six main steps that are described in detail in this chapter: (1) Cell cultivation, (2) Generation of cell extracts, (3) GUS enzyme assay, (4) Preparation of standards, (5) Fluorometric determination of GUS enzyme activity, and (6) Data analysis and evaluation. Of note, Gus activity can also be visualized in plate assays using either the colorimetric substrate 5-bromo-4-chloro-3-indolyl-β-d-glucuronide (X-Gluc) or MUG (Fig. 1). In concert with the well-established liquid assays described in this chapter, this feature is of broad applicability, for instance in genetic screens for secretion-deficient or so-called super-secretor mutants. Hence, establishing the Gus reporter system for assaying UPS is the basis to further characterization of the respective underlying export machinery and will hence find broad applicability also in other organisms.

2   Materials

All solutions should be prepared using ultrapure water ($H_2O_{bid.}$) and per analysis (*p.a.*) grade reagents. Prepare and store all reagents and solutions at room temperature unless stated otherwise.

2.1 Cell Culture

1. Complete medium (CM-glc): 0.25% (w/v) casamino acids, 0.1% (w/v) yeast extract, 1.0% (v/v) vitamin solution, 6.25% (v/v) salt solution, 0.05% (w/v) herring sperm DNA , and 0.15% (w/v) NH_4NO_3. Dissolve in $H_2O_{bid.}$. Adjust to pH 7.0 with 5 M NaOH. Autoclave 5 min 121 °C for sterilization and add 1% (w/v) glucose using a 50% (w/v) stock solution (see **Note 1**).

2. Vitamin solution: 0.1% (w/v) thiamine hydrochloride, 0.05% (w/v) riboflavin, 0.05% (w/v) pyridoxine, 0.2% (w/v) calcium pantothenate, 0.05% (w/v) p-aminobenzoic acid, 0.2% (w/v) nicotinic acid, 0.2% (w/v) choline chloride, 1% (w/v) myo-inositol. Dissolve in $H_2O_{bid.}$ and sterilize the solution using filters (0.2 μm pore size). Store aliquots at −20 °C.

3. Salt solution: 16% (w/v) KH_2PO_4, 4% (w/v) Na_2SO_4, 8% (w/v) KCl, 1.32% (w/v) $CaCl_2 \times 2H_2O$, 8% (v/v) trace elements, 1% (w/v) $MgSO_4$ (water free) (see **Note 1**). Dissolve in $H_2O_{bid.}$ and sterilize the solution using filters (0.2 μm pore size).

4. Trace elements: 0.06% (w/v) H_3BO_3, 0.14% (w/v) $MnCl_2 \times 4H_2O$, 0.4% (w/v) $ZnCl_2$, 0.4% (w/v) $Na_2MoO_4 \times 2H_2O$, 0.1% (w/v) $FeCl_3 \times 6H_2O$, 0.04% (w/v) $CuSO_4 \times 5H_2O$. Dissolve in $H_2O_{bid.}$ and sterilize the solution using filters (0.2 μm pore size).

5. Cell cultivation: Glass reaction tubes, tube rotator, 250 mL baffled flasks, benchtop orbital shaker running at 200 rpm.

2.2 Preparation of Cell Extracts and Supernatants

1. Phosphate buffered saline (PBS): 0.89% (w/v) Na_2HPO_4, 1.97% (w/v) KH_2PO_4, 0.09% (w/v) $MgCl_2 \times 6H_2O$, 0.02% (w/v) KCl, 0.8% (w/v) NaCl. Eventually adjust the solution to pH 7.4 using a 5 M NaOH stock solution. Autoclave 5 min at 121 °C.

2. Lysis buffer for native cell extracts (see **Note 2**): 0.01 mM phenylmethylsulfonylfluorid (PMSF), 12.5 μM benzamidine hydrochloride hydrate, 100 μL protease inhibitor cocktail 50×. Dissolve in PBS pH 7.4.

3. Ball mill (we use a MM400 from Retsch) equipped with PTFE adapter for five reaction tubes (1.5 and 2.0 mL) and 0.25–0.5 mm glass beads (see **Note 3**).

4. Bradford solution (store at 4 °C).

5. Protein determination: Tecan fluorescence reader infinite M200 equipped with Microtest Plate 96 Well F.

6. Cell harvest: Centrifuge suited for 50 mL PET tubes.

7. Supernatant preparation: funnel, folded filter paper (Particle Size: 5–8 μm Thickness: 0.16 mm Pore Size: 5–8 μm).

2.3 GUS Assay Components

1. 2× GUS assay buffer for supernatant (see **Note 4**): 0.028 mM β-mercaptoethanol, 0.8 mM Na_2-EDTA pH 8.0, 10 mM $NaPO_4$ buffer pH 7.0, 0.0042% (v/v) lauroyl-sarcosine,

0.004% (v/v) triton-X-100, 1/50 vol bovine serum albumin fraction V (BSA), 2 mM 4-methylumbelliferyl-β-d-glucuronide × 3H_2O (MUG).

2. 1× GUS assay buffer for cell extracts (*see* **Note 4**): 1× GUS extraction buffer, 1/50 vol bovine serum albumin fraction V (BSA), 2 mM 4-methylumbelliferyl-β-d-glucuronide × 3H_2O (MUG).

3. 2× GUS extraction buffer: 0.028 mM β-mercaptoethanol, 0.8 mM Na$_2$-EDTA pH 8.0, 10 mM NaPO$_4$ buffer pH 7.0, 0.0042% (v/v) lauroyl-sarcosine, 0.004% (v/v) triton-X-100. Dissolve in $H_2O_{bid.}$.

4. Sodium phosphate buffer (NaPO$_4$ buffer pH 7.0): Prepare 1 M NaPO$_4$ buffer stock solutions (solution 1: 14.2% (w/v) Na$_2$HPO$_4$; solution 2: 13.8% (w/v) Na$_2$HPO$_4$; both dissolved in $H_2O_{bid.}$ each). Use 1 L solution 1 and titrate with solution 2 until pH 7.0 is reached (about 300 mL solution 2 needed). Store at 37 °C to prevent salt precipitation.

5. EDTA stock solution: Dissolve 0.5 M Na2-EDTA × 2H_2O stock in H_2O_{bid}, adjust to pH 8.0. Autoclave 5 min at 121 °C to sterilize.

6. GUS stop buffer: 0.2 M Na$_2$CO$_3$. Dissolve in H_2O_{bid}.

7. X-Gluc stock solution (100 mg/mL, solved in DMSO).

8. 4-Methylumbelliferone (MU).

9. Fluorescence measurements: Tecan fluorescence reader infinite M200 equipped with black 96 Well Microplates (96 Well, PS, F-Bottom (chimney well), μCLEAR®, black, CELLSTAR®, TC, sterile).

3   Methods

Here we describe a liquid assay (*see* **Note 5**) based on the fluorescent substrate MUG that can be applied to determine if a protein is secreted unconventionally in yeast-like growing cells (*see* **Note 6**) by using protein fusions with the GUS reporter enzyme. Of note, we are using *U. maydis* AB33 derivatives that have been generated by homologous recombination and thus harbor the respective fusion genes upstream of a constitutive promoter (*see* **Note 7**) stably integrated in their genomes. In this fungus the preferred locus for insertion of expression constructs is the *ips*locus. Importantly, the copy number in this locus can vary such that the insertion of only a single copy needs to be verified by Southern blot analysis resulting in comparable strains [9]. The full-length expression of the fusion proteins which is crucial to allow drawing conclusions from the reporter assay should furthermore be documented by performing Western Blot analyses using antibodies against an epitope tag that is added to the fusion protein (*see* **Note 8**).

3.1 Growing Cultures

1. Inoculate a preculture of each strain in glass tubes using 3 mL medium supplemented with 1% (w/v) glucose (CM-glc). Use *U. maydis* cultures on agar plates for inoculation (*see* **Note 9**). Incubate at 28 °C for about 20–24 h on a rotating wheel incubator.

2. Use 6 µL of the densely grown precultures to inoculate main cultures of 20 mL CM-glc, and incubate in sterile 100 mL baffled flasks shaking at 200 rpm and 28 °C (*see* **Note 10**).

3. Grow the cells until they reach an OD_{600} of about 0.5. This will take approximately 14–16 h. Determine the volume of cells needed to adjust an OD_{600} of 0.2 and harvest this volume by centrifugation at $3500 \times g$ for 5 min. Resuspend the cells in sterile $H_2O_{bid.}$ for washing and repeat the centrifugation step. Resuspend the cells of each strain in 20 mL CM-glc (*see* **Note 6**) and incubate in sterile 100 mL baffled flasks for 6 h at 200 rpm and 28 °C.

4. After 6 h determine the OD_{600} of each culture. These values will be needed for calculating enzyme activities later on.

5. To harvest cells, use 2 mL reaction tubes and centrifuge 1 mL of each culture at $16,000 \times g$ for 10 min. Discard the supernatants, resuspend the cell pellets in 1 mL 1× GUS extraction buffer, and repeat the centrifugation step for washing. Discard the supernatants and use the cell pellets for the generation of cell extracts (*see* **Note 11**).

6. To harvest culture supernatants, use 2 mL reaction tubes and centrifuge 2 mL of the CM-glc cultures at $16,000 \times g$ for 10 min. Filter the supernatants into a fresh tube using filter papers. Keep at least 1 mL of each culture supernatant for the GUS activity assay (*see* **Note 12**).

3.2 Preparing Cell Extracts

Work on 4 °C or ice during the whole procedure.

1. Dissolve the cell pellet in 200 µL 1× GUS extraction buffer and transfer the suspension into 2 mL reaction tubes. Add a volume of about 100 µL glass beads (*see* **Note 13**) and crack the cells in the ball mill for 5 min and 30 Hz.

2. Transfer the suspension into a fresh 1.5 mL reaction tube and centrifuge for 5 min at $8000 \times g$.

3. Transfer 100 µL supernatant into a fresh 1.5 mL reaction tube. Store on ice.

4. Determine the protein concentration in the extracts (*see* **Note 14**).

3.3 GUS Enzyme Assay

1. Prewarm appropriate amounts of 1× GUS assay buffer for cell extracts and 2× GUS assay buffer for supernatants to 37 °C (calculate needed volumes from this subheading **steps 3** and **4**, respectively).

2. For the enzyme assay using cell extracts, mix 10 μL cell extract with 990 μL 1× GUS extraction buffer and add 1 mL prewarmed 1× GUS assay buffer for cell extracts.

3. For the enzyme assay using culture supernatants, mix 1 mL prewarmed 2× GUS assay buffer for supernatants and 1 mL cell-free culture supernatant.

4. After 0, 2, 3, and 4.5 h (*see* **Note 15**) transfer 200 μL of the reaction mixture to a fresh 1.5 mL reaction tube containing 800 μL GUS stop buffer to stop the reaction.

5. Store stopped samples at 4 °C in the dark after addition of GUS stop buffer until all reactions are stopped and then proceed to Subheading 3.5 (Fluorescence determination of GUS enzyme activity).

3.4 Generation of MU Standards

1. Prepare a 100 mM stock solution by dissolving 17.6 mg 4-methylumbelliferone (MU) in 1 mL GUS stop buffer (*see* **Note 16**).

2. Use the 100 mM MU stock solution to generate dilutions with a final concentration of 0.1, 1, 10, and 100 μM in a volume of 1 mL in reaction tubes (*see* **Note 17**). Pure stop buffer is used for determining background fluorescence at 0 μM MU.

3. To prepare the fluorescence measurements of the MU standards for cell extracts, mix 200 μL of each solution (0–100 μM MU in stop buffer) with 600 μL GUS stop buffer and 200 μL 1× GUS extraction buffer.

4. To prepare the fluorescence measurements of the MU standards for cell-free culture supernatants, mix 200 μL of each solution (0–100 μM MU in stop buffer) with 600 μL GUS stop buffer, 100 μL 2× GUS extraction buffer, and 100 μL CM-gluc.

5. Store standards at 4 °C in the dark until all reactions are stopped and then proceed to Subheading 3.5.

3.5 Fluorometric Determination of GUS Enzyme Activity

1. Use a black 96-well plate (*see* Subheading 2.3) and for each sample generated in Subheading 3.3, add 200 μL in technical triplicates.

2. On the same plate, add 200 μL of each MU standard generated in Subheading 3.4 in triplicates.

3. Determine the fluorescence of the samples using a fluorometer. The excitation and emission wavelengths are set to 365 and 465 nm, respectively, and the gain to 60 (*see* **Note 18**) at ambient temperature.

4. Use raw data (relative fluorescence units) in the output file for data analysis using Microsoft Excel or a similar program (Fig. 2).

3.6 Data Analysis and Evaluation

1. Generate the arithmetical means of the three technical replicates (Fig. 2(1); *see* **Note 19**).

2. Subtract the blank. This is the value obtained from the arithmetical mean of the values obtained for the 0 µM MU standard solution (Fig. 2(2)).

3. Generate the MU standard curve. Therefore, the arithmetic means of the values obtained for the standard solutions need to be determined (Fig. 2(3)) and represented as a graph, containing the RFU values on the *y*-axis and the MU concentration [µM] on the *x*-axis (Fig. 2(3), Graph). The different data points can be connected in a straight line. Display the equation of the linear curve and use the slopes for further calculations.

4. Convert the relative fluorescence units into µM MU by dividing the values by the slope of the linear standard curve (Fig. 2(4)).

GUS activities supernatants

strain	h	RFU (3 replicates)			average	w/o blank	µM	µM/h	µM/min	OD$_{600}$	U/OD$_{600}$
AB33	0	37	36	40	38	12	0.07				
	2	35	34	35	35	9	0.05	0.002	0.00003	1.555	0.00004
	3	33	34	36	34	8	0.05				
	4.5	36	47	37	40	14	0.08				
AB33 GUS-Cts1	0	35	32	34	34	8	0.05				
	2	2508	2529	2518	2518	2492	14.49	6.553	0.109	1.665	0.13119
	3	3513	3421	3557	3497	3471	20.19				
	4.5	5134	5123	5101	5119	5093	29.63				

MU standard

µM MU	RFU (3 replicates)			average	w/o blank
0	24	25	29	26	0
0.1	44	44	43	44	18
1	230	225	232	229	203
10	2053	2076	2082	2070	2044
100	17160	17309	17383	17284	17258

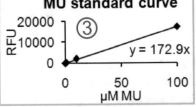

MU standard curve — y = 172.9x

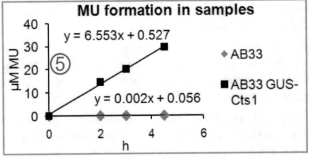

MU formation in samples — y = 6.553x + 0.527 (AB33 GUS-Cts1); y = 0.002x + 0.056 (AB33)

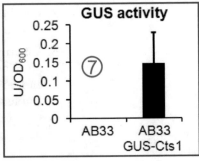

GUS activity

Fig. 2 Data analysis. Details are explained in the text (Subheading 3.6)

5. Similar to the standard curve, generate graphs of the values obtained for 0, 2, 3, and 4.5 h of each sample (Fig. 2(5), Graph). Connect the data points of each sample and display the equations of the linear curves.

6. Divide each slope by 60 to obtain μM/min (Fig. 2(6)). This corresponds to the MU concentration in the reaction volume of 2 mL per minute.

7. To finally determine the respective GUS activities for cell extracts, all values are related to the protein amounts in the samples (in mg; determined in **step 4** of Subheading 3.2) used in the assay (not shown).

8. To finally determine the respective GUS activities for supernatants, all values need to be multiplied by a factor of 2, because the samples have been diluted (*see* **step 4** in Subheading 3.3). Then, take the OD_{600} (documented in **step 4** of Subheading 3.1) into account by subtracting the values for each sample. This provides the GUS activities in 1 L culture supernatants in U/OD_{600} (Fig. 2(7), Table).

9. Use the U/OD_{600} values of biological triplicates of all samples (*see* **Note 20**) to generate the final figure, in which the arithmetical mean of the values is displayed together with the respective standard deviation (Fig. 2(7), Graph).

4   Notes

1. The quality of the CM-glc medium is of fundamental importance. For example, we observed variations (much lower GUS activities) when we replace water-free $MgSO_4$ by alternative salts containing water.

2. Prepare fresh for each experiment.

3. If large cell culture volumes need to be processed, the bead mill can alternatively be equipped with 35 mL stainless steel cups and 12 mm stainless steel beads.

4. Prepare freshly and keep the solution in the dark. It is stable for at least 1 h on 4 °C.

5. Indicator plates containing the colorimetric substrate X-Gluc can be used to verify and complement the liquid assay. Plates should be prepared using 100 μL of an X-Gluc stock solution. Let the plates dry for at least 1 h to allow soaking with the substrate and evaporation of DMSO. Cells should be grown to an OD_{600} of 0.5. Then, spot 4 μL onto a plate and incubate for at least 24 h. It is essential to add a control strain expressing intracellular GUS to evaluate cell lysis.

6. For assaying filamentously growing cells of *U. maydis*, AB33 cultures need to be shifted to nitrogen minimal (NM) medium supplemented with glucose [9, 15].

7. The choice of the promoter may be crucial to the experiment. If possible (meaning if there is sufficient expression to allow a detection of the gene product), we recommend to stick to the native promoter of the candidate gene. However, if expression levels are low, strong constitutive promoters may need to be introduced [9, 11].

8. To allow for checking the integrity of the fusion proteins in Western blot analyses, an epitope tag should be added to the reporter proteins. Using a fluorescence tag like eGFP furthermore provides the possibility to localize the proteins in the cell. In cases where fusion proteins show strong degradation in Western blot analyses, it may be dangerous to draw conclusions from the reporter assay. In our hands, multiple protease deletions were helpful in reducing this problem [11].

9. We recommend using freshly grown agar plate cultures of similar age for inoculation of liquid cultures. *U. maydis* strains can be stored for long term in glycerol stocks at –80 °C. A small portion of such frozen stock should be plated on agar plates and incubated for 2 days at 28 °C to ensure proper growth of precultures in liquid. Do not keep these plates for more than 2 weeks.

10. Be cautious when handling liquid cultures of *U. maydis*. We observe that cultures start growing worse when shaking is abolished for short periods. Thus, make sure that cultures are always well aerated, even in times when you have to remove them from the shaker. In these periods, mix by shaking the flasks by hand from time to time and place them back onto the shaker as soon as possible. Do not keep liquid cultures on the bench for longer periods without agitation.

11. Cell pellets can be frozen at –20 °C without losing GUS activity.

12. Cell-free culture supernatants can be frozen at –20 °C without losing GUS activity for at least 1 month.

13. To add an approximate volume of 100 μL of glass beads to our samples, we prepare self-made measuring spoons. Therefore, we mark the filling height of 100 μL in a 1.5 mL reaction tube using water and cut off this part with a scalpel. The section is then connected to a 1 mL plastic pipet tip by melting the narrow end of the latter with the bunsen burner and subsequent gluing to form a small spoon. This procedure easily provides a convenient device for aliquoting equal amounts of glass beads.

14. Protein determination should be performed according to standard protocols.

15. The assay may also be performed as an enzyme kinetic. To this end, the assay volume should be reduced to 200 μL and continuous reading should be performed in the fluorescent reader, determining fluorescence emission every 10 min at 37 °C for about 90 min. The MU standards need to be adapted to this new setup.

16. MU standards need to be prepared in GUS stop buffer because the powder is insoluble in H_2O.

17. The MU standard solutions can be stored at –20 °C without showing variations in fluorescence for at least 1 month.

18. We established gain 60 for our assays. However, depending on the experimental conditions and the specific type of plate reader, it may be necessary to determine the optimal gain for the conducted assay anew. However, after determining the optimal gain, this value should be fixed for all biological replicates.

19. Preparing a calculation spreadsheet in Microsoft Excel or a comparable program will facilitate data processing.

20. The values of different biological replicates tend to differ quite a bit in the reporter assay. Hence, to make sure that the results are significant and that the standard deviation is in a satisfactory range, more than three biological replicates may need to be conducted.

Acknowledgements

We are grateful to Drs. Janine Koepke, Saskia Kreibich, Thomas Brefort, and Regine Kahmann for their input and support, and to Dr. Michael Feldbrügge for critical discussions, comments on the manuscript, and general support. J.S. and M.T. received a fellowship of the Graduate School of the Cluster Industrial Biotechnology 2021 (CLIB-GC) and K.S. was supported by the Strategic Research Fund of the Heinrich-Heine University Düsseldorf. The scientific activities of the Bioeconomy Science Center were supported financially by the Ministry of Innovation, Science and Research within the framework of the NRW Strategieprojekt BioSC (No. 313/323-400-002 13).

References

1. Rothman JE (1994) Mechanisms of intracellular protein transport. Nature 372(6501):55–63. doi:10.1038/372055a0

2. Rapoport TA (2007) Protein translocation across the eukaryotic endoplasmic reticulum and bacterial plasma membranes. Nature 450(7170):663–669. doi:10.1038/nature06384

3. Malhotra V (2013) Unconventional protein secretion: an evolving mechanism. EMBO J 32(12):1660–1664. doi:10.1038/emboj.2013.104

4. Rabouille C, Malhotra V, Nickel W (2012) Diversity in unconventional protein secretion. J Cell Sci 125(Pt 22):5251–5255. doi:10.1242/jcs.103630

5. De Marchis F, Bellucci M, Pompa A (2013) Unconventional pathways of secretory plant proteins from the endoplasmic reticulum to the vacuole bypassing the Golgi complex. Plant Signal Behav 8(8):pii:e25129. doi:10.4161/psb.25129

6. Monteleone M, Stow JL, Schroder K (2015) Mechanisms of unconventional secretion of IL-1 family cytokines. Cytokine 74(2):213–218. doi:10.1016/j.cyto.2015.03.022

7. La Venuta G, Zeitler M, Steringer JP, Müller HM, Nickel W (2015) The startling properties of fibroblast growth factor 2: how to exit mammalian cells without a signal peptide at hand? J Biol Chem. doi:10.1074/jbc.R115.689257

8. McGrath JP, Varshavsky A (1989) The yeast STE6 gene encodes a homologue of the mammalian multidrug resistance P-glycoprotein. Nature 340(6232):400–404. doi:10.1038/340400a0

9. Stock J, Sarkari P, Kreibich S, Brefort T, Feldbrügge M, Schipper K (2012) Applying unconventional secretion of the endochitinase Cts1 to export heterologous proteins in *Ustilago maydis*. J Biotechnol 161(2):80–91. doi:10.1016/j.jbiotec.2012.03.004

10. Koepke J, Kaffarnik F, Haag C, Zarnack K, Luscombe NM, König J, Ule J, Kellner R, Begerow D, Feldbrügge M (2011) The RNA-binding protein Rrm4 is essential for efficient secretion of endochitinase Cts1. Mol Cell Proteomics 10(12):111–011213. doi:10.1074/mcp.M111.011213

11. Sarkari P, Reindl M, Stock J, Müller O, Kahmann R, Feldbrügge M, Schipper K (2014) Improved expression of single-chain antibodies in *Ustilago maydis*. J Biotechnol 191:165–175. doi:10.1016/j.jbiotec.2014.06.028

12. Feldbrügge M, Kellner R, Schipper K (2013) The biotechnological use and potential of plant pathogenic smut fungi. Appl Microbiol Biotechnol 97(8):3253–3265. doi:10.1007/s00253-013-4777-1

13. Vollmeister E, Schipper K, Baumann S, Haag C, Pohlmann T, Stock J, Feldbrügge M (2012) Fungal development of the plant pathogen *Ustilago maydis*. FEMS Microbiol Rev 36(1):59–77. doi:10.1111/j.1574-6976.2011.00296.x

14. Kämper J, Kahmann R, Bölker M, Ma LJ, Brefort T, Saville BJ, Banuett F, Kronstad JW, Gold SE, Müller O, Perlin MH, Wösten HA, de Vries R, Ruiz-Herrera J, Reynaga-Pena CG, Snetselaar K, McCann M, Pérez-Martin J,

Feldbrügge M, Basse CW, Steinberg G, Ibeas JI, Holloman W, Guzman P, Farman M, Stajich JE, Sentandreu R, Gonzalez-Prieto JM, Kennell JC, Molina L, Schirawski J, Mendoza-Mendoza A, Greilinger D, Münch K, Rössel N, Scherer M, Vranes M, Ladendorf O, Vincon V, Fuchs U, Sandrock B, Meng S, Ho EC, Cahill MJ, Boyce KJ, Klose J, Klosterman SJ, Deelstra HJ, Ortiz-Castellanos L, Li W, Sanchez-Alonso P, Schreier PH, Hauser-Hahn I, Vaupel M, Koopmann E, Friedrich G, Voss H, Schlüter T, Margolis J, Platt D, Swimmer C, Gnirke A, Chen F, Vysotskaia V, Mannhaupt G, Güldener U, Münsterkotter M, Haase D, Oesterheld M, Mewes HW, Mauceli EW, DeCaprio D, Wade CM, Butler J, Young S, Jaffe DB, Calvo S, Nusbaum C, Galagan J, Birren BW (2006) Insights from the genome of the biotrophic fungal plant pathogen *Ustilago maydis*. Nature 444(7115):97–101. doi:10.1038/nature05248

15. Brachmann A, Weinzierl G, Kämper J, Kahmann R (2001) Identification of genes in the bW/bE regulatory cascade in *Ustilago maydis*. Mol Microbiol 42(4):1047–1063

16. Langner T, Özturk M, Hartmann S, Cord-Landwehr S, Moerschbacher B, Walton JD, Göhre V (2015) Chitinases are essential for cell separation in *Ustilago maydis*. Eukaryot Cell 14(9):846–857. doi:10.1128/EC.00022-15

17. Langner T, Göhre V (2016) Fungal chitinases: function, regulation, and potential roles in plant/pathogen interactions. Curr Genet 62:243–254

18. Haag C, Steuten B, Feldbrügge M (2015) Membrane-coupled mRNA trafficking in fungi. Annu Rev Microbiol 69:265–281. doi:10.1146/annurev-micro-091014-104242

19. Göhre V, Vollmeister E, Bölker M, Feldbrügge M (2012) Microtubule-dependent membrane dynamics in *Ustilago maydis*: trafficking and function of Rab5a-positive endosomes. Commun Integr Biol 5(5):485–490

20. Hull GA, Devic M (1995) The beta-glucuronidase (gus) reporter gene system. Gene fusions; spectrophotometric, fluorometric, and histochemical detection. Methods Mol Biol 49:125–141. doi:10.1385/0-89603-321-X:125

21. Iturriaga G, Jefferson RA, Bevan MW (1989) Endoplasmic reticulum targeting and glycosylation of hybrid proteins in transgenic tobacco. Plant Cell 1(3):381–390. doi:10.1105/tpc.1.3.381

Chapter 11

Stress-Inducible Protein 1 (STI1): Extracellular Vesicle Analysis and Quantification

Marcos Vinicios Salles Dias, Vilma Regina Martins, and Glaucia Noeli Maroso Hajj

Abstract

This chapter is derived from our experience in the study of stress-Inducible Protein 1 (STI1) in extracellular vesicles. We used different techniques to isolate, explore, and characterize the extracellular vesicles that contained this protein. Ultracentrifugation and gel chromatography were used to isolate extracellular vesicles of different sizes, nanotracking particle analysis (NTA) determined number and size of vesicles, while flow cytometry and ELISA were used to determine the specific protein content of vesicles.

Key words Stress-Inducible Protein 1 (STI1), Extracellular vesicles, Ultracentrifugation, Nanoparticle tracking analysis (NTA), Gel chromatography

1 Introduction

This chapter is derived from our experience in the study of Stress-Inducible Protein 1 (STI1) in extracellular vesicles. STI1 is a cytoplasmic co-chaperone that acts in the transfer of client proteins to Hsp70 and Hsp90 [1]. However, several studies indicate that STI1 can bind the Cellular Prion Protein (PrP^C) in the outer leaflet of the cell surface, which elicits several neurotrophic properties [2–4]. To bind PrP^C in the outer leaflet of neurons, STI1 is secreted by astrocytes in extracellular vesicles [5]. Our methods were then originally standardized to the detection of STI1 in such vesicles, but can be applied to the research of other proteins of interest.

This protocol for the isolation of extracellular vesicles was standardized for cultures of mouse primary astrocytes. To identify secreted proteins, Western blot is frequently a method of choice. However, some technical difficulties may appear (*see* **Note 1**). For example, it is a challenge to compare levels of specific proteins between different cell types or among treatments. Since the protein composition of extracellular vesicles is very variable depending

Andrea Pompa and Francesca De Marchis (eds.), *Unconventional Protein Secretion: Methods and Protocols*, Methods in Molecular Biology, vol. 1459, DOI 10.1007/978-1-4939-3804-9_11, © Springer Science+Business Media New York 2016

on the cellular source, the choice of a standard protein for normalization is complicated. Another issue is protein quantification; depending on the volume of conditioned media used to isolate vesicles, the protein amount may be too low to be determined by standard protein quantification methods like Bradford assay. More accurate results can be achieved by counting the number of extracellular vesicles in samples by using nanoparticle tracking analysis (NTA). NTA methodology has been used in the literature to characterize extracellular vesicle population [5–9]. In the NTA method, particles are visualized by light scattering using a light microscope. A video is taken and the NTA software tracks the Brownian motion of individual particles, calculating their size and total concentration. Our protocol is fitted for the Nanosisht LM10 equipped with the NTA 2.3 software (Malvern Instruments). Other brands may be used but adaptations to the protocol may be required.

To identify specific proteins of the surface of extracellular vesicles, two techniques can be applied: flow cytometry and proteinase K digestion. Because of its reduced size (50–150 nm), extracellular vesicles cannot be analyzed by standard flow cytometry methods. To circumvent this problem, it is possible to couple the extracellular vesicles in beads of a size that is in the detection range of the flow cytometer. This methodology has been used by different authors to semiquantitatively detect surface proteins of exosomes [10–12]. The bead-vesicle complexes are then labeled with the antibody of interest and followed by fluorophore-conjugated secondary antibody and analysis. Alternatively, intact vesicles can be exposed to proteinase K (PK) treatment. This treatment will only digest proteins present at the surface [13, 14], leaving intact proteins present inside the vesicles.

2   Materials

All solutions should be prepared using ultrapure water and analytical grade reagents. All reagents should be prepared and stored at room temperature unless indicated otherwise. Diligently follow all waste disposal regulations when disposing waste materials.

2.1   Conditioned Media Collection

1. 75 cm² Tissue culture flask with a confluent monolayer of primary astrocytes.

2. 0.45 μm Syringe filter units.

3. Syringe.

4. 15 ml Tubes.

5. Amicon Ultra 15 centrifugal filters (EMD Millipore UFC901024).

6. Benchtop centrifuge equipped with a rotor for 15 ml tubes.

7. Sterile phosphate-buffered saline (PBS): 137 mM NaCl, 2.7 mM KCl, 10 mM Na$_2$HPO$_4$, 2 mM KH$_2$PO$_4$.

8. Dulbecco's modified Eagle's medium—high glucose (DMEM) supplemented with antibiotics (gentamicin 40 mg/ml).

9. Fetal calf serum (FCS) previously cleared by ultracentrifugation at $100,000 \times g$ for 16 h (*see* **item 3**).

2.2 Nanoparticle Tracking Analysis to Size and Count Extracellular Vesicles

1. NanoSight LM10 (Malvern Instruments).

2. 3 ml Syringe.

3. Polystyrene latex microspheres 100, 200, and 300 nm (Sigma-Aldrich cat. 43302).

4. Fresh PBS sterile filtered in a 0.2 μm filter.

2.3 Separation by Ultracentrifugation

1. Beckman ultracentrifuge or similar.

2. SW-41Ti rotor for Beckman ultracentrifuge or similar.

3. Open top ultra-clear or polyallomer tubes for rotor SW-41Ti.

4. Phosphate-buffered saline (PBS).

2.4 Separation by Gel Filtration Chromatography

1. Glass low-pressure column 2.5 × 30 cm (Bio-Rad Econo-Column® Chromatography Columns, 2.5 × 30 cm 7374253 or similar).

2. Laboratory stand.

3. 140 ml Superose 12 prep grade (GE Healthcare 17-0536-01).

4. Peristaltic pump (Bio-Rad Model EP-1 Econo Pump or similar).

5. Fraction collector (Bio-Rad Model 2110 Fraction Collector or similar).

6. 0.8 mm Silicone Tubing (Bio-Rad 7318210 or similar).

7. Low-Pressure System Fittings Kit (Bio-Rad 7318220).

8. 13 × 100 mm Tubes for fraction collection and storage (Bio-Rad 2239750).

9. PBS.

10. 40% Glycerol.

2.5 Trichloroacetic Acid Precipitation of Proteins

1. Refrigerated microcentrifuge.

2. 100% Trichloroacetic acid solution (w/v): Solution of TCA can be made by the addition of 227 ml of water to 500 g of TCA. The resulting clear solution will have a density of approximately 1.45 g/ml. Store refrigerated.

3. Pure acetone stored at −20 °C.

4. Urea buffer: 1% SDS (w/v), 36% urea (w/v), 50 mM Tris pH 7.2.

5. 4× Laemmli buffer: 195 mM Tris–HCl, pH 6.8, 10% (w/v) SDS, 30% (w/v) glycerol, 15% (v/v) β-mercaptoethanol, 0.01% (w/v) bromophenol blue.

2.6 ELISA	1. Polystyrene 96-well high-binding ELISA plate.
	2. Sealing tape for 96-well plates.
	3. Antibody against the protein of interest.
	4. Anti-STI1 Antibody (StressMarq SPC-203D).
	5. HRP-conjugated secondary antibody.
	6. Microplate reader with a 490 nm filter.
	7. PBS.
	8. Recombinant protein of interest.
	9. Blocking buffer: 5 % Nonfat dry milk in PBS.
	10. PBS+ 0.3 % TritonX-100.
	11. OPD substrate solution: 1 mg/ml of OPD (Sigma-Aldrich 3175) in 0.05 M of phosphate-citrate buffer (pH 5) containing 30 % hydrogen peroxide.
	12. 4 N H_2SO_4.

2.7 Flow Cytometry to Detect Proteins in the Extracellular Vesicles Surface

1. Flow cytometer.
2. 4 μm Aldehyde/sulfate latex beads (Invitrogen, A37304).
3. Refrigerated Microcentrifuge (Eppendorf 5427R or similar).
4. Anti-STI1 Antibody (StressMarq SPC-203D).
5. Anti-Flotillin-1 antibody 1:100 (ABCAM ab41927).
6. Anti-Hsp90 antibody 1:50 (ABCAM ab13492).
7. Alexa 546 anti-rabbit 1:200 (Invitrogen, A-11035).
8. Sterile PBS.
9. 1 M Glycine.
10. BSA 0.5 % (w/v) in PBS.

2.8 Proteinase K Digestion of Proteins from the Surface of Extracellular Vesicles

1. Proteinase K (GIBCO 25530-015).
2. PBS.
3. PK buffer: 10 mM EDTA, 5 mM *N*-ethylmaleimide.
4. 0.5 % (v/v) Triton X-100.
5. 4× Laemmli buffer.

3   Methods

3.1 Conditioned Media Collection

1. Culture primary astrocytes in a 75 cm² flask under normal growth culture media (DMEM supplemented with antibiotics and 10 % FCS) until confluent (*see* **Note 2**).
2. Aspirate off the normal growth medium, wash monolayer 3× with sterile PBS, then replenish cells with 5 ml of serum-free DMEM, and incubate for 48 h.

3. Connect one syringe filter in a syringe with no embolus and place it over a 15 ml tube. Collect the conditioned media and pour over the filter. Do not apply pressure; filter the medium only by gravity, as pressure may lead to breakage of extracellular vesicles or cell debris.

4. Next, the conditioned media should also be cleared of cell debris by sequential centrifugation. Pour conditioned media at a 15 ml tube and centrifuge at $1500 \times g$ for 10 min. Remove supernatant and pour over another 15 ml tube, centrifuging at $4500 \times g$ for 10 min. Remove supernatant to another 15 ml tube and centrifuge at $10,000 \times g$ for 30 min [15]. The conditioned media can now be used for Western blot, ELISA, and nano-tracking analysis or separated by ultracentrifugation or gel chromatography.

5. For many applications, the conditioned media cleared of cell debris may be concentrated by the use of Amicon Ultra 15 centrifugal filters. Depending on the size of the desired protein or vesicle, different membrane cutoffs may be used. A molecular weight cutoff of 10 kDa will retain most of the proteins and all vesicles. Add up to 15 ml of cleared conditioned media (12 ml if using a fixed-angle rotor) to the Amicon® Ultra filter device. Place capped filter device into centrifuge rotor and spin at $4000 \times g$ (swinging-bucket rotor) or $5000 \times g$ (fixed-angle rotor) maximum for approximately 20 min or until volume reaches 200 µl. To recover the concentrated solute, insert a pipette into the bottom of the filter device and withdraw the sample.

3.2 Nanoparticle Tracking Analysis to Size and Count Extracellular Vesicles

1. Use isolate extracellular vesicles by your choice method (ultracentrifugation or gel chromatography). A volume of at least 500 µl is needed. If needed, dilute the sample in fresh sterile filtered PBS. Use the same PBS aliquot to dilute all your samples. Always try to analyze fresh samples, since freeze/thaw cycles could break exosomal membranes.

2. The temperature for extracellular vesicle measurement should not exceed 37 °C. Turn on the NTA 2.3 software. Apply ~500 µl of the same PBS used to dilute the sample, in a sample chamber of the NanoSight LM10. In this PBS sample, you should not be able to detect any particles. If particles are detected in this step, use this value to subtract the value of sample quantification. Avoid making the measurements close to sources of vibration at the bench (e.g., centrifuges).

3. Use a clean syringe of 3 ml to apply the samples in the chamber. After allowing a brief time to equilibrate (approximately 5–30 s), set the NTA mode to capture. The sample should be analyzed within 15 min of the initial dilution to achieve acceptable levels of precision. Localize the thumbprint area and

ensure that all the measurements are being made in the correct region. The best imaging position is as close as possible to the thumbprint. Set up the camera gain and camera shutter speed for the first sample. Gradually increase the camera level (which alters the shutter speed and gain) until the image is close to saturated, and then slowly reduce the level until particles are observed as single bright points. Adjust the focus if necessary.

4. Dilute your sample (if required) in order to achieve 40–60 particles on the screen at one time and then capture a 30-s video. Introduce a fresh volume of sample into the chamber and make another recording. Clean the chamber between samples and between different dilutions of the same sample by introducing clean PBS in order to avoid the risk of presence of residual sample. Repeat until three videos have been captured. For very low particle counts, it may be necessary to increase the number of videos captured in order to achieve an acceptable number of tracked events. Concentration measurements decrease with time due to adherence to the sample chamber. This can be minimized by making several short measurements (e.g., 3×30 s) rather than one long measurement. Avoid presence of air bubbles inside the chamber. Air bubbles can deflect the laser, and cause background scattering events. Reintroduce the sample to remove them.

5. Analyze the videos using the NTA software adjusting the detection threshold and the minimum expected particle size (should be adjusted for 30 nm). The other parameters (blur, minimum tracking length, and extracted background) should be selected. At least 1000 events in total should be tracked. A PDF report is automatically generated, showing the concentration at each vesicle size.

6. Regular quality control measurements of microsphere beads of similar size of extracellular vesicles (100, 200, and 300 nm) should be made for calibration purposes.

3.3 Separation by Ultracentrifugation

1. Place conditioned media cleared of cellular debris (not concentrated) in the ultracentrifugation tube. The tube must be filled up to the top or it will collapse during centrifugation; this will require approximately 12 ml of solution. The conditioned media may be diluted with PBS to achieve the desired volume or conditioned media from more than one 75 cm^2 flask may be used. Balance the tubes perfectly.

2. Centrifuge at $100,000 \times g$ for 1 h. Carefully remove the supernatant to another centrifugation tube and resuspend the pellet in 100 µl of PBS. Usually, pellets are not visible.

3. Balance the tubes perfectly and centrifuge again for $100,000 \times g$ for 2 h. Carefully remove the supernatant to another centrifugation tube and resuspend the pellet in 100 µl of PBS.

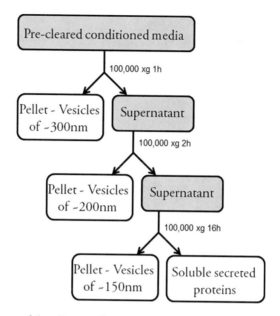

Fig. 1 Scheme of the ultracentrifugation strategy for the purification of extracellular vesicles of different size

4. Balance the tubes perfectly and centrifuge again for $100,000 \times g$ for 16 h. Collect supernatant to a 15 ml tube and resuspend the pellet in 100 μl of PBS (Fig. 1). The concentration and size of vesicles may be checked by nanoparticle tracking analysis (NanoSight LM10, Malvern Instruments) (described in Subheading 3.2) (*see* **Note 3**). Vesicles are ready to be used for Western blot, ELISA, or functional assays. The supernatant may also be probed for soluble secreted proteins. For this application, if concentration is needed, it can be done by the use of the Amicon concentration unit (described in Subheading 3.1).

3.4 Separation by Gel Filtration Chromatography

1. Sepharose beads are supplied fully hydrated and contain preservatives. To remove preservatives, place gel beads in at least three bead volumes of ultrapure water and gently stirr for 10 min. Let the beads settle and remove excess supernatant with the aid of a pipette.

2. Beads are then equilibrated in PBS by adding three bead volumes in PBS and stirring gently for about 10 min. Let the slurry settle and remove excess buffer to one bead volume.

3. The slurry is degased by placing the beads under vacuum for about 5 min. Occasional swirling of the container will help to fully degas the slurry. Do not use a stir bar as beads may be crushed. Give preference to manual swirling.

4. Mount the column vertically on a laboratory stand. Connect approximately 50 cm of tubing to the outlet of the column using a two-way stopcock (contained in the Bio-Rad Low-

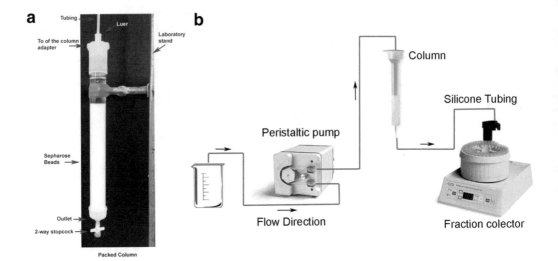

Fig. 2 (**a**) Picture of the mounted gel chromatography column. (**b**) Scheme of the setting for the gel filtration chromatography

Pressure System Fittings Kit). Using a syringe, inject PBS into the column outlet tubing until the empty column is filled with buffer to just above the bed support screen. Close the two-way stopcock (this procedure removes trapped air from below the support screen (Fig. 2a, b).

5. Pour the 50:50 bead/buffer slurry gently to avoid air bubbles using a funnel in order to completely fill the column. Open the stopcock and allow buffer to flow out. As the buffer is drained, pour the remaining of the 50:50 bead/buffer until all slurry is used and the column is completely filled with the Sepharose (*see* **Note 4**). After the column is completely packed, close the stopcock.

6. Connect the top of the column adapter to 50 cm of tubing using an appropriate luer and connect the other extremity of the tubing to the peristaltic pump. Fill the tubing and pump with PBS making sure that they are free of air bubbles. Pump 4–5 bead bed volumes of PBS through the column at a flow rate of 1 ml/min (Fig. 2b) (*see* **Note 5**).

7. Connect the outlet tubing of the column to the fraction collector using an appropriate luer (contained in the Bio-Rad Low-Pressure System Fittings Kit). Fill the fraction collector with 40 collection tubes.

8. Open the top of the column adapter. Mix 500 μl of concentrated conditioned media with 200 μl of 40% glycerol. Pipette carefully the mix in the column using a micropipette placed in close proximity to the top of the gel bed. Make a thin, uniform layer. Replace the top of the column adapter.

9. Connect outlet tubing to the fraction collector using an appropriate luer (contained in the Bio-Rad Low-Pressure System Fittings Kit). Open stopcock and turn on the pump allowing continuous PBS flow at 1 ml/min rate. In the fraction collector set volume of fraction to 3.5 ml. Collect fractions until the phenol red from the concentrated media runs through the entire column (typically 30 fractions). Fractions are ready to use for ELISA or proteins can be precipitated with trichloroacetic acid for Western Blots. Fraction 2 contains vesicles of approximately 300 nm in size, fractions 5–6 contain vesicles of approximately 200 nm, and fractions 11–12 contain vesicles of approximately 70 nm.

3.5 Trichloroacetic Acid Precipitation of Proteins

1. Chill the pre-cleared fractions collected from gel filtration chromatography by incubating on ice.

2. Add the 100% TCA solution to each sample to achieve the final concentration of 15% TCA. Agitate by inverting the tube. Incubate for 5 min at −20 °C and 1 h on ice.

3. Distribute the solution into 1.5 ml tubes (in cases of the 3.5 ml fractions collected from the gel filtration chromatography three tubes will be necessary) and centrifuge at maximum speed for 30 min at 4 °C. Remove as much supernatant as possible carefully so as not to disturb the pellet. In cases where low amounts of protein are present pellet may be unvisible.

4. Fill the tube with pre-chilled acetone and incubate on ice for 15 min. Centrifuge at maximum speed for 30 min at 4 °C. Remove as much supernatant as possible carefully so as not to disturb the pellet. Leave tubes open until completely dry (no smell of acetone should be present).

5. Add 0.75 μl of 1 M Tris and 6.75 μl of urea buffer to each tube and resuspend well. Add Laemmli buffer to 1×.

3.6 ELISA

1. A standard curve should be prepared using purified recombinant protein of interest. In the case of STI1 a range of 23.4–1500 pM can be used. Freshly prepare a standard curve by making serial dilutions of the protein stock solution in sterile PBS. For STI1 measurements, the following standard solutions were used: 23.4, 46.8, 93.7, 187.5, 375, 750, and 1500 pM. Coat the three wells of an ELISA plate with 100 μl of each of the standard solutions (triplicate measurement). Include a background, negative control sample containing sterile PBS. Ensure that no bubbles are present during the incubation steps and prior to reading the plate.

2. Coat the remaining wells of the plate with 100 μl of the sample of interest in triplicates. Seal the plate and incubate overnight at 4 °C. Gently remove the coating solution and wash the plate

three times by filling the wells with 200 µl PBS. After each washing step, make sure that the wells are totally dry before starting the next step.

3. Block the remaining binding sites in the coated wells by adding 200 µl blocking buffer. Seal the plate and incubate for 2 h at 37 °C. Wash the plate three times by filling the wells with 200 µl of PBS+ 0.3% TritonX-100 and once with 400 µl of PBS (this amount ensures that any excess of detergent is removed).

4. Dilute the antibody of interest in sterile PBS to the appropriate concentration. The information regarding appropriate concentration is generally supplied by the antibody manufacturer; otherwise it should be empirically determined. Anti-STI1 antibody was diluted to a concentration of 8 ng/µl. Add 100 µl of diluted antibody to each well. Seal the plate and incubate overnight at 4 °C. Wash the plate three times by filling the wells with 200 µl of PBS+ 0.3% TritonX-100 and once with 400 µl of PBS.

5. Dilute the HRP-conjugated secondary antibody in PBS to the appropriate concentration. In the case of STI1, anti-rabbit-HRP was diluted 1:1000. Add 100 µl of diluted antibody to each well, seal the plate, and incubate for 1 h at 37 °C. Wash the plate three times by filling the wells with 200 µl of PBS+ 0.3% TritonX-100 and once with 400 µl of PBS.

6. Add 50 µl of OPD substrate solution in each well and incubate the plate in the dark, until the color development (approximately 15 min) at room temperature. Stop the reaction by the addition of 50 µl of 4 N H_2SO_4. Read the absorbance at 490 nm.

7. Once the intensity of each well has been measured on the plate reader, calculate the average absorbance values for each triplicate sample. Then generate a standard curve by graphing the mean absorbance for each sample (x-axis) vs. the standard concentration (y-axis). Draw the trendline for the data to generate the equation of the line (i.e., $y = mx + b$; m = slope of the line and $b = y$ intercept). Calculate the concentration of each sample by using the average of the triplicate samples for x in the equation (see **Note 6**).

3.7 Flow Cytometry to Detect Proteins in the Extracellular Vesicles Surface

1. Prepare EVs by ultracentrifugation method, and resuspend the pellet in 200 µl of fresh sterile PBS. For example, one preparation starting from 35 ml of astrocyte-conditioned media yields sufficient material for the analysis of five samples.

2. Mix the vesicle preparation with 20 µl of aldehyde/sulfate latex beads for 30 min at room temperature. Add PBS to a final volume of 1 ml, and incubate overnight at 4 °C under agitation.

3. Prepare a set of beads with no vesicles to be used as negative controls in the flow cytometry. Incubate 20 µl of aldehyde/

sulfate latex beads with 1 ml of PBS and follow the same procedures indicated for the vesicle/bead preparation.

4. To block remaining binding sites, incubate vesicle-coated beads for 30 min at room temperature with 110 µl of 1M glycine. Wash the vesicle/bead complexes by centrifuging for 3 min at $1600 \times g$, room temperature, discard the supernatant, and resuspend the bead pellet in 1 ml PBS/BSA 0.5%. Repeat this step twice (a pellet will become visible in the side of the tube after the second wash). Resuspend the pellet in 1 ml 0.5% BSA in PBS and split the sample in five tubes (200 µl each).

5. Use a sample of 200 µl to each analysis. One sample must be used as negative control stained with isotype-matched antibody. Incubate the remained samples with the desired antibodies for 1 h at 4 °C. The information regarding appropriate concentration is generally supplied by the antibody manufacturer; otherwise it should be empirically determined. In the case of anti-STI1, incubation was done with 20 ng of antibody. Use other surface exosomal proteins as a positive control (Flotillin-1 and CD-9 for example). Use one sample of an exosomal internal protein as a negative control (Hsp90).

6. Wash the vesicle/bead complexes by centrifuging for 3 min at $1600 \times g$, room temperature, discard the supernatant, and resuspend the bead pellet in 1 ml BSA 0.5% in PBS.

7. Incubate all the samples with secondary antibody (fluorophore-conjugated secondary antibody) to the appropriate concentration. In case of STI1, Alexa 546 anti-rabbit was used at 1:200 concentration at 4 °C for 40 min. Analyze antibody-stained vesicle/bead complexes on a flow cytometer. Adjust the forward scatter (FSC) and side scatter (SSC) to see both single beads and doublet beads. Compare fluorescence obtained with specific antibody and with irrelevant isotype control (Fig. 3).

3.8 Proteinase K Digestion of Proteins from the Surface of Extracellular Vesicles

1. Isolate EVs by ultracentrifugation method. For primary astrocytes cultures we use 35 ml of conditioned media. Resuspend the pellet in 300 µl of PK buffer and split it into three tubes (100 µl each). One tube will be the control (no PK treatment), one tube will be PK treated, and one tube PK treated + Triton X-100 0.5% (this is a positive control for the PK digestion; since the Triton will solubilize the membrane all proteins should be digested).

2. In tube 1 add 400 ng of PK, and in tube 2 add 400 ng PK and 10 µl of 5% Triton X-100. Incubate the reactions at 37 °C for 10 min. Add Laemmli buffer to the final concentration of 1× and perform Western blot against the protein of interest.

3. Compare the signal intensity in each lane. The lane with no PK represents the total protein. The lane treated with PK + Triton

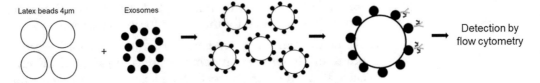

Fig. 3 Scheme of the preparation steps for the flow cytometry of extracellular vesicles

X-100 should present no signal. If there is still a visible band in this conditions the digestion was incomplete and should be repeated with an increased concentration of PK. If the extracellular vesicle contains the protein of interest at surface, the intensity of the PK-treated band should be lower than the control with no PK.

4   Notes

1. Western blot of conditioned media can be tricky due to the large amounts of albumin that are contained in the serum used to grow cells, which will distort the electrophoretic run, especially around 60 kDa. One alternative to overcome this difficulty is the use of serum-free media for the conditioning. However cells can be sensitive to serum starvation, so this option is not always viable. Additionally, serum starvation may alter secretion pathways, so this also has to be considered in the analysis. In cases where serum is used, it must be previously cleared of extracellular vesicles by ultracentrifugation at $100,000 \times g$ for 16 h. Other alternative is the isolation of the extracellular vesicles by ultracentrifugation or gel filtration chromatography, which removes the albumin. In the cases where extracellular vesicles are isolated by ultracentrifugation, pellets can be resuspended in volumes sufficient for loading on the SDS-PAGE gel. However, separation by gel filtration chromatography yields fractions of 3.5 ml that will need to be concentrated in order to reach adequate volumes for an SDS-PAGE. Concentration of the volume may be achieved by TCA precipitation.

2. Astrocytes can be maintained in complete confluence and a 75 cm^2 bottle of confluent astrocytes yields approximately 3×10^6 vesicles/ml in a total of 5 ml of media, which is sufficient for most applications [5]. However, different cells may have different yields of vesicles and thus escalation may be needed if working with other types of cells. Additionally, since some types of cells cannot be grown to confluence, the optimal condition for each cell must be preserved. In those cases, it is preferable to grow multiple plates with fewer amounts of cells. High apoptotic rates should also be avoided, since these cells

secrete apoptotic bodies that could be mistaken for extracellular vesicles. We do not recommend freezing the conditioned media or separated extracellular vesicles, since a rupture of the vesicles can occur with freezing.

3. In astrocyte-conditioned media, the 1-h centrifugation separates vesicles of average size of 300 nm, while the 2-h centrifugation separates vesicles of average size of 200 nm and the 16-h centrifugation separates vesicles of an average size of 150 nm.

4. As the column gets packed, the speed of packing reduces considerably; do this procedure in advance as it may take several hours. To improve speed, when the slurry settles in the column, the excess PBS can be removed with a pipette and more 50:50 bead/buffer can be added. Do not leave the column unattended for long periods when the stopcock is open as the columns may run out of buffer and overdry. If the surface of the beads turns an opaque white, beads are dehydrating, which may be prejudicial to the chromatography.

5. Columns may be stored mounted at room temperature just by turning off the pump and closing the column outlet. For preservation during long periods of inactivity pump 4–5 bead bed volumes of PBS containing 20% ethanol or 0.1% sodium azide. For use after long-term storage wash out preservative by pumping 4–5 bead bed volumes with PBS.

6. If the concentration of the sample exceeds the highest point of the curve or does not lie within the linear range of the curve, then dilute the sample prior to measurement. If a diluted sample is used, remember to multiply by the dilution factor to obtain the final concentration value.

References

1. Blatch GL, Lässle M, Zetter BR, Kundra V (1997) Isolation of a mouse cDNA encoding mSTI1, a stress-inducible protein containing the TPR motif. Gene 194:277–282
2. Zanata SM, Lopes MH, Mercadante AF et al (2002) Stress-inducible protein 1 is a cell surface ligand for cellular prion that triggers neuroprotection. EMBO J 21:3307–3316
3. Lopes MH, Hajj GNM, Muras AG et al (2005) Interaction of cellular prion and stress-inducible protein 1 promotes neuritogenesis and neuroprotection by distinct signaling pathways. J Neurosci 25:11330–11339
4. Roffé M, Beraldo FH, Bester R et al (2010) Prion protein interaction with stress-inducible protein 1 enhances neuronal protein synthesis via mTOR. Proc Natl Acad Sci U S A 107:13147–13152
5. Hajj GNM, Arantes CP, Dias MVS et al (2013) The unconventional secretion of stress-inducible protein 1 by a heterogeneous population of extracellular vesicles. Cell Mol Life Sci 70:3211–3227
6. Costa-Silva B, Aiello NM, Ocean AJ et al (2015) Pancreatic cancer exosomes initiate pre-metastatic niche formation in the liver. Nat Cell Biol 17:1–7
7. Sokolova V, Ludwig AK, Hornung S, Rotan O, Horn PA, Epple M, Giebel B (2011) Characterisation of exosomes derived from human cells by nanoparticle tracking analysis and scanning electron microscopy. Colloids Surfaces B Biointerfaces 87:146–150
8. Dragovic RA, Gardiner C, Brooks AS et al (2011) Sizing and phenotyping of cellular vesicles using nanoparticle tracking analysis. Nanomedicine 7:780–788
9. Gardiner C, Ferreira YJ, Dragovic RA, Redman CWG, Sargent IL (2013) Extracellular vesicle sizing and enumeration by nanoparticle tracking analysis. J Extracell vesicles 2:1–11

10. Kim JW, Wieckowski E, Taylor DD, Reichert TE, Watkins S, Whiteside TL (2005) Fas ligand-positive membranous vesicles isolated from sera of patients with oral cancer induce apoptosis of activated T lymphocytes. Clin Cancer Res 11(3):1010–1020

11. Welton JL, Khanna S, Giles PJ, Brennan P, Brewism IA, Staffurth J, Mason MD, Clayton A (2010) Proteomics analysis of bladder cancer exosomes. Mol Cell Proteomics 9:1324–1338

12. Wahlgren J, Karlson TDL, Brisslert M, Vaziri Sani F, Telemo E, Sunnerhagen P, Valadi H (2012) Plasma exosomes can deliver exogenous short interfering RNA to monocytes and lymphocytes. Nucleic Acids Res 40:e130

13. Hardy KM, Hoffman EA, Gonzalez P, McKay BS, Stamer WD (2005) Extracellular trafficking of myocilin in human trabecular meshwork cells. J Biol Chem 280:28917–28926

14. Gupta S, Knowlton AA (2007) HSP60 trafficking in adult cardiac myocytes: role of the exosomal pathway. Am J Physiol Heart Circ Physiol 292:H3052–H3056

15. Théry C, Amigorena S, Raposo G, Clayton A (2006) Isolation and characterization of exosomes from cell culture supernatants and biological fluids. Curr Protoc cell Biol 3:322

Chapter 12

Analysis of Yeast Extracellular Vesicles

Marcio L. Rodrigues, Debora L. Oliveira, Gabriele Vargas,
Wendell Girard-Dias, Anderson J. Franzen, Susana Frasés,
Kildare Miranda, and Leonardo Nimrichter

Abstract

Extracellular vesicles (EV) are important carriers of biologically active components in a number of organisms, including fungal cells. Experimental characterization of fungal EVs suggested that these membranous compartments are likely involved in the regulation of several biological events. In fungal pathogens, these events include mechanisms of disease progression and/or control, suggesting potential targets for therapeutic intervention or disease prophylaxis. In this manuscript we describe methods that have been used in the last 10 years for the characterization of EVs produced by yeast forms of several fungal species. Experimental approaches detailed in this chapter include ultracentrifugation methods for EV fractionation, chromatographic approaches for analysis of EV lipids, microscopy techniques for analysis of both intracellular and extracellular vesicular compartments, interaction of EVs with host cells, and physical chemical analysis of EVs by dynamic light scattering.

Key words Fungi, Yeast, Extracellular vesicles, *Cryptococcus*, *Candida*, *Saccharomyces*

1 Introduction

Organisms of the three domains of life shed extracellular vesicles (EVs) into their microenvironment [1]. During the last decade, it has been demonstrated that EVs play fundamental roles in cancer, infectious diseases, and neurodegenerative disorders, suggesting potential targets for diagnosis, prognosis, and therapeutic intervention in different syndromes [2]. EVs are spherical bilayered compartments ranging in diameter from 20 to 1000 nm [1]. Due to their complex molecular composition, characterization of EVs in different biological systems requires a combination of several experimental approaches, including centrifugation-based protocols, microscopy techniques, chromatographic analysis, proteomics, glycomics, lipidomics, and nucleic acid sequencing, among others.

Andrea Pompa and Francesca De Marchis (eds.), *Unconventional Protein Secretion: Methods and Protocols*, Methods in Molecular Biology, vol. 1459, DOI 10.1007/978-1-4939-3804-9_12, © Springer Science+Business Media New York 2016

In the fungi, EVs were firstly described in the yeast pathogen *Cryptococcus neoformans* [3]. Since then, EVs have been characterized in yeast forms a number of fungal species, including *Candida albicans* [4–7], *C. parapsilosis* [8], *Histoplasma capsulatum* [8], *Malassezia sympodialis* [9], *Paracoccidioides brasiliensis* [7, 10–13], *Saccharomyces cerevisiae* [14], and *Sporothrix schenckii* [8]. Remarkably, studies with *Alternaria infectoria* provided the only experimental evidence demonstrating that filamentous fungal forms also produce EVs [15].

Fundamental biological properties have been suggested for most of the fungal species listed above [3, 4, 7–10, 12, 13, 16–19]. However, many questions about fungal EVs remain still unanswered. In a context of high biological importance combined with still obscure properties, it is clear that improving the methods for analysis of fungal EVs will be determinant for advancement of the field. In this manuscript, we describe protocols that have been used by our group for the analysis of EVs produced by yeast forms of the pathogens *C. neoformans* and *C. albicans* and by the model organism *S. cerevisiae*.

2   Materials

Prepare all solutions with ultrapure water (18.2 MΩ cm at 25 °C) and analytical grade reagents.

2.1 Media for Storage and Growth of Yeast Cultures

1. Media: Yeast cells can be stored in standard media used for fungal growth, including Sabouraud's medium and brain-heart infusion (BHI).

2. Sabouraud's broth (2 % dextrose and 1 % peptone): Weight d-glucose (20 g) and meat peptone (10 g) and make up to 1000 ml with water. Sterilize through autoclaving.

3. BHI medium: Dissolve 37 g of BHI in 1000 ml water. Sterilize through autoclaving.

4. Solid media: Prepare Sabouraud's or BHI broth as described above and supplement with 2 % agar (20 g per liter) (*see* **Note 1**).

5. Minimal medium (MM): 15 mM d-Glucose, 10 mM $MgSO_4$, 29.4 mM KH_2PO_4, 13 mM glycine, 3 µM thiamine-HCl. Dissolve d-glucose (2.7 g), $MgSO_4$ (1.2 g), KH_2PO_4 (4 g), glycine (1 g), and 10 µl of thiamine solution in 800 ml water. Stir, adjust pH to 5.5, and make up to 1000 ml with water in a volumetric cylinder.

6. 0.3 M Thiamine solution: Dissolve 100 mg of the vitamin in 1 ml of water and filter through sterile 0.22 µm membranes. Add 10 µl of the thiamine stock solution for each liter of MM.

2.2 EV Isolation	1. Phosphate-buffered saline 0.01 M (PBS): Dissolve 0.26 g KH_2PO_4, 1.25 g K_2HPO_4, and 8.71 g NaCl in a beaker containing 700 ml of water. Stir, adjust pH to 7.4, and make up to 1000 ml with water in a volumetric cylinder. Filter through sterile 0.22 μm membranes. Store at 4 °C.

2. Membranes for filtration and ultrafiltration, respectively: 0.4 μm Polycarbonate membrane and 100 kDa polyethersulfone membrane for stirred ultrafiltration cells.

3. Density gradient: Stock solution of 60% iodixanol in water. Gradient fractions (6%, 7.2%, 8.4%, 9.6%, 10.8%, 12%, 13.2%, 14.4%, 15.6%, 16.8%, 18%) are diluted in water.

2.3 Lipid and Protein Content of EVs

1. HPTLC silica gel $60F_{254}$ plate: Glass plates with 5 cm × 10 cm dimension.

2. Separation solvent: Combine hexane, diethyl ether, and glacial acetic acid to form a 40:20:1 (vol:vol:vol) mixture (*see* **Note 2**).

3. Sterol detection reagent: 0.05% $FeCl_3$, 5% acetic acid, and 5% sulfuric acid. Weight 0.05 g $FeCl_3$ and dissolve in 90 ml water in a glass beaker. Add 5 ml of glacial acetic acid. Slowly add 5 ml of concentrated sulfuric acid. Stock at room temperature and protect from light.

4. Sterol quantification: Amplex Red cholesterol assay kit (Molecular probes).

5. Dissolving buffer: Chloroform:methanol:0.75% KCl (8:4:3, v/v/v).

6. Ergosterol.

7. Protein quantification: BCA protein assay kit (Pierce).

2.4 EV Staining

1. 1,1′-Dioctadecyl-3,3,3′,3′-tetramethylindocarbocyanine solution (DiI) staining solution: Add 5 μl of stock solution (1 mM) to 995 μl of PBS. Mix well by gently pipetting (*see* **Note 3**).

2.5 Host Cells

1. Macrophage-like cultures: RAW 264.7 murine macrophages (American Type Culture Collection).

2. Sodium pyruvate.

3. L-Glutamine.

4. Gentamicin.

5. 4-(2-Hydroxyethyl)-1-piperazineethanesulfonic acid (HEPES).

6. 2-Beta-mercaptoethanol.

7. Complete Dulbecco's minimal essential medium (DMEM) supplemented with 10% fetal calf serum (FCS): Add 10 ml of filter-sterilized FCS to 90 ml of DMEM. Supplement with 2 mM L-glutamine, 1 mM sodium pyruvate, 10 mg/ml gentamicin, 10 mM HEPES, and 50 mM 2-beta-mercaptoethanol.

2.6 Transmission Electron Microscopy

1. Cellulose capillaries for high-pressure freezing (Leica Microsystems): 200 µm inner diameter, wall thickness 8 µm.

2. High-pressure freezing-carries filling: 1-Hexadecen.

3. Osmium tetroxide.

4. Acetone.

5. Glutaraldehyde.

6. Uranyl acetate.

7. Lead citrate.

8. Sodium citrate.

9. Freeze substitution medium: To prepare the osmium tetroxide stock solution, dilute 1 g of the crystal in 25 ml of pure acetone. For the working solution, mix 500 µl of 4% osmium tetroxide in acetone, 1.4 µl of 70% glutaraldehyde, and 9.6 µl of deionized water and add pure acetone to complete 1 ml. Keep the solution frozen in liquid nitrogen.

10. Epoxide resin embedding: Epon-812 or Spurr (EMS).

11. Electron microscopy copper grids: 200 mesh grid and slot grid.

12. Post-staining solutions: To prepare 5% uranyl acetate, add 2.5 g of the solid to 50 ml of distilled water. Store at room temperature protected from light; solution can be used 24 h after preparation. To prepare Reynold's solution, add 1.33 g of lead citrate, 1.76 g of sodium citrate, and 5 ml of 1 N NaOH to 30 ml of distilled water. Stir for 10 min and make up to 50 ml with water. Store at 4 °C.

13. Substitution medium: 2% Osmium tetroxide, 0.1% glutaraldehyde, and 1% of water in acetone.

2.7 Cryoultramicrotomy and Immunogold Electron Microscopy

1. 0.2 M Sodium cacodylate buffer: Add 21.4 g of the solid to 400 ml of distilled water. Adjust pH to 7.2–7.4 and add distilled water to complete 500 ml.

2. Formaldehyde.

3. Gelatin.

4. Polyvinylpyrrolidone.

5. Methylcellulose.

6. Fixation medium: Mix 4 µl of 25% glutaraldehyde, 250 µl of 16% formaldehyde, 500 µl of sodium cacodylate buffer, and 246 µl of distilled water. Adjust pH to 7.2.

7. Sample preparation and cryoprotectant solutions: For 10% gelatin, add 1 g of gelatin powder to 5 ml of warmed distilled water under shaking. After dissolution, make up to 10 ml with water. For 2.3 M sucrose/PVP solution, add 2.5 g of polyvinylpyrrolidone (PVP) to 5 ml of 2.3 M sucrose (prepared previously by adding 78.7 g of the solid to 100 ml phos-

phate buffer 0.1 M, pH 7.2). Leave the solution overnight under shaking and add 2.3 M sucrose to complete 10 ml. Store at −20 °C.

8. Thawing and staining solutions: To prepare 3% methylcellulose, add 0.3 g of methylcellulose to 10 ml of water. For staining solution, prepare a 9:1 (v/v) mixture of 3% polyvinyl alcohol (PVA) (previously prepared by adding 0.3 g of PVA to 10 ml of water) and 5% uranyl acetate (prepared as mentioned above).

9. Grid washing solution: 3 g BSA per 100 ml PBS, pH 8.0.

10. Quenching solution: Add 26 mg of NH_4Cl to 10 ml of PBS, to form 50 mM NH_4Cl.

3 Methods

3.1 EV Isolation

1. Pick up a single colony from stock cultures and inoculate into an Erlenmeyer flask containing 20 ml of MM (Subheading 2.1). Incubate for 48 h at 30 or 37 °C under shaking (150 rpm).

2. Transfer the culture to 600 ml of liquid media contained in 2 l Erlenmeyer flasks. Cultivate yeast cells up to stationary phase (cell density of approximately 1×10^8 cells/ml) at 30 or 37 °C with shaking (150 rpm).

3. Remove fungal cells and debris by sequential centrifugations at 4000, 10,000, and 15,000 $\times g$ for 15 min, at 4 °C. Discard pellets after each centrifugation step. Vacuum-filter cell-free supernatants through 0.4–0.8 µm polycarbonate membranes to remove possible cellular contaminants.

4. Concentrate supernatants approximately 20-fold using the ultrafiltration system (100 kDa cutoff membrane). Ice baths may be used to prevent microbial contamination.

5. Ultracentrifuge concentrated supernatants at 100,000 $\times g$ for 1 h at 4 °C; resulting pellets are usually not visible. Discard ultracentrifugation supernatants and gently suspend pellets in PBS. Repeat the ultracentrifugation protocol twice, always discarding supernatants. Suspend pellets in 200 µl PBS (crude EV fractions) or, alternatively, use dry pellets for lipid extraction. For *C. neoformans* EVs, removal of contaminant glucuronoxylomannan (GXM) is required (*see* **Note 4**). For further EV fractionation, iodixanol density gradient may be used (*see* **Note 5**).

3.2 Analysis of Sterol and Protein EV Content

1. Add 300 µl methanol and 600 µl chloroform to a 120 µl suspension of crude EVs. Alternatively, the 900 µl methanol-chloroform mixture can be added to dry ultracentrifugation pellets. Vortex vigorously for 10 s. White precipitates (protein aggregates or polysaccharides) are usually formed.

2. Spin down for 10 s at $2000 \times g$ at 4 °C; a two-phase system will be formed. Discard the upper phase and concentrate the lower phase to dryness under a N_2 stream.

3. Fractionate the dry lower phase according to the Folch partition method [20] by dissolving the sample in 1 ml of dissolving buffer. Vortex vigorously and collect the lower phase enriched with neutral lipids. Concentrate to dryness under a N_2 stream.

4. Dissolve the sample in 60 μl of a 2:1 (v/v) chloroform:methanol mixture. Using a microliter syringe, load 20 μl of the lipid mixture into an HPTLC silica gel plate to form band-shaped loading areas 0.5–1 cm wide, 1 cm above the bottom of plate. Sterol standard (ergosterol, 2 μg) is required for retention factor (Rf) comparisons between samples.

5. Place the HPTLC plate in a chromatography chamber under the conditions described in Subheading 2.3. Remove plates from the chamber when the distance between the solvent front and the plate's edge is approximately 1 cm. Wait for natural solvent evaporation in a chemical hood (10–15 min, room temperature).

6. Spray the dry plate with the sterol detection reagent (subheading 2.3) and heat it at 100 °C for 5 min (Fig. 1). Sterols develop as purple bands. For densitometry analysis of the spots, we recommend ImageJ software (imagej.nih.gov/ij/).

7. Quantify sterols with the Amplex Red Cholesterol assay kit, following the manufacturer's instructions.

8. Quantify proteins with the BCA protein assay kit [5] (*see* **Note 6**). Protein identification can be performed through LC-MS/MS proteomics, as previously described [4, 14]. Proteomics protocols are out of the scope of this chapter and will not be described here.

3.3 Sample Preparation by High-Pressure Freezing and Freeze Substitution for Routine TEM Observation and/or Electron Tomography

1. Prepare yeast cell or EV pellets as detailed in Subheading 3.1. Place pellets between two types of aluminum carriers, protecting the biological material within the 200 μm cavity on one carrier (fill the whole cavity). Alternatively, cells or EVs may be placed in 200 μm wide cellulose capillaries previously cut into 2 mm pieces (for fitness into 3 mm HPF carriers) (*see* **Note 7**).

2. Mount the sandwiched samples in the HPF holder and freeze it using high-pressure freezing equipment (HPM 010 or 100, Bal-Tec, Corp., Liechtenstein). The machine should be previously loaded with liquid nitrogen. After freezing, carriers are removed from the holder and stored in liquid nitrogen.

3. Remove samples carefully from liquid nitrogen and immerse them in the substitution medium (*see* **Note 8**).

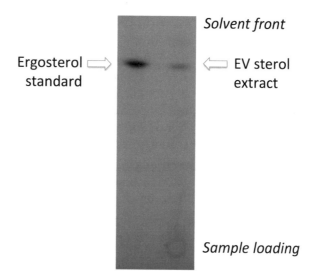

Fig. 1 Illustrative HPTLC analysis of a sterol extract obtained from crude EV fractions

4. Apply a substitution-heating curve consisting of keeping the samples for 72 h at 90 °C, 15 h at –20 °C, and 2 h at 4 °C.

5. After substitution, wash the samples three times with acetone at room temperature. Embed samples in epoxide resins (Epon or Spurr). Polymerize at 60 °C for 72 h.

6. Using a ultramicrotome, obtain 70 nm thick sections for conventional transmission electron microscopy or 200 nm thick sections for electron tomography and collect the sections in 200 mesh copper grids. Alternatively, series of sections may be obtained and collected on formvar-coated slot grids.

7. Post-stain the grids for 20 min in 5 % uranyl acetate in water and 5 min in 1 % lead citrate (post-staining solutions, 2.7). For electron tomography, incubate both sides of the grids in 10 nm colloidal gold for 5 min and wash with distilled water.

8. Transfer the grids to a transmission electron microscope and acquire images using either a film or a digital acquisition system (CCD camera).

9. For electron tomography, transfer the grids containing the 200 nm thick sections to a transmission electron microscope equipped with electron tomography capabilities, operating at 200 kV.

10. Position the grid to ensure that the grid's bars are parallel to the axis of the tomography holder to minimize obstruction of the electron beam at high tilt angles. Similarly, for slot grids containing serial sections, position the slot (assuming that the ribbon of serial sections is placed along the slot axis) parallel to the holder axis.

11. After defining the region of interest, perform the corrections of eccentric height and focus, and record the tilt series from –70° to +70° with an angular increment of 1° or 2°.

12. At the end of the tomography acquisition, a single file comprising the tilt series will be generated (MRC file). Align the tilt series based on fiducial markers (colloidal gold applied as mentioned above) with appropriate software (for example, IMOD software package—University of Colorado, USA).

13. After alignment of the tilt series, perform the three-dimensional (3D) reconstruction using, for example, weighted back projection or SIRT. We suggest the IMOD software package (University of Colorado, USA). A 3D model can be generated automatically by manual segmentation of the structures of interest in the 3D volume (Fig. 2). Different software is available for this purpose, including IMOD software package (University of Colorado, USA), AMIRA (Visage Imaging, USA), and Cytoseg (National Center for Microscopy and Image Research, USA).

3.4 Cryoultra-microtomy and Immunogold Electron Microscopy

1. For immunogold labeling, fungal cells in the fixation medium (Subheading 2.8) for 1 h at 4 °C.

2. Wash the cells twice in 0.1 M sodium cacodylate buffer and spin down to obtain strongly tight pellets. If firm pellets are not formed, alternatively embed sediments in 10% gelatin. With a toothpick, transfer the pellet and the surrounding buffer to a parafilm-covered Petri dish. Section the sediment in 2 mm wide cubes using a razor blade. Transfer pellet pieces to the polyvinyl-pyrrolidone/sucrose cryoprotectant solution (Subheading 2.8) and incubate for 2 h. Mount the pellet in cryoultramicrotomy holders and plunge it into liquid nitrogen, making a fast circular movement to avoid freezing by nitrogen gases formed around the sample (Leidenfrost effect).

3. Transfer the sample to a previously cooled cryoultramicrotome loaded with a glass or cryodiamond knife. Sample and chamber temperatures should be adjusted to –90 °C; knife temperature must be –70 °C. Obtain ultrathin cryosections, arrange them with an eyelash, and collect them in 2.3 M sucrose or 3% methylcellulose (Subheading 2.8), using a wire loop.

4. Thaw the sections over formvar-coated 300 mesh nickel grids and transfer them to the grid washing solution, cells facing down. Incubate the grids for 30 min in quenching solution (Subheading 2.8).

5. Wash the grids three times for 5 min with the grid washing solution.

6. Incubate the grids with the primary antibody for 1–3 h.

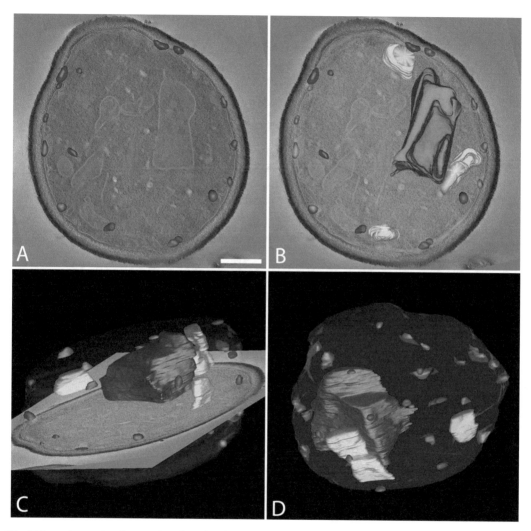

Fig. 2 Three-dimensional reconstruction of a whole *S. cerevisiae* cell by serial electron tomography. Cells were prepared using the high-pressure freezing-freeze substitution protocol described here. A ribbon of 200 nm sections was obtained by ultramicrotomy and collected in formvar-coated slot grids. Tilt series of ±70° at a 1° interval was obtained from each profile of the cell in each section and reconstructed. Segmentation was performed in IMOD, as described before. (**a**) Virtual section of a tomogram showing the general structure of the cell. Budding extracellular vesicles are shown in *blue*. (**b, c**) Projection of virtual sections and reconstructed models of intracellular structures at different angles. B—Orthogonal projection, C—tilted view. (**d**) 3D model of the reconstructed volume (whole cell) showing the segmentation of a few surface and intracellular structures. *Blue*: Extracellular vesicles. *Red*: Nucleus. *Yellow*: mitochondria (partial segmentation). *Green*: Plasma membrane-intracellular space. Bar: 1 μm

7. Wash the grids three times for 10 min with the grid washing solution (Subheading 2.8). Incubate the grids with the secondary antibody conjugated with gold particles of varying sizes, depending on the structure that should be labeled (usually between 5 and 15 nm).

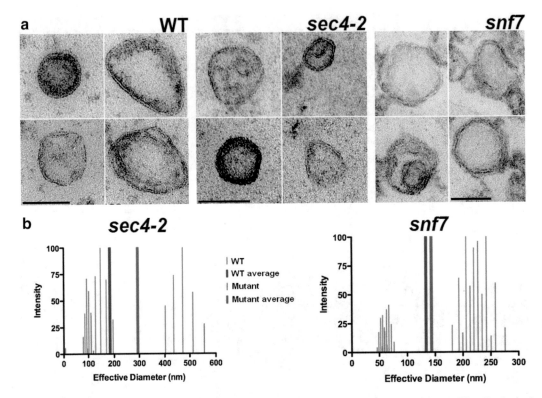

Fig. 3 TEM (**a**) and DLS (**b**) analyses of *S. cerevisiae* EVs. In (**a**), each individual panel exemplifies the typical vesicle morphology for wild-type (WT) and secretion mutants (*sec4.2* and *snf7*). Scale bar, 100 nm. B. Light scattering analysis showing diameter distribution and average values of vesicles obtained from WT or mutant (*sec4-2* and *snf7*) cells. For strain details, *see* ref. 14, from which this figure has been reproduced

8. Wash the grids three times for 10 min with grid washing solution, and then quickly wash in water (30 s each wash). Place the grids in a mixture of PVA-uranyl acetate staining solution for 10 min. Drain off the excess of liquids and let the grids naturally dry at room temperature before observation in a transmission electron microscope (Fig. 3a).

3.5 Physical Chemical Analysis of EVs s by Dynamic Light Scattering (See Note 9)

1. Analyses of fungal EVs by dynamic light scattering (DLS) in our laboratory during the past 5 years have been performed using a quasi-elastic light scattering (QELS) in a 90Plus/BI-MAS Multi Angle Particle Sizing analyzer (Brookhaven Instruments Corp., Holtsville, NY). Suspend purified vesicles in PBS until current count rate is between 10,000 and 90,000 counts per second. DLS analysis should be performed at 22 °C with an equilibrium time of 2 min. A minimum of six replicates with five measurements per replicate should be used.

2. Check the raw correlation data to ensure that the amplitude (*y*-intercept) is stable and the correlograms are smooth (i.e., decay exponentially to a flat baseline) (*see* **Note 10**). Discard

the sample if sediment is visible at the bottom of the cuvette following measurements (*see* **Note 11**).

3. Report the mean *z*-average diameter and mean polydispersity index along with their standard deviations based on the replicates.

4. Obtain the multimodal size distributions of particles (Fig. 3b) by a non-negatively constrained least squares algorithm (NNLS) based on the intensity of light scattered by each particle.

3.6 DiI staining of EVs

1. Prepare 995 µl of an EV suspension with total sterol content adjusted to 2 µg/ml. Add 5 µl of DiI (stock solution, Subheading 2, **item 5**) (*see* **Note 12**). Incubate the suspension for 15 min at 25 °C protecting from light.

2. Ultracentrifuge DiI-stained EVs at $100,000 \times g$ for 1 h, 4 °C, and discard the supernatant. Wash stained EVs with PBS three times as described in Subheading 3.2.

3. Suspend DiI-stained EVs to generate samples with sterol concentrations corresponding to 2 µg/ml. For EV suspension, use media suitable for interaction with host cells [4, 8, 17] as mentioned in Subheading 2.4.

4. To prepare negative controls, repeat the procedures above with fractions containing no vesicles.

3.7 Interaction of EVs DiI Stained with Host Cells

1. Place RAW 264.7 cells previously suspended in suitable media (Subheading 2.6) into glass cover slip-covered wells of 24-well plates. Initial cell density will correspond to 2×10^5 cells/well. Incubate at 37 °C in a 5 % CO_2 atm until macrophages firmly adhere to the plate (approximately 2 h).

2. Replace culture media with fresh media (500 µl) containing DiI-labeled EV suspensions with sterol concentrations varying from 1 to 0.001 µg/ml. Incubate at 37 °C (5 % CO_2) for 15 min, 1 h, 16 h, and 24 h. After each time point wash monolayers three times with 500 µl PBS and fix the cells with 4 % paraformaldehyde (in PBS) for 1 h at 25 °C or overnight at 4 °C.

3. Transfer glass cover slips to glass slides containing 3 µl of PROLONG GOLD antifade (Life Technologies P36934). Seal with uncolored nail polish and let cover slips dry for 60 min.

4. Observe slides under fluorescence microscopy (excitation, 546 nm; emission, 590 nm). If available, microscopes equipped with deconvolution systems will allow simpler interpretation of the intracellular distribution of EVs, as illustrated in Fig. 4.

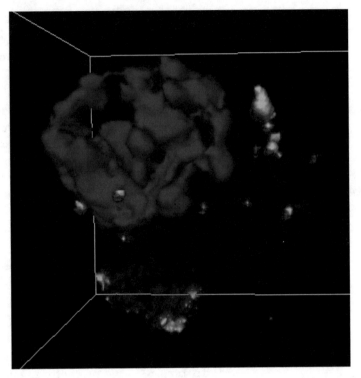

Fig. 4 Tridimensional analysis showing cell surface binding and internalization of DiI-stained *C. albicans* EVs in a dendritic cell model. Z-stacks and deconvolution profiles were obtained on an AxioVision 4.8 (Carl Zeiss International) inverted fluorescence microscope. EVs are shown in *red*; nucleus is stained in *blue* (DAPI) and the cell surface is stained in *green* (β-cholera toxin subunit-FITC; details in [4])

4   Notes

1. Autoclave and distribute media into Petri dishes (20 ml per 13 mm plates) before the temperature reaches 50 °C. Keep plates for 5 min at room temperature for agar solidification.

2. Pre-saturate the chromatographic chamber with the solvent mixture by adding the separation solvent to the chamber's bottom part and incubating for 10 min in a gas hood. The mixture must be freshly prepared and used only once to avoid changes in concentration due to evaporation. Volumes will vary depending on chamber dimensions; 0.5 cm solvent height is acceptable.

3. Alternative long-chain lipophilic carbocyanine analogs, such as DiO, can be similarly used for EV staining.

4. To remove contaminant polysaccharide from preparations of *C. neoformans* EVs, suspend pellets from the $100{,}000 \times g$ centrifugation in 50 μl PBS and add to the wells of a 96-well enzyme-linked immunosorbent assay (ELISA) plate previously coated with a monoclonal antibody to the major capsular polysaccharide

of *C. neoformans* (MAb 18B7 to GXM; coating with 10 μg/ml solution, for 1 h at room temperature; antibody provided by Dr. Arturo Casadevall for the studies performed so far). Block with PBS containing 1% BSA (1 h at 37 °C). Incubate the plates for 1 h at room temperature, and collect the unbound fraction, containing vesicles free of soluble polysaccharides. Filter through 0.4 μm membranes to remove any potential aggregate or contaminating cells.

5. Fractionate EVs by loading vesicle samples into an iodixanol gradient (Optiprep). Prepare solutions of iodixanol ranging from 6 to 18% in PBS, with 1.2% concentration increments between each solution. Load iodixanol fractions sequentially from higher (bottom) to lower (top) concentrations into the centrifuge tube. Each fraction will have 300 μl and the final volume in the tube will be 3.3 ml. Gently dispense the crude EV suspension at the top of the 6% gradient fraction and ultracentrifuge at 250,000×*g* for 75 min. In our experience, a Beckman swing-bucket rotor (50.1Ti) has been successfully used. Keep ultracentrifuge brake off. Collect 11 fractions of 300 μl each from top to bottom. Fractions can be tested for morphological features, biochemical composition, and biological activity [21, 22].

6. The BCA-based method is not efficient for protein detection in *Cryptococcus* EVs, likely because of polysaccharide interference.

7. Samples are taken into the void by capillarity, so it is important to ensure that the space between the capillaries is filled up with 1-hexadecene, allowing pressure and heat transfer to take place.

8. The substitution medium should be precooled to –90 °C using a computer-controlled freeze substitution apparatus.

9. DLS measures the hydrodynamic diameter. In analysis of EV suspensions, light is scattered by vesicles; hence variations in the intensity of the scattered light can be detected. The reason explaining these characteristics relies on differences in the phases of waves scattered by different particles. The DLS experimental time-averaged autocorrelation data contains information on all the diffusional timescales present in the system. The data is usually fitted based on the cumulate approach with a least number of exponential decays sufficient to reproduce the correlation curve. The following equations are used for data interpretation:

$$\Gamma = Dq^2 \qquad (1)$$

with *D* being the diffusion coefficient of the nanoparticles and *q* the scattering vector given by [23]

$$q = 4\pi n\lambda^{-1} \sin(\theta / 2) \qquad (2)$$

where λ is the wavelength of the incident light, n is the refractive index of the suspension medium, and θ is the scattering angle. By using the "Stokes-Einstein" relationship the hydrodynamic diameter d of the particles can then be calculated on the assumption of a log normal size distribution [23]:

$$d = kT / 3\pi\eta D \qquad (3)$$

where k is the Boltzmann constant, T is the absolute temperature, and η is the viscosity of the dispersing medium. This technique is extremely precise when there is a simple distribution of decay times, e.g., for scattering from spherical nanoparticles with a narrow distribution of sizes about a single mean, although issues arise in accurately determining the size distribution.

10. Noisy correlograms and/or fluctuating amplitudes for a given sample can be attributed to the presence of dust/foreign particles in the sample, concentration variations from sample precipitation or aggregation, solvent evaporation, or dirty cuvettes.

11. Sediment indicates that the sample either contains a significant portion of large (micrometer) size particles or the target particles are unstable during the time frame of the experiment.

12. DiI and DiO are long-chain lipophilic carbocyanines that diffuse laterally within the plasma membrane. Therefore, staining with DiI can modify EV dimensions, which has been in fact observed in our laboratory.

Acknowledgement

We are grateful to Jorge José-Jó Bastos Ferreira for helpful suggestions. This work was supported by grants from the Brazilian agencies CNPq, CAPES, and FAPERJ. The authors also acknowledge support from the Instituto Nacional de Ciência e Tecnologia de Inovação em Doenças Negligenciadas (INCT-IDN). M.L.R. is the recipient of a Pathfinder Award from the Wellcome Trust (UK).

References

1. Colombo M, Raposo G, Thery C (2014) Biogenesis, secretion, and intercellular interactions of exosomes and other extracellular vesicles. Annu Rev Cell Dev Biol 30:255–289. doi:10.1146/annurev-cellbio-101512-122326

2. Lo Cicero A, Stahl PD, Raposo G (2015) Extracellular vesicles shuffling intercellular messages: for good or for bad. Curr Opin Cell Biol 35:69–77. doi:10.1016/j.ceb.2015.04.013

3. Rodrigues ML, Nimrichter L, Oliveira DL, Frases S, Miranda K, Zaragoza O, Alvarez M, Nakouzi A, Feldmesser M, Casadevall A (2007) Vesicular polysaccharide export in Cryptococcus neoformans is a eukaryotic solution to the problem of fungal trans-cell wall transport. Eukaryot Cell 6(1):48–59. doi:10.1128/EC.00318-06

4. Vargas G, Rocha JD, Oliveira DL, Albuquerque PC, Frases S, Santos SS, Nosanchuk JD,

Gomes AM, Medeiros LC, Miranda K, Sobreira TJ, Nakayasu ES, Arigi EA, Casadevall A, Guimaraes AJ, Rodrigues ML, Freire-de-Lima CG, Almeida IC, Nimrichter L (2015) Compositional and immunobiological analyses of extracellular vesicles released by Candida albicans. Cell Microbiol 17(3):389–407. doi:10.1111/cmi.12374

5. Wolf JM, Espadas J, Luque-Garcia J, Reynolds T, Casadevall A (2015) Lipid biosynthetic genes affect Candida albicans extracellular vesicle morphology, cargo, and immunostimulatory properties. Eukaryot Cell 14(8):745–754. doi:10.1128/EC.00054-15

6. Gil-Bona A, Llama-Palacios A, Parra CM, Vivanco F, Nombela C, Monteoliva L, Gil C (2015) Proteomics unravels extracellular vesicles as carriers of classical cytoplasmic proteins in Candida albicans. J Proteome Res 14(1):142–153. doi:10.1021/pr5007944

7. Peres da Silva R, Puccia R, Rodrigues ML, Oliveira DL, Joffe LS, Cesar GV, Nimrichter L, Goldenberg S, Alves LR (2015) Extracellular vesicle-mediated export of fungal RNA. Sci Rep 5:7763. doi:10.1038/srep07763

8. Albuquerque PC, Nakayasu ES, Rodrigues ML, Frases S, Casadevall A, Zancope-Oliveira RM, Almeida IC, Nosanchuk JD (2008) Vesicular transport in Histoplasma capsulatum: an effective mechanism for trans-cell wall transfer of proteins and lipids in ascomycetes. Cell Microbiol 10(8):1695–1710. doi:10.1111/j.1462-5822.2008.01160.x

9. Gehrmann U, Qazi KR, Johansson C, Hultenby K, Karlsson M, Lundeberg L, Gabrielsson S, Scheynius A (2011) Nanovesicles from Malassezia sympodialis and host exosomes induce cytokine responses--novel mechanisms for host-microbe interactions in atopic eczema. PLoS One 6(7), e21480. doi:10.1371/journal.pone.0021480

10. Albuquerque PC, Cordero RJ, Fonseca FL, Peres da Silva R, Ramos CL, Miranda KR, Casadevall A, Puccia R, Nosanchuk JD, Nimrichter L, Guimaraes AJ, Rodrigues ML (2012) A Paracoccidioides brasiliensis glycan shares serologic and functional properties with cryptococcal glucuronoxylomannan. Fungal Genet Biol 49(11):943–954. doi:10.1016/j.fgb.2012.09.002

11. Vallejo MC, Nakayasu ES, Longo LV, Ganiko L, Lopes FG, Matsuo AL, Almeida IC, Puccia R (2012) Lipidomic analysis of extracellular vesicles from the pathogenic phase of Paracoccidioides brasiliensis. PLoS One 7(6), e39463. doi:10.1371/journal.pone.0039463

12. Vallejo MC, Nakayasu ES, Matsuo AL, Sobreira TJ, Longo LV, Ganiko L, Almeida IC, Puccia R (2012) Vesicle and vesicle-free extracellular proteome of Paracoccidioides brasiliensis: comparative analysis with other pathogenic fungi. J Proteome Res 11(3):1676–1685. doi:10.1021/pr200872s

13. Vallejo MC, Matsuo AL, Ganiko L, Medeiros LC, Miranda K, Silva LS, Freymuller-Haapalainen E, Sinigaglia-Coimbra R, Almeida IC, Puccia R (2011) The pathogenic fungus Paracoccidioides brasiliensis exports extracellular vesicles containing highly immunogenic alpha-galactosyl epitopes. Eukaryot Cell 10(3):343–351. doi:10.1128/EC.00227-10

14. Oliveira DL, Nakayasu ES, Joffe LS, Guimaraes AJ, Sobreira TJ, Nosanchuk JD, Cordero RJ, Frases S, Casadevall A, Almeida IC, Nimrichter L, Rodrigues ML (2010) Characterization of yeast extracellular vesicles: evidence for the participation of different pathways of cellular traffic in vesicle biogenesis. PLoS One 5(6), e11113. doi:10.1371/journal.pone.0011113

15. Silva BM, Prados-Rosales R, Espadas-Moreno J, Wolf JM, Luque-Garcia JL, Goncalves T, Casadevall A (2014) Characterization of Alternaria infectoria extracellular vesicles. Med Mycol 52(2):202–210. doi:10.1093/mmy/myt003

16. Rizzo J, Oliveira DL, Joffe LS, Hu G, Gazos-Lopes F, Fonseca FL, Almeida IC, Frases S, Kronstad JW, Rodrigues ML (2014) Role of the Apt1 protein in polysaccharide secretion by Cryptococcus neoformans. Eukaryot Cell 13(6):715–726. doi:10.1128/EC.00273-13

17. Oliveira DL, Freire-de-Lima CG, Nosanchuk JD, Casadevall A, Rodrigues ML, Nimrichter L (2010) Extracellular vesicles from Cryptococcus neoformans modulate macrophage functions. Infect Immun 78(4):1601–1609. doi:10.1128/IAI.01171-09

18. Rodrigues ML, Nakayasu ES, Oliveira DL, Nimrichter L, Nosanchuk JD, Almeida IC, Casadevall A (2008) Extracellular vesicles produced by Cryptococcus neoformans contain protein components associated with virulence. Eukaryot Cell 7(1):58–67. doi:10.1128/EC.00370-07

19. Rodrigues ML, Nimrichter L, Oliveira DL, Nosanchuk JD, Casadevall A (2008) Vesicular trans-cell wall transport in fungi: a mechanism for the delivery of virulence-associated macromolecules? Lipid Insights 2:27–40

20. Folch J, Lees M, Sloane Stanley GH (1957) A simple method for the isolation and purification of total lipids from animal tissues. J Biol Chem 226(1):497–509

21. Cantin R, Diou J, Belanger D, Tremblay AM, Gilbert C (2008) Discrimination between exosomes and HIV-1: purification of both vesicles from cell-free supernatants. J Immunol Methods 338(1-2):21–30. doi:10.1016/j.jim.2008.07.007

22. Oliveira DL, Nimrichter L, Miranda K, Frases S, Faull KF, Casadevall A, Rodrigues ML

(2009) Cryptococcus neoformans cryoultra-microtomy and vesicle fractionation reveals an intimate association between membrane lipids and glucuronoxylomannan. Fungal Genet Biol 46(12):956–963. doi:10.1016/j.fgb.2009.09.001

23. Meli F, Klein T, Buhr E, Frase CG, Gleber G, Krumrey M, Duta A, Duta S, Korpelainen V, Bellotti R, Picotto GB, Boyd RD, Cuenat A (2012) Traceable size determination of nanoparticles, a comparison among European metrology institutes. Meas Sci Technol 23(12)

Chapter 13

Exploring the *Leishmania* Hydrophilic Acylated Surface Protein B (HASPB) Export Pathway by Live Cell Imaging Methods

Lorna MacLean, Helen Price, and Peter O'Toole

Abstract

Leishmania major is a human-infective protozoan parasite transmitted by the bite of the female phlebotomine sand fly. The *L. major* hydrophilic acylated surface protein B (HASPB) is only expressed in infective parasite stages suggesting a role in parasite virulence. HASPB is a "nonclassically" secreted protein that lacks a conventional signal peptide, reaching the cell surface by an alternative route to the classical ER-Golgi pathway. Instead HASPB trafficking to and exposure on the parasite plasma membrane requires dual N-terminal acylation. Here, we use live cell imaging methods to further explore this pathway allowing visualization of key events in real time at the individual cell level. These methods include live cell imaging using fluorescent reporters to determine the subcellular localization of wild type and acylation site mutation HASPB18-GFP fusion proteins, fluorescence recovery after photobleaching (FRAP) to analyze the dynamics of HASPB in live cells, and live antibody staining to detect surface exposure of HASPB by confocal microscopy.

Key words Live cell imaging, *Leishmania*, Nonclassical protein secretion, FRAP (fluorescence recovery after photobleaching), HASPB (hydrophilic acylated surface protein B), Acylation

1 Introduction

The *Leishmania*-specific HASPs are lifecycle stage-regulated, with protein expression restricted to infective extracellular metacyclic promastigotes and intracellular amastigotes. The protein displays both inter- and intraspecific variation, mainly in the central amino acid repeat regions [1–6]. The *L. major* cDNA16 locus on chromosome 23 contains genes encoding HASPB, the closely related HASPA, and another stage-regulated protein, small hydrophilic endoplasmic reticulum-associated protein (SHERP). The cDNA16 locus is essential for differentiation of host-infective parasites (a process termed metacyclogenesis) in the midgut of the sand fly vector [7]. HASPB protein is immunogenic in the host [8, 9] and low-dose vaccination with recombinant *Leishmania donovani*

Andrea Pompa and Francesca De Marchis (eds.), *Unconventional Protein Secretion: Methods and Protocols*, Methods in Molecular Biology, vol. 1459, DOI 10.1007/978-1-4939-3804-9_13, © Springer Science+Business Media New York 2016

HASPB can induce long-term T cell-mediated protection when administered in the absence of exogenous antigen in mice [10–12]. Although HASPB is a promising vaccine candidate, the function of this protein and its export pathway to the parasite plasma membrane are not fully understood. It has been shown previously, using green fluorescent protein (GFP) reporter cell lines, that the N-terminal 18 amino acids of *Leishmania major* HASPB (termed here as HASPB18) are sufficient for targeting to the promastigote plasma membrane [2]. Key to this, N-terminal myristoylation and palmitoylation are known to be essential co- and post-translational modifications that facilitate trafficking of the protein to the plasma membrane. While the wild type HASPB18-GFP reporter protein is transported to the plasma membrane, disruption of the N-myristoylation site (G2A mutation) prevents both acylation events, resulting in retention of the HASPB18-GFP G2A reporter protein in the cytosol [2]. In contrast, disruption of the N-terminal palmitoylation site (C3S mutation) leads to accumulation of the HASPB18-GFP C3S reporter protein in the vicinity of the Golgi. At the plasma membrane, surface exposure of the HASPB-GFP reporter protein has been detected biochemically by surface biotinylation [2]. Although HASPB is a *Leishmania*-specific protein, the HASPB trafficking pathway is conserved in higher eukaryotes [13, 14]. Here, we describe the use of live cell imaging techniques to further investigate the route by which HASPB is transported to the plasma membrane and flagellum. We show using Lysotracker RED and FM4-64 co-localization studies that HASPB18-GFP CS3 protein is found not only at the Golgi but also in the lysosome and in small acidic structures at the posterior end of the parasite, not previously detected in fixed cells. The latter have been identified as a subset of autophagosomes by co-localization with monomeric red fluorescent protein (mRFP)-ATG8 (Fig. 1) [15–17], implicating these subcellular organelles in the transport pathway of HASPB. Autophagosomes have also been implicated in the trafficking pathway of Acb1, an unconventionally secreted protein in yeast [18, 19]. We use fluorescence recovery after photobleaching (FRAP) to analyze the dynamics of HASPB in live cells, demonstrating that once localized at the plasma membrane, HASPB18-GFP can undergo bidirectional movement within the inner leaflet of the membrane and on the flagellum (Fig. 2a, c). We also show that transfer of HASPB18-GFP between the flagellum and the plasma membrane is compromised (Fig. 2d), leading to the hypothesis that *Leishmania* parasites have a diffusion barrier that separates the cell body membrane and flagellum, a mechanism that may regulate the molecular composition of the flagellar membrane. Lastly we describe live cell antibody labeling to detect surface exposure of HASPB18-GFP (Fig. 3a, b), the metacyclic-specific surface glycoconjugate, lipophosphoglycan (using mAb 3 F12), and native full-length HASPB in wild type *L. major* metacyclic parasites, showing that surface exposure of full-length HASPB

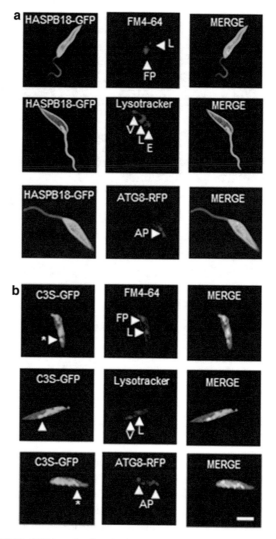

Fig. 1 HASPB18-GFP localization in *L. major* promastigotes. Live cell imaging methods were used to localize the HASPB18-GFP (**a**) and HASPB18-GFP C3S fusion proteins (**b**) in transfected *L. major* parasites. These images were originally published in [17]. Top two panels: Live *L. major* expressing either GFP fusion protein was imaged after immobilization in PBS-primed CyGEL and labeling with Lysotracker RED DND-99 or FM4-64 for 90 min. * small vesicular structures of unknown origin. Bottom panel: Live cell staining of HASPB18-GFP and C3S parasites that are also expressing the autophagosomal marker, RFP-ATG8. *Arrowheads* indicate subcellular compartments as follows: L, lysosome; FP, flagellar pocket; E, endosomes; V, vesicle; AP, autophagosome. Size bar, 5 μm (all images at the same magnification)

correlates with entry into the mammalian-infective phase of the extracellular parasite growth cycle (Fig. 3c, d). In order to produce accurate data with live cell imaging techniques, an effective and reproducible method was required for total immobilization of the

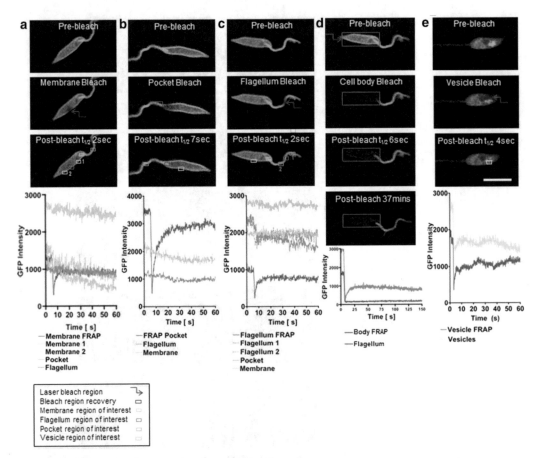

Fig. 2 FRAP analysis of HASPB18-GFP intracellular movement. The dynamics and direction of GFP fusion protein movement in live CyGEL-immobilized HASPB18-GFP (A-D) and HASPB18-GFP C3S (E) *L. major* parasites were investigated by FRAP (fluorescence recovery after photobleaching) analysis. This data was originally published in [17]. Pre-bleach, bleach, and post-bleach images are shown of a typical FRAP; the bleached region of interest (ROI) and recovery at other parts of the cell over time are graphically represented below each set of images, plotting GFP intensity versus time. (**a**). FRAP (*red* ROI) at the plasma membrane and analysis of two adjacent plasma membrane ROIs (*grey*), flagellar pocket ROI (*cyan*), and flagellum ROI (*pink*). (**b**). FRAP (*red* ROI) at the flagellar pocket and simultaneous analysis of plasma membrane ROI (*grey*) and flagellar pocket ROI (*pink*). (**c**). FRAP (red ROI) at the flagellum and analysis of two flagellar ROIs (*pink*), flagellar pocket ROI (*cyan*), and plasma membrane ROI (*grey*). (**d**). FRAP (*red* ROI) of the whole cell body and analysis of this ROI up to 37 min post-bleach. (**e**). FRAP (*red* ROI) of vesicles in metacyclic HASPB18-GFP C3S and analysis of the adjacent vesicle ROI (*yellow*). Size bar, 10 μm (all images presented at the same magnification)

Fig. 3 (continued) and metacyclic-specific LPG (EC LPG 3 F12, using monoclonal 3 F12 antibody and AlexaFluor-647 conjugated secondary antibody). Surface exposure of full-length HASPB was confirmed across the entire cell body and flagellum in a punctate pattern. The DAPI-stained kinetoplast and nucleus are visible and their relative positions identify this cell as a metacyclic parasite; this identification is also verified by staining with the metacyclic-specific antibody, 3 F12. EC HASPB co-localized with EC metacyclic LPG in this parasite, which was intact at the time of labeling as indicated by the absence of Sulfo-NHS-AMCA staining. Size bar, 5 μm. (**d**). Flow cytometry analysis of early passage (*p*=3) *L. major* FVI wild type parasites sampled from early log phase (Day 2), late log phase (Day 5), and stationary, metacyclic-rich phase (Day 7). Sulfo-NHS-AMCA staining was used to distinguish live/dead cells to allow gating on live parasites only. Live parasites were labeled for detection of EC HASPB and metacyclic-specific LPG as above

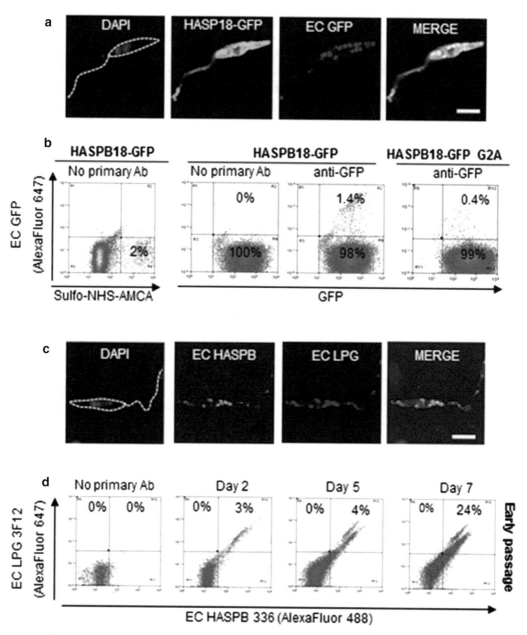

Fig. 3 HASPB18-GFP and full-length HASPB are exposed on the surface of live metacyclic *L. major*. (This data was originally published in [17].) (**a**). Confocal microscopic analysis of HASPB18-GFP surface exposure in metacyclic *L. major* (following labeling with Sulfo-NHS-AMCA to distinguish between live and dead cells), using mouse anti-GFP and detection by AlexaFluor-647-conjugated goat anti-mouse IgG. Extracellular HASPB18-GFP (EC GFP) decorating the surface of the cell body and flagellum in a punctate distribution, while intracellular protein was detected by GFP fluorescence. Size bar, 5 μm. (**b**). Surface HASPB18-GFP exposure was determined by FACS analysis of live HASPB18-GFP *L. major* using the protocol in A (following labeling with Sulfo-NHS-AMCA to distinguish between live and dead cells) using mouse anti-GFP and detection by AlexaFluor-647-conjugated goat anti-mouse IgG. Non-*N*-myristoylated HASPB18-GFP G2A parasites were used as the control for nonsurface exposure. The live/dead analysis (*left hand panel*) detected 2 % dead cells in the HASPB18-GFP parasite population; the remaining analyses (*center* and *right hand panels*) are gated on live cells only. EC GFP, extracellular GFP. (**c**). Confocal microscopy of live labeled early passage (p = 3) *L. major* for detection of EC HASPB (using polyclonal antibody 336 and AlexaFluor-488 conjugated secondary antibody)

highly motile *Leishmania* promastigote parasites while maintaining cell viability. Here we employed a technique previously validated using a thermoreversible gel CyGEL (Biostatus Ltd, UK) which allows reproducible live cell imaging in *L. major*, immobilizing cells while maintaining a high level of cell viability (>95%) following 2 h incubation in CyGEL [20].

2   Materials

2.1 *L. major* Parasites (Summarized in Tables 1a–c)

1. *L. major* wild type (MHOM/IL/81/Friedlin, FVI strain).

2. Homozygous null (knock out, KO) *L. major* line, ΔcDNA16::HYG/ΔcDNA16::PAC, deleted for the diploid LmcDNA16 locus encoding the SHERP and HASP genes.

3. *L. major* expressing the N-terminal 18 amino acids of HASPB as a C-terminal fusion with GFP (HASPB18-GFP) from an episomal pX NEO plasmid.

4. *L. major* myristoylation-minus mutation of the HASPB18-GFP transgene described above (HASPB18-GFP G2A).

5. *L. major* palmitoylation mutant of the HASPB18-GFP transgene described above (HASPB18-GFP C3S).

6. *L. major* HASPB18-GFP transfected with N-terminal RFP-tagged ATG8 (subcloned from ATG8-GFP into pNUS-HnRFP).

7. *L. major* HASPB18-GFP C3S transfected with N-terminal RFP-tagged ATG8.

2.2 *L. major* Growth Media

1. 500 ml of 5× Medium 199 (M199) supplemented with Earle's Salts: mix 27.45 g M199 powder, 0.825 g $NaHCO_3$, make up with H_2O, adjust to pH 7.4 with NaOH and filter sterilize. This can be stored at 4 °C for up to 6 months.

2. 500 ml of 1× M199: use 100 ml 5× M199 plus the following supplements: 10% (v/v) fetal calf serum (FCS), 40 mM HEPES pH 7.4, 100 µM Adenine hemisulfate salt, 0.005% (w/v) Hemin (from a stock solution of 0.25% (w/v) in 50% (v/v) triethanolamine), 100 U penicillin/100 µg Streptomycin, adjusted up to 500 ml with sterile water and filter sterilized. This can be stored at 4 °C for up to 3 months.

3. *L. major transfected parasite lines*: Transgenic *L. major* were maintained in M199 under appropriate drug selection. *L. major* expressing HASPB18-GFP and acylation mutants as well as those expressing full-length HASPB-GFP were grown in media supplemented with 100–500 ng/ml neomycin. Parasites also expressing ATG8-GFP were additionally supplemented with 10 µg/ml blasticidin.

Table 1
***L. major* transgenic parasite lines**

(a)

Transgene	N-Terminal HASPB sequence	Localization	Drug selection
HASPB18-GFP	MGSSCTKDSAKEPQKSAD	Flagellar pocket, plasma membrane	100–500 ng/ml neomycin
HASPB18-GFP G2A	MASSCTKDSAKEPQKSAD	Cytosol	100–500 ng/ml neomycin
HASPB18-GFP C3S	MGSSSTKDSAKEPQKSAD	Golgi, lysosome, acidic vesicles	100–500 ng/ml neomycin

(b)

Transgene	Description	RFP-ATG8 localization	Drug selection
HASPB18-GFP/ RFP-ATG8	N terminal 18aa HASPB/ RFP-tagged ATG8	Autophagosomes	100–500 ng/ml neomycin and 10 µg/ml blasticidin
HASPB18-GFP C3S/ RFP-ATG8	Palmitoylation site mutant N terminal 18aa HASPB/ RFP-tagged ATG8	Autophagosomes	100–500 ng/ml neomycin and 10 µg/ml blasticidin

(c)

Cell line	Description	Genotype	Drug selection
cDNA16 KO	Homozygous null, deleted for diploid LmcDNA16 locus encoding SHERP, HASPA and HASPB genes	ΔcDNA16::HYG/ ΔcDNA16::PAC	100–500 ng/ml neomycin

2.3 Live Cell Imaging

1. Fluorescent reporters: 40 µM FM4-64 (N-(3-Triethylammoniumpropyl)-4-(6-(4-(Diethylamino) Phenyl) Hexatrienyl) Pyridinium Dibromide) emission/excitation maxima 515/640 nm; 50 nM Lysotracker Red DND-99 emission/excitation 577/590 nm.

2. Thermoreversible immobilization gel: CyGEL (Biostatus Ltd, UK).

2.4 Live Cell Antibody Labeling

1. Live/dead stain: 1 mM Sulfo-succinimidyl-7-amino-4-methylcoumarin-3-acetic acid (Sulfo-NHS-AMCA).

2. Sulfo-NHS-AMCA termination reagent: 10 mM Tris–HCl in water, adjust to pH 8.5, filter sterilize, store at room temperature.

3. Primary Antibodies (all diluted in blocking solution): Rabbit polyclonal anti-HASPB336 (1:300); Mouse monoclonal anti-

3F12 (undiluted); Mouse monoclonal anti-GFP (1:200; Invitrogen).

4. Secondary Antibodies: AlexaFluor-488-conjugated goat anti-rabbit IgG (1:250; Invitrogen); AlexaFluor-647-conjugated goat anti-mouse IgG (1:250; Invitrogen).

5. Antibody blocking solution: 1% (w/v) fatty acid-free bovine serum albumin (BSA) in PBS, filter sterilize, and store 4 °C.

6. Fixing solution: 4% (w/v) paraformaldehyde in PBS, stored as 8% (w/v) at –20 °C.

7. Mounting solution: Vectashield with Dapi (Vector laboratories).

8. Phosphate buffered saline (PBS).

3   Methods

3.1   L. major Cell Culture

L. major promastigotes were maintained at 26 °C in vitro, inoculated into culture medium M199 at 10^5 ml^{-1} and grown from early logarithmic procyclic parasites (day 2) to stationary phase metacyclic parasites (day 7) (Subheading 2.1; Table 1a–c) (*see* **Note 1**).

3.2   Live Cell Imaging

1. Fluorescent reporters: To analyze the HASPB pathway in detail, procyclic and metacyclic *L. major* HASPB18-GFP and C3S palmitoylation mutant parasites were incubated live with either Lysotracker RED or FM4-64 fluorescent tracers (subheading 2.3, **item 1**) and immobilized in CyGEL (subheading 2.3, **item 2**) for imaging (*see* **Note 2**).

2. Labeling: collect 1×10^7 cells in microcentrifuge tubes by centrifugation ($800 \times g$ for 10 min at RT), wash in PBS, and resuspend in 5 μl PBS.

3. Treat parasites with 40 μM FM4-64 or 50 nM Lysotracker Red DND-99 for 90 min at RT.

4. Live Leishmania parasites were then immobilized for imaging using CyGEL. This is an optically clear compound which is liquid when ice cold but forms a solid matrix upon warming to 15 °C and above (*see* **Note 3**) Live labeled parasites (<10 μl) were immobilized by adding to 200 μl of ice-cold PBS-primed CyGEL (CyGEL is supplied with 40× PBS) (*see* **Notes 4** and **5**).

5. Aliquot a volume of 50–100 μl of each suspension (using pre-cooled pipette tips) onto a glass coverslip (No. 1.5, 22 × 40 mm) laid on several layers of tissue paper on top of a flat cold block.

6. Place a glass slide on top of the coverslip, allowing the mixture to spread out briefly.

7. Lift the samples off the cold block using the layers of tissue and incubate at 20 °C for 5 min to allow the CyGEL to solidify.

8. Invert the slides, seal by painting nail varnish around the coverslip/slide edge and leave to dry for 5 min.

9. Parasites were then immediately imaged by confocal microscopy using a Zeiss LSM510 meta confocal microscope with a Plan-Apochromat 63×/1.4 Oil DIC III objective. GFP transgene expression was visualized using the 488 nm laser for excitation with emission collected through a 505 LP filter; and FM4-64 and Lysotracker RED were excited at 543 nm and emission collected using a 560 LP filter.

10. The *L. major* HASPB18-GFP and C3S parasite lines expressing the autophagosomal marker mRFP-ATG8 (Table 1b) were immobilized and imaged as above (unlabeled), GFP excitation/emission as above, mRFP excited at 561 nm and emission collected using a 590 LP filter.

11. All images were taken within 2 h of immobilization (*see* **Note 6**).

3.3 FRAP Analysis

FRAP analysis was performed on live mid to late log phase *L. major* episomally expressing HASPB18-GFP and palmitoylation mutant HASPB18-GFP C3S to investigate the dynamics and direction of GFP fusion protein movement. Using the 488 nm laser to rapidly photobleach a small region of interest (ROI) within the HASPB18-GFP fluorescence, the ROI is repopulated by mobile unbleached HASPB18-GFP molecules as the bleached molecules diffuse away until equilibrium is reached (*see* **Note 7**).

1. Immobilize parasites in PBS-primed CyGEL on slides and visualize GFP transgene expression by confocal microscopy as described above in subheading 3.2, **step 1** (*see* **Note 8**).

2. The laser scanned the selected ROI with 100 iterations at an elevated laser power. Pre- and post-bleach images were collected as part of a time series up to 156 s.

3. A further 37 min time point was included when the whole cell body ROI bleach was performed.

4. Analysis was performed using SigmaPlot11 and data fitted according to a single exponential.

3.4 Live Cell Antibody Labeling and Imaging

Labeling was performed on live parasites to detect surface proteins, using methods designed to minimize antibody capping but maximize signal. Prior to live cell staining, the amine-reactive fluorophore Sulfo-NHS-AMCA was used to confirm cell viability; dead cells stained with this reagent emit a strong blue fluorescence throughout the cell and these were removed from further analysis [1, 21].

1. Analyze approximately 2×10^7 *L. major* FVI wild type parasites, LmcDNA16 locus KO (Table 1a, c), and those expressing HASPB18-GFP and HASPB18-GFP G2A (*see* **Note 9**).

2. Collect cells by centrifugation at $800 \times g$ for 5 min, remove supernatant.

3. Wash cell pellet in 1 ml ice-cold PBS (*see* **Note 10**).

4. Incubate cells with 90 µl Sulfo-NHS-AMCA (1 mM; Subheading 2.4, **item 1**) on ice for 10 min, before termination of the reaction by the addition of 10 mM Tris–HCl pH 8.5 to a concentration of 10 mM (addition of 10 µl of 100 mM Tris–HCl pH 8.5; Subheading 2.4, **item 2**) [22].

5. Wash cells three times with 1 ml cold 1% (w/v) fatty acid-free BSA blocking solution (filter sterilized) (*see* Subheading 2.4, **item 5**) and resuspend in 200 µl of blocking solution for 20 min at 20 °C.

6. Centrifuge cells at $800 \times g$ at 10 min, remove blocking solution before addition of 100 µl primary antibodies (Subheading 2.4, **item 3**) for 30 min at 20 °C. Incubate Live *L. major* wild type parasites with rabbit polyclonal anti-HASPB336 (1:300) [6] and mouse monoclonal anti-3F12 (undiluted) [23] to detect native extracellular HASPB and metacyclic-specific form of the parasite surface glycoconjugate lipophosphoglycan (LPG) respectively.

7. Wash parasites three times in PBS and fix in 4% (w/v) paraformaldehyde/PBS (PFA; Subheading 2.4, **item 6**) for 20 min on ice before being washed in PBS and secondary antibody (*see* Subheading 2.4).

8. Detect with 100 µl (1:250 in blocking solution) AlexaFluor-488-conjugated goat anti-rabbit IgG and AlexaFluor-647-conjugated goat anti-mouse IgG respectively for 60 min at 20 °C.

9. Detect extracellular GFP in HASPB18-GFP and HASPB18-GFP G2A parasites by incubation of live cells with mouse anti-GFP (1:200) followed by fixation and detection with AlexaFluor-647-conjugated goat anti-mouse IgG as above.

10. Wash parasites three times in PBS, and resuspend in 200 µl PBS.

11. Place 100 µl of this sample onto poly-lysine slides, allow to adhere for 20 min, remove liquid and mount with 10 µl of Vectashield containing DAPI (*see* Subheading 2.4 **item 7**) to detect the parasite nucleus and kinetoplast, place coverslip on top, and seal with nail polish (*see* **Note 11**).

12. HASPB18-GFP and G2A parasites were imaged by confocal microscopy as described above, detecting Sulfo-NHS-AMCA (dead cells) and DAPI (nuclei and kinetoplasts) using the 405 nm laser for excitation and emission collected through a 420–480 nm filter, GFP transgene using the 488 nm laser for excitation with emission collected through a 505 LP filter and extracellular HASPB18-GFP detected using anti-GFP/AlexaFluor-647-conjugated goat anti-mouse IgG excitation 633 nm and emission collected using a 650 nm LP filter.

13. Wild type parasites were imaged, detecting dead cells and DAPI as above, extracellular 3F12 by anti-3F12/AlexaFluor-647-

conjugated goat anti-mouse IgG excitation 633 nm and emission collected using a 650 nm LP filter, and extracellular full-length endogenous HASPB detected with anti-HASPB336/AlexaFluor-488-conjugated goat anti-rabbit IgG using the 488 nm laser for excitation with emission collected through a 505 LP filter (*see* **Note 12**).

4   Notes

1. All transfected parasite lines were grown under appropriate drug selection (*see* Subheading 2.2). All parasites categorized as early passage cells were </= p6 relative to the time of extraction from the lymph nodes of experimentally infected BALB/c mice.

2. The lipophilic probe FM4-64 is transported through the endosomal system in a time- and temperature-dependent manner. Under the labeling conditions described below, FM4-64 signal appeared primarily in the flagellar pocket/early endosomes and lysosome of the parasite. Lysotracker probes accumulate in low pH compartments, therefore targeting the late endosomes, lysosome, and acidocalcisomes in *L. major*.

3. CyGEL matrix can also act as a controlled delivery system for fluorescent probes for subcellular localization studies rather than pretreatment although the initial uptake of dye is slightly delayed using this method.

4. This was carefully mixed by flicking the tube avoiding the generation of bubbles (it is important for imaging that the slide has no bubbles). To remove any bubbles, gently tap the coverslip with the end of a pipette tip while CyGEL is still in liquid form.

5. It is essential that parasites are well dispersed through CyGEL. This can be aided by flicking the tube or gently pipetting the cell suspension up and down using precooled pipette tips. If the parasites form clumps they are impossible to image with any precision.

6. The most technically difficult part of setting up live cell imaging is that timing is crucial. It is best if the time between mounting CyGEL-immobilized sample and imaging is very short (about 10–15 min), allowing enough time to carry out imaging within 2 h when cell viability is >95 %. Initially the *Leishmania* flagella are still motile, which can act as an indication that the cell being imaged is alive. This then slows and eventually ceases once the CyGEL has fully set, to allow successful imaging. It would also be possible to detect live cells by using propidium iodide exclusion to ensure that only live cells are being imaged.

7. The rate of mobility and percentage recovery can be analyzed, giving an indication of the speed of movement and the percentage of mobile versus immobile molecules. Furthermore, by recording

the corresponding fluorescence loss induced by photobleaching (FLIP) in another ROI outside of the bleached ROI, the exchange of bleached for unbleached HASPB18-GFP in the cell will allow analysis of the direction of movement of HASPB18-GFP.

8. ROI of a fixed dimension chosen for HASPB18-GFP included the plasma membrane, flagellum, flagellar pocket (FRAP performed ≥5 times) and a larger ROI was also chosen to bleach the whole cell body. The small vesicle-like structures identified in HASPB18-GFP C3S expressing cells were also chosen as a ROI.

9. The LmcDNA16 locus KO and myristoylation mutant were selected as controls for surface exposed GFP and HASPB and a no primary antibody control was also used.

10. The pellet was resuspended in filter sterilized PBS, pipetted up and down, centrifuged $800 \times g$ 5 min, and supernatant removed.

11. Slides can be stored briefly in the dark at 4 °C before imaging.

12. As labeling was carried out in microcentrifuge tubes the remaining 100 µl of each sample was used for flow cytometry. Cells (without DAPI) were analyzed on a Beckman Coulter CyAn ADP and data evaluated using Summit 4.3 Software to give quantitative information on cell viability using Sulfo-NHS-AMCA staining, then live cells were gated to determine the population with surface exposed HASPB18-GFP and native HASPB. This also allowed us to demonstrate that early passage *L. major* expressing extracellular 3F12 also stains with HASPB, confirming this double positive population to be metacyclic promastigotes.

Acknowledgements

We thank Jeremy Mottram (University of Glasgow) for the gift of RFP-ATG8 plasmid, and Barbara Smith, Michael Hodgkinson, and Ian Morrison for technical assistance. We also thank Deborah Smith for her support. This work was funded by Wellcome Trust programme grant 077503 awarded to Deborah F. Smith (University of York).

References

1. Depledge DP, MacLean LM, Hodgkinson MR, Smith BA, Jackson AP, Ma S, Uliana SR, Smith DF (2010) Leishmania-specific surface antigens show sub-genus sequence variation and immune recognition. PLoS Negl Trop Dis 4(9), e829. doi:10.1371/journal.pntd.0000829

2. Denny PW, Gokool S, Russell DG, Field MC, Smith DF (2000) Acylation-dependent protein export in Leishmania. J Biol Chem 275(15):11017–11025

3. Alce TM, Gokool S, McGhie D, Stager S, Smith DF (1999) Expression of hydrophilic surface proteins in infective stages of Leishmania donovani. Mol Biochem Parasitol 102(1):191–196

4. McKean PG, Trenholme KR, Rangarajan D, Keen JK, Smith DF (1997) Diversity in repeat-containing surface proteins of Leishmania major. Mol Biochem Parasitol 86(2):225–235

5. Rangarajan D, Gokool S, McCrossan MV, Smith DF (1995) The gene B protein localises to the surface of Leishmania major parasites in the absence of metacyclic stage lipophosphoglycan. J Cell Sci 108(Pt 11):3359–3366

6. Flinn HM, Rangarajan D, Smith DF (1994) Expression of a hydrophilic surface protein in infective stages of Leishmania major. Mol Biochem Parasitol 65(2):259–270

7. Sadlova J, Price HP, Smith BA, Votypka J, Volf P, Smith DF (2010) The stage-regulated HASPB and SHERP proteins are essential for differentiation of the protozoan parasite Leishmania major in its sand fly vector, Phlebotomus papatasi. Cell Microbiol 12(12):1765–1779. doi:10.1111/j.1462-5822.2010.01507.x

8. Jensen AT, Gasim S, Moller T, Ismail A, Gaafar A, Kemp M, el Hassan AM, Kharazmi A, Alce TM, Smith DF, Theander TG (1999) Serodiagnosis of Leishmania donovani infections: assessment of enzyme-linked immunosorbent assays using recombinant L. donovani gene B protein (GBP) and a peptide sequence of L. donovani GBP. Trans R Soc Trop Med Hyg 93(2):157–160

9. Jensen AT, Gaafar A, Ismail A, Christensen CB, Kemp M, Hassan AM, Kharazmi A, Theander TG (1996) Serodiagnosis of cutaneous leishmaniasis: assessment of an enzyme-linked immunosorbent assay using a peptide sequence from gene B protein. Am J Trop Med Hyg 55(5):490–495

10. Stager S, Alexander J, Kirby AC, Botto M, Rooijen NV, Smith DF, Brombacher F, Kaye PM (2003) Natural antibodies and complement are endogenous adjuvants for vaccine-induced CD8+ T-cell responses. Nat Med 9(10):1287–1292. doi:10.1038/nm933

11. Stager S, Smith DF, Kaye PM (2000) Immunization with a recombinant stage-regulated surface protein from Leishmania donovani induces protection against visceral leishmaniasis. J Immunol 165(12):7064–7071

12. Maroof A, Brown N, Smith B, Hodgkinson MR, Maxwell A, Losch FO, Fritz U, Walden P, Lacey CN, Smith DF, Aebischer T, Kaye PM (2012) Therapeutic vaccination with recombinant adenovirus reduces splenic parasite burden in experimental visceral leishmaniasis. J Infect Dis 205(5):853–863. doi:10.1093/infdis/jir842

13. Stegmayer C, Kehlenbach A, Tournaviti S, Wegehingel S, Zehe C, Denny P, Smith DF, Schwappach B, Nickel W (2005) Direct transport across the plasma membrane of mammalian cells of Leishmania HASPB as revealed by a CHO export mutant. J Cell Sci 118(Pt 3):517–527. doi:10.1242/jcs.01645

14. Tournaviti S, Pietro ES, Terjung S, Schafmeier T, Wegehingel S, Ritzerfeld J, Schulz J, Smith DF, Pepperkok R, Nickel W (2009) Reversible phosphorylation as a molecular switch to regulate plasma membrane targeting of acylated SH4 domain proteins. Traffic 10(8):1047–1060. doi:10.1111/j.1600-0854.2009.00921.x

15. Besteiro S, Williams RA, Morrison LS, Coombs GH, Mottram JC (2006) Endosome sorting and autophagy are essential for differentiation and virulence of Leishmania major. J Biol Chem 281(16):11384–11396. doi:10.1074/jbc.M512307200

16. Williams RA, Tetley L, Mottram JC, Coombs GH (2006) Cysteine peptidases CPA and CPB are vital for autophagy and differentiation in Leishmania mexicana. Mol Microbiol 61(3):655–674. doi:10.1111/j.1365-2958.2006.05274.x

17. Maclean LM, O'Toole PJ, Stark M, Marrison J, Seelenmeyer C, Nickel W, Smith DF (2012) Trafficking and release of Leishmania metacyclic HASPB on macrophage invasion. Cell Microbiol 14(5):740–761. doi:10.1111/j.1462-5822.2012.01756.x

18. Duran JM, Anjard C, Stefan C, Loomis WF, Malhotra V (2010) Unconventional secretion of Acb1 is mediated by autophagosomes. J Cell Biol 188(4):527–536. doi:10.1083/jcb.200911154

19. Manjithaya R, Anjard C, Loomis WF, Subramani S (2010) Unconventional secretion of Pichia pastoris Acb1 is dependent on GRASP protein, peroxisomal functions, and autophagosome formation. J Cell Biol 188(4):537–546. doi:10.1083/jcb.200911149

20. Price HP, MacLean L, Marrison J, O'Toole PJ, Smith DF (2010) Validation of a new method for immobilising kinetoplastid parasites for live cell imaging. Mol Biochem Parasitol 169(1):66–69. doi:10.1016/j.molbiopara.2009.09.008

21. Engstler M, Pfohl T, Herminghaus S, Boshart M, Wiegertjes G, Heddergott N, Overath P (2007) Hydrodynamic flow-mediated protein sorting on the cell surface of trypanosomes. Cell 131(3):505–515. doi:10.1016/j.cell.2007.08.046

22. Grunfelder CG, Engstler M, Weise F, Schwarz H, Stierhof YD, Morgan GW, Field MC, Overath P (2003) Endocytosis of a glycosylphosphatidylinositol-anchored protein via clathrin-coated vesicles, sorting by default in endosomes, and exocytosis via RAB11-positive carriers. Mol Biol Cell 14(5):2029–2040. doi:10.1091/mbc.E02-10-0640

23. Sacks DL, da Silva RP (1987) The generation of infective stage Leishmania major promastigotes is associated with the cell-surface expression and release of a developmentally regulated glycolipid. J Immunol 139(9):3099–3106

Chapter 14

Characterization of the Unconventional Secretion of the Ebola Matrix Protein VP40

Olivier Reynard and Mathieu Mateo

Abstract

While most secreted proteins use the classical endoplasmic reticulum (ER)-Golgi secretion pathway to reach the extracellular medium, a few proteins are secreted through unconventional secretary pathways. Viral proteins can be secreted through unconventional secretion pathways. Here, we describe how we have recently demonstrated that the Ebola virus (EBOV) matrix protein VP40 is released from transfected and infected cells in a soluble form through an unconventional secretion pathway.

Key words Ebola, Ebola virus, VP40, Virus, Unconventional secretion

1 Introduction

Ebola virus (EBOV) and Marburg virus form the Filoviridae family, a group of enveloped negative-strand RNA viruses responsible for severe disease in humans. The structural EBOV matrix protein VP40 is essential for assembly and budding by allowing incorporation of viral ribonucleocapsids in budding virus particles [1]. When expressed alone in mammalian cells, VP40 promotes the formation of virus-like particles (VLPs) resembling filamentous virions [2]. Several crystal structures of VP40 have been solved. The monomeric VP40 is formed by two functionally interrelated domains: an N-terminal oligomerization domain and a C-terminal membrane-binding domain [3]. VP40 does not seem to exist as a monomer in cells and several oligomerization states have been described that may be associated with different functions [4, 5]. Dimers of VP40 are formed in the cytosol and transported to the plasma membrane. Initial steps of VP40 intracellular trafficking have been proposed to be associated with the COPII vesicular transport system and notably through an interaction of VP40 with Sec24C [6, 7]. Binding of VP40 to the plasma membrane through a proline-rich domain induces the formation of VP40 hexamers [8]. The formation of the

Andrea Pompa and Francesca De Marchis (eds.), *Unconventional Protein Secretion: Methods and Protocols*, Methods in Molecular Biology, vol. 1459, DOI 10.1007/978-1-4939-3804-9_14, © Springer Science+Business Media New York 2016

filamentous EBOV particle depends on VP40 hexamerization at the plasma membrane [9, 10]. There, VP40 recruits the components of the ESCRT machinery TSG101 and Nedd4 through two overlapping late domains PTAPPEY to allow budding and release of the virus particle [11]. In addition, VP40 can assemble octamers that bind RNA in a sequence-specific manner [12].

In a recent publication, we demonstrated that VP40 is also released from transfected or infected cells as a monomeric soluble form (sVP40) via an unconventional secretion pathway [13]. While producing VP40 VLPs, we constantly noticed the presence of VLP-free VP40 in the supernatant. In silico analysis using secretomeP software suggested that VP40 could be released with fairly high probability through an unconventional secretion [14]. We tested this hypothesis by developing an experimental approach allowing discriminating the release of VP40 as a soluble protein from the release of VP40 in the form of enveloped VLPs and that is described herein in details. Using a similar approach, we also demonstrated that the matrix protein of human metapneumovirus is secreted from infected cells through an unconventional secretion pathway [15]. Such method could therefore be applied to any other viral matrix proteins of interest.

Initially samples from supernatant, ultrasupernatant, VLPs, and cell lysates were analyzed by nonreducing and low SDS western blot (0.1%). This analysis displayed in Fig. 1a revealed the absence of oligomeric VP40 in the ultrasupernatant. Next, we confirmed that the presence of the protein in the ultrasupernatant does not result from the lysis of virus-like particles by checking the stability of particles for 3 days at 37 °C. Western blot analysis of the VLPs fraction showed no loss in VLP stability (Fig. 2b). Then to prove that sVP40 is not released in a membrane-associated form, we performed a flotation assay through ultracentrifugation of the supernatant in a discontinued sucrose gradient (40–30–10%). Figure 2c shows that the major part of the VP40 is found in the bottom fractions (40%) containing soluble proteins while VLPs accumulate at the 30–10% interface. Next we showed that secretion of sVP40 occurs in the absence of cytotoxicity and apoptosis as demonstrated by both LDH cytotoxicity assay and PI/DIOC6 staining of VP40-transfected cells, respectively, when compared to mock-transfected cells.

Finally, we demonstrated that VP40 secretion is independent of the standard ER/golgi secretion pathway. Classical secretion of cellular proteins from mammalian cells requires a signal peptide sequence and occurs via the ER-Golgi network. We could not detect the presence of a classical signal peptide in the VP40 sequence. Using two inhibitors of ER-to-Golgi transport only or both trans-Golgi-network-to-cell-surface transport and ER-to-Golgi transport, brefeldin A (BFA) and H89, respectively, we demonstrated that VP40 is not secreted through a conventional pathway (Fig. 3).

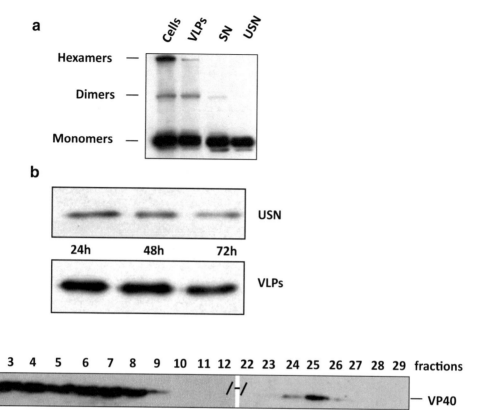

Fig. 1 Presence of soluble VP40 in the supernatant. (**a**) Immunoblot analysis of cell lysate (cells), supernatant (SN), virus-like particles (VLPs), and ultrasupernatant (USN). (**b**) Immunoblots showing the stability of VP40 VLPs at 37 °C in the USN (*upper panel*) and in the VLPs (*lower panel*). (**c**) Immunoblot analysis of the fractions resulting from the flotation assay. For **a**, **b**, and **c**, immunoblot was realized using mouse anti-VP40 antibodies

2   Materials

2.1   Cells and Transfections

1. HEK 293T cells.
2. Dulbecco's modified Eagle's medium (DMEM).
3. Glutamine.
4. Fetal calf serum.
5. Transfection reagent.
6. Eukaryotic expression vector for VP40 and a conventionally secreted protein (Ebola sGP).

2.2   VLP Release and Ultra-centrifugation

1. PBS-CM: PBS, 0.1 mM $CaCl_2$, 1 mM $MgCl_2$.
2. Sucrose: 10, 20, 30, 40 % solution in PBS-CM (w/v).
3. Tabletop low-speed centrifuge.

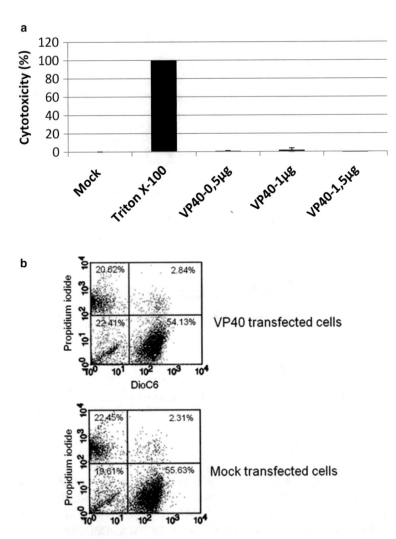

Fig. 2 Cytotoxicity of VP40 expression. (**a**) LDH cytotoxicity detection assay. Triton X100 treated condition was used to normalize value at 100 % LDH release and LDH release was also determined in cells expressing increasing amounts of VP40 (0.5, 1, and 1.5 μg). (**b**) Apoptosis measurement on VP40-transfected HEK 293 T using DIOC6 and PI staining

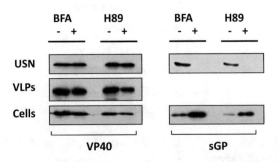

Fig. 3 Blockage of conventional secretion pathways. Immunoblots showing no blockage of VP40 budding (VLPs) or secretion (USN) by BFA and H89 (*left panels*). Both BFA and H89 block the secretion of sGP (*right panels*)

4. Ultracentrifuge with swinging rotor (for 38 ml tube as SW32 rotor).

5. Brefeldin A solution (γ,4-dihydroxy-2-(6-hydroxy-1-heptenyl)-4-cyclopentanecrotonic acid λ-lactone), mother solution at 1 mg/ml in ethanol.

6. H89 solution (N-[2-(p-bromocinnamylamino)ethyl]-5-isoquinolinesulfonamide dihydrochloride), mother solution at 5 mg/ml in PBS-CM.

2.3 CO-IP Buffer

1. CO-IP buffer: 20 mM Tris–HCl, pH 7.0, 100 mM NaCl, 5 mM EDTA, 0,4 % sodium deoxycholate (v/v), 1 % Nonidet P40 (v/v).

2.4 Cytotoxicity Assay

1. LDH cytotoxicity detection kit.

2. Trypsin 0.25 % solution in DMEM.

3. 1 % Triton X100 in PBS-CM.

4. Anti mouse phycoerythrin secondary antibody (10–20 μg/ml).

5. Propidium iodide 1 mg/ml.

6. DIOC6 1 mM.

7. Flow cytometer.

2.5 Western Blot

1. Precast gel 4–20 % gel or in-house prepared with 4 % (w/v) stacking and 8 % (w/v) separating gel, prepared with acrylamide:bisacrylamide mixture (37.5:1).

2. Precast transfer sandwich.

3. Laemmli buffer: 4 % SDS (w/v), 20 % glycerol (w/v), 10 % 2-mercaptoethanol (w/v), 0.004 % bromphenol blue (w/v), and 0.125 M Tris–HCl, pH approx. 6.8.

4. Mini sodium dodecyl sulfate-polyacrylamide-gel (SDS-PAGE) gel system.

5. SDS-PAGE molecular weight standards.

6. 0.1 % (v/v) Tween 20 in phosphate-buffered saline, PBS (PBST).

7. 10 % (w/v) skimmed milk in PBST.

8. Anti-VP40 monoclonal antibody (in-house production, commercial antibodies available at IBT Bioservice).

9. Anti-mouse IgG horseradish-peroxidase-conjugated secondary antibody: Working solution is 1:25,000 in 1 % milk-PBST.

10. Western chemiluminescent substrate system.

11. Digital imager for chemiluminescence.

3   Methods

3.1   Cell Cultures and Transfection

1. HEK 293T cells were cultured at 37 °C in Dulbecco's modified Eagle's medium supplemented with 10% fetal calf serum (FCS).

2. Transfection of cells was performed using Turbofect (Thermo) according to the manufacturer's instructions.

3.2   VLP Assays

1. Plate 5×10^5 HEK293T cells per well of 6-well plates.

2. On the following day, transfect cells with 3 μg of plasmid expressing VP40 under the control of a CMV promoter (*see* **Note 1**).

3. At 14 h post-transfection, clarify the culture supernatant (SN) by centrifugation at $5000 \times g$ for 5 min.

4. Keep an aliquot of the SN for further analysis.

5. Lyse the cells in 800 μl of CO-IP buffer.

6. Load 35 ml of the supernatant on 3 ml of a 20% sucrose cushion in 38 ml tubes.

7. Centrifuge at $250,000 \times g$ for 2 h at 4 °C in a ultracentrifuge.

8. Keep an aliquot of the ultrasupernatant (USN) for further analysis and resuspend the VLP pellet in 1000 μl of PBS.

9. Resuspend the different fractions in Laemmli buffer containing 0.1% SDS and no β-mercaptoethanol.

10. Separate the fractions on an 8% SDS-PAGE.

11. Transfer the protein to a PVDF membrane.

12. Block the membrane for at least 1 h in 10% PBS-nonfat milk and then stain the membrane using anti-VP40 antibodies and HRP-conjugated secondary antibodies.

13. Reveal using an HRP substrate such as chemiluminescence (ECL) HRP substrate with stable light output for mid-femtogram-level detection.

3.3   VLP Stability

1. Resuspend the VLP pellet in a given volume and separate in equal volume fractions (one fraction per time point).

2. Incubate at 37 °C for a given period (24, 48, and 72 h).

3. Load the VLPs on a 20% sucrose cushion and centrifuge at $250,000 \times g$ for 2 h at 4 °C in an ultracentrifuge.

4. Resuspend the pellet in Laemmli buffer and analyze by SDS-PAGE followed by western blot using anti-VP40 antibodies.

3.4   Floatation Assay

1. Clarify the SN of VP40-transfected cells as in subheading 3.1, **step 2**.

2. Adjust the clarified SN to 40% sucrose.

3. Load 12 ml of SN adjusted to 40 % sucrose at the bottom of a 38 ml ultracentrifuge tube and layer 21 ml of 30 % sucrose and then 5 ml of 10 % sucrose solutions in PBS-CM on top of it (*see* **Note 2**).

4. Centrifuge the samples at $130,000 \times g$ for 20 h.

5. Collect gradient fractions from the bottom to the top (*see* **Note 3**) and analyze by western blot using anti-VP40 antibodies. Bottom fractions contain soluble proteins, and top fractions contain membrane-associated proteins.

3.5 Measuring Cytotoxicity by Lactate Dehydrogenase Cytotoxicity Detection Assay

1. Transfect 2.5×10^5 293T cells per well of 12-well plate with increasing amounts of plasmid expressing VP40.

2. At 14 h post-transfection, as a positive control for cytotoxicity, pretreat a well with a PBS 1 % triton X-100 solution during 10 min.

3. Harvest all supernatant and perform LDH assay following the manufacturer's recommendation (LDH cytotoxicity detection kit).

4. Measure absorbance at 490 nm using a spectrophotometer.

3.6 Measuring Cell Death by Dioc6/Propidium Iodide Flow Cytometry Analysis

1. Transfect 5×10^5 293T cells in a well of 6-well plate with 1 μg of phCMV-VP40 or phCMV-empty.

2. At 24 h post-transfection, wash the cells and add trypsin (0.25 % solution) for 5 min to detach the cells and separate them into two fractions.

3. Stain one fraction with VP40 antibodies (primary anti-VP40 10F6 for 30 min and secondary anti-mouse PE for 30 min) to determine the percentage of transfected cells.

4. Stain the other fraction with DIOC6 (3,3′-dihexyloxacarbocyanine iodide) at 40 nM for 20 min at 37 °C, wash once, and add propidium iodide just prior analysis.

5. Analyzse samples by flow cytometry using FL1 (emission wavelength = 488 nm) and FL3 (emission wavelength = 600 nm).

3.7 Blocking Conventional Secretion Pathways

1. Transfect 5×10^5 HEK293T cells in a well of 6-well plate with 3 μg of plasmid expressing VP40 or the soluble form of the Ebola glycoprotein sGP as a control.

2. At 14 h post-transfection, rinse the cells with fresh culture medium and incubate for 7 h in the presence or absence of 3.6 μM BFA, an inhibitor of ER-to-Golgi transport, or 50 μM H89 (Sigma), an inhibitor of both trans-Golgi-network-to-cell-surface transport and ER-to-Golgi transport (*see* **Note 4**).

3. Isolate soluble proteins by ultracentrifugation on a 20 % sucrose cushion as described in Subheading 3.1.

4. Analyze the cells, VLP, and USN fractions by western blot using anti-VP40 or anti-sGP proteins.

4 Notes

1. Transfections were performed using Turbofect reagent (Thermo) using a DNA:turbofect ratio of 1:3 (1 μg plasmids: 3 μl turbofect) and following the manufacturer's recommendation. All reagents must be at room temperature at the time of mixing for efficient transfection.

2. Collecting fractions is tricky. For reproducible fractionation, it is better to use a fraction collector device monitored by a peristaltic pump. If not available, the bottom of the tube can be perforated perpendicularly to the tube with a red-hot 21 G needle, in such a way that the needle will finally be placed inside the tube at 1 or 2 mm above the bottom of the tube with the needle bevel orientated top. This may allow a drip collection of the sample.

3. The 40 % could be loaded first as described or alternatively 30 and 10 % sucrose solutions could be added first and the 40 % sucrose sample solution added through the first two layers with a fine pipette (glass Pasteur pipette). Proceeding this way will generate very clean solutions interface that will result in cleaner results at analysis.

4. H89 and brefeldin A are toxic compounds that cannot be used for a long time period. The proper concentration is cell type specific and should be adjusted using an appropriate control, for instance a conventionally secreted protein.

References

1. Harty RN, Brown ME, Wang G, Huibregtse J, Hayes FP (2000) A PPxY motif within the VP40 protein of Ebola virus interacts physically and functionally with a ubiquitin ligase: implications for filovirus budding. Proc Natl Acad Sci U S A 97(25):13871–13876

2. Timmins J, Scianimanico S, Schoehn G, Weissenhorn W (2001) Vesicular release of Ebola virus matrix protein VP40. Virology 283(1):1–6

3. Ruigrok RW, Schoehn G, Dessen A, Forest E, Volchkov V, Dolnik O, Klenk HD, Weissenhorn W (2000) Structural characterization and membrane binding properties of the matrix protein VP40 of Ebola virus. J Mol Biol 300(1):103–112

4. Dolnik O, Volchkova VA, Escudero-Perez B, Lawrence P, Klenk HD, Volchkov VE (2015) Shedding of Ebola virus surface glycoprotein is a mechanism of self-regulation of cellular cytotoxicity and has a direct effect on virus infectivity. J Infect Dis 212(Suppl 2):S322–S328. doi:10.1093/infdis/jiv268

5. Warfield KL, Bradfute SB, Wells J, Lofts L, Cooper MT, Alves DA, Reed DK, VanTongeren SA, Mech CA, Bavari S (2009) Development and characterization of a mouse model for Marburg hemorrhagic fever. J Virol 83(13):6404–6415. doi:10.1128/JVI.00126-09

6. Yamayoshi S, Noda T, Ebihara H, Goto H, Morikawa Y, Lukashevich IS, Neumann G, Feldmann H, Kawaoka Y (2008) Ebola virus matrix protein VP40 uses the COPII transport system for its intracellular transport. Cell Host Microbe 3(3):168–177

7. Reynard O, Nemirov K, Page A, Mateo M, Raoul H, Weissenhorn W, Volchkov VE (2011) Conserved proline-rich region of Ebola virus matrix protein VP40 is essential for plasma membrane targeting and virus-like particle release. J Infect Dis 204(Suppl 3):S884–S891. doi:10.1093/infdis/jir359

8. Scianimanico S, Schoehn G, Timmins J, Ruigrok RH, Klenk HD, Weissenhorn W (2000) Membrane association induces a conformational change in the Ebola virus matrix protein. EMBO J 19(24):6732–6741

9. Volchkova VA, Dolnik O, Martinez MJ, Reynard O, Volchkov VE (2015) RNA editing of the GP gene of Ebola virus is an important

pathogenicity factor. J Infect Dis 212(Suppl 2):S226–S233. doi:10.1093/infdis/jiv309

10. Kaletsky RL, Simmons G, Bates P (2007) Proteolysis of the Ebola virus glycoproteins enhances virus binding and infectivity. J Virol 81(24):13378–13384. doi:10.1128/ JVI.01170-07

11. Timmins J, Schoehn G, Ricard-Blum S, Scianimanico S, Vernet T, Ruigrok RW, Weissenhorn W (2003) Ebola virus matrix protein VP40 interaction with human cellular factors Tsg101 and Nedd4. J Mol Biol 326(2):493–502

12. Hoenen T, Volchkov V, Kolesnikova L, Mittler E, Timmins J, Ottmann M, Reynard O, Becker S, Weissenhorn W (2005) VP40 octamers are essential for Ebola virus replication. J Virol 79(3):1898–1905

13. Unconventional secretion of Ebolavirus matrix protein VP40 Olivier Reynard, St. Patrick Reid, Audrey Page, Mathieu Mateo, Nathalie Alazard-Dany, Christopher F. Basler and Viktor E. Volchkov. J Infectious Diseases 2011 Nov, 204 (Suppl 3) : S757–S1097

14. Bendtsen JD, Jensen LJ, Blom N, Von Heijne G, Brunak S (2004) Feature-based prediction of non- classical and leaderless protein secretion. Protein Eng Des Sel 17(4):349–356

15. Bagnaud-Baule A, Reynard O, Perret M, Berland JL, Maache M, Peyrefitte C, Vernet G, Volchkov V, Paranhos-Baccala G (2011) The human metapneumovirus matrix protein stimulates the inflammatory immune response in vitro. PLoS One 6(3):e17818. doi:10.1371/ journal.pone.0017818

Chapter 15

Role and Characterization of Synuclein-γ Unconventional Protein Secretion in Cancer Cells

Caiyun Liu, Like Qu, and Chengchao Shou

Abstract

Synuclein-γ (SNCG), the third member of synuclein family, is implicated in both neurodegenerative diseases and cancer. Overexpression of SNCG in cancer cells is linked to tumor progression and chemoresistance. Without any known signal sequence required for conventional protein secretion, SNCG is elevated in the serum of cancer patients and the medium of cultured cancer cells. SNCG actively secretes from cancer cells and extracellular SNCG promotes malignant phenotypes of cancer cells. Here, we describe methods for the characterization of SNCG as an unconventional secretion protein from cancer cells and investigation of the effect of extracellular SNCG on the phenotypes of cancer cells.

Key words Synuclein-γ, Unconventional secretion, Cancer cell, Migration, Invasion

1 Introduction

The synucleins are small proteins (127–140 amino acids) consisting of α-synuclein (SNCA), β-synuclein (SNCB), and γ-synuclein (SNCG). Overexpressed SNCG is associated with cancer cell metastasis, chemoresistance, and correlates with adverse outcome in several cancers. SNCG participates in estrogen receptor-alpha36-mediated estrogen signaling [1], Akt/mTOR signaling [2], insulin-like growth factor (IGF-1) signaling [3], mitogen-activated protein kinases (MAPK) pathways [4], and microtubule regulation [5]. Elevated serum SNCG levels are associated with pancreatic [6], gastrointestinal, esophageal, and colorectal [7, 8] cancers. Recent reports have uncovered that SNCG is actively released into the extracellular microenvironment by unconventional secretion pathway and the extracellular SNCG promotes cancer cell migration and invasion [9].

Secretomes include proteins secreted using a signal peptide through the classical ER-Golgi pathway, proteins secreted through different nonclassical pathways (e.g., exosomes, microvesicles), and extracellular domains of plasma membrane proteins generated

Andrea Pompa and Francesca De Marchis (eds.), *Unconventional Protein Secretion: Methods and Protocols*, Methods in Molecular Biology, vol. 1459, DOI 10.1007/978-1-4939-3804-9_15, © Springer Science+Business Media New York 2016

by protease shedding [10]. Accumulating evidence indicates that a vast number of intracellular proteins without any known signal sequence required for conventional protein secretion are secreted from cells by unconventional secretion pathway [11], but a unified picture of the pathways and mechanism of unconventional secretion remain elusive [12]. Unconventional secretion is a major contributor of cancer cell secretomes and these proteins play important roles during tumorigenesis [9, 11].

However, secretome proteins can be contaminated by proteins from bovine serum or cell lysis. A number of false positives coming from serum and several proteins are both in serum and being secreted from cancer cells [11]; therefore the viability of cell lines should be monitored as a quality control for the generation of secretomes. Here, we show that SNCG is naturally secreted in the medium during the culture of cancer cells, and describe the methods used to characterize SNCG as an unconventional secretion protein, and investigate the biological role of extracellular SNCG for cancer cells.

2   Materials

All solutions are prepared using ultrapure water (prepared by purifying deionized water to attain a sensitivity of 18.2 MΩ cm at 25 °C) and analytical grade reagents. All concentrations are final concentrations unless stated otherwise.

2.1   Cell Lines and Cell Cultures

1. Cancer cell lines: HT-29, HCT-116, and Lovo (*see* **Note 1**).

2. Cell culture medium: RPMI 1640 supplemented with 10% (v/v) fetal bovine serum.

3. Cell culture dishes, 24-well culture plates (Costar, Cambridge, MA, USA).

4. Amicon Ultra-3 centrifugal filters, 3 kDa (UFC500324, Millipore, Billerica, MA, USA).

5. Humidified tissue culture incubator at 37 °C with 5% CO_2.

6. Ethylenediaminetetraacetic acid (EDTA): Dissolve 0.2 g EDTA in 500 mL PBS to obtain a 0.04% (w/v) EDTA and sterilize by autoclaving. Store at room temperature.

7. Trypan blue dye: Dissolve 1 mg Trypan blue dye in 10 mL of methanol to obtain a 1% solution and store at room temperature.

8. A hemocytometer is used to count cells.

9. CytoTox 96 Non-Radioactive Cytotoxicity Assay kit from Promega (Madison, WI, USA) is used to assess the viability of cells.

2.2 Chemical Reagents and Inhibitors

1. Brefeldin A (BFA): An inhibitor of the classical ER-Golgi dependent pathway. Dissolve BFA in dimethyl sulfoxide (DMSO) at 1 mg/mL. Aliquot to avoid multiple freeze/thaw cycles. Store at –20 °C.

2. Glyburide: An inhibitor of ABC transporter. Dissolve glyburide at 100 mM in DMSO. Aliquot and store at –20 °C.

3. 5,5-(N-N-Dimethyl)-amiloride hydrochloride (DMA): An inhibitor of exosome secretion pathway (*see* **Note 2**). Dissolve DMA at 5 mM in methanol. Aliquot and store at –20 °C.

4. Ammonium chloride (NH$_4$Cl): An inhibitor of lysosomal-associated pathway. Dissolve NH$_4$Cl at 5 M in ultrapure water. Store at 4 °C.

5. Akt inhibitor (1L6-Hydroxymethyl-chiro-inositol-2-(R)-2-O-methyl-3-O-octadecyl-*sn*-glycerocarbonate). Dissolve Akt inhibitor at 10 mM in DMSO. Aliquot and store at –20 °C.

6. LY294002 (Cell Signaling, Danvers, MA, USA): An inhibitor of PI3K. Dissolve LY294002 at 10 mM in DMSO. Aliquot and store at –20 °C.

7. U0126 (Cell Signaling): An inhibitor of MAPK. Dissolve U0126 at 10 mM stock in DMSO. Aliquot and store at –20 °C.

8. SP600125 (Cell Signaling): An inhibitor of JNK. Prepare as 50 mM stock in DMSO. Aliquot and store at –20 °C.

2.3 The Sandwich ELISA

1. 96-well Microtiter plates.

2. Anti-SNCG mAb 42#.

3. Horseradish peroxidase (HRP)-linked mAb 1# conjugate.

4. Purified SNCG proteins.

5. 3,3′,5,5′-tetramethylbenzidine (TMB) substrate solution.

6. Stop solution (2 M H$_2$SO$_4$).

7. Microplate Reader.

2.4 Sodium Dodecylsulfate Polyacrylamide Gel Electrophoresis (SDS-PAGE) and Western Blot

1. Thirty percent acrylamide/Bis solution: 29.2% acrylamide monomer and 0.8% Bis (cross-linker) in ultrapure water. Filtered through a 0.2 μm nitrocellulose filter (Millipore) and stored at 4 °C.

2. 1.5 M Tris–HCl, pH 8.8. Filtered through a 0.2 μm nitrocellulose filter and stored at room temperature.

3. 1 M Tris–HCl, pH 6.8. Filtered the solution through a 0.2 μm nitrocellulose filter and stored at room temperature.

4. Ammonium persulfate solution (10% w/v). Filtered the solution through a 0.2 μm nitrocellulose filter and stored at –20 °C.

5. N,N,N′,N′-Tetramethylene diamine (TEMED) (Amresco, Solon, OH, USA), stored at 4 °C.

6. Sodium Dodecyl Sulfate (SDS).

7. Resolving gel (10 mL, 15%): Mix 2.2 mL of ultrapure water, 5 mL of 30% Acrylamide mixture, 2.6 mL of 1.5 M Tris–HCl, pH 8.8, 100 μL of 10% SDS, 100 μL of 10% APS, and 4 μL of TEMED.

8. Stacking gel (5 mL): Mix 2.8 mL of ultrapure water, 0.66 mL of 30% acrylamide mixture, 0.5 mL of 1.0 M Tris–HCl, pH 6.8, 40 μL of 10% SDS, 40 μL of 10% APS, and 4 μL of TEMED.

9. Reducing sample buffer (2×): Mix 2 mL of 0.5 M Tris–HCl (pH 6.8), 4 mL of 10% SDS, 2 mL of glycerol, 2 mL of 2-mercaptoethanol, and 1 mg bromophenol blue. Store at room temperature.

10. SDS-PAGE running buffer (10 x stock solution): Weigh 30.3 g Tris, 144 g glycine, and 10 g SDS in ultrapure water and make it to 1 L. Store at room temperature.

11. Transferring buffer: Containing 25 mM Tris-base, 0.2 M glycine, and 20% (v/v) methanol in ultrapure water.

12. Rainbow molecular weight marker proteins.

13. Nitrocellulose membranes, Whatman 3 mm filter paper, and Electroblotting apparatus.

14. Phosphate buffered saline (PBS) solution with Tween-20 (PBST): 150 mM NaCl, 10 mM disodium hydrogen phosphate, 10 mM sodium dihydrogen phosphate, pH 7.2, 0.1% Tween-20.

15. Protease inhibitor cocktail.

16. Blocking solution: 4% (w/v) skim milk in PBST (see Note 3).

17. Antibody diluent: 4% (w/v) skim milk in PBST (see Notes 3 and 4).

18. Primary antibodies: Mouse anti-Annexin-2 (Novus, Littleton, CO, USA), rabbit anti-HSP-70 (Epitomics, Burlingame, CA, USA) (see Note 5); Rabbit anti-THBS4 (Abcam, Cambridge, MA, USA). Mouse anti-SNCG mAb 1# was generated and characterized as in Reference 7 (see Note 6); mouse Anti-GAPDH, marker for housekeeping protein.

19. Secondary antibodies: Anti-mouse and -rabbit conjugated to horse radish peroxidase (HRP).

20. Western Chemiluminescent HRP Substrate detection system.

2.5 Transwell Migration Assay, Invasion Assay

1. Cancer cell lines and cell culture (see Subheading 2.1).

2. Purified recombinant GST and GST-SNCG proteins (see Note 7).

3. Chamber with polycarbonate membrane inserts (BD Falcon, 8 μm pore size).

4. 24-well plate for the base of the housing chamber.

5. Sterile forceps.

6. Matrigel gel solution (see Note 8).

7. Chilled pipette tips and 1.5-mL tubes in the refrigerator for handling Matrigel.

8. Precooled methanol.

9. Crystal violet staining solution: dissolve 10 g crystal violet in 100 mL methanol, store at room temperature. Dilute the stock solution by 100× in PBS to obtain 0.1% crystal violet solution.

10. Cotton swab.

11. Resin-based mounting medium.

12. Cover slip.

2.6   Adhesion Assay

1. Cancer cell lines and cell culture (*see* Subheading 2.1).

2. Purified recombinant GST and GST-SNCG proteins (*see* Subheading 2.5, **item 2**).

3. 96-well flat-bottom plates.

4. Matrigel (*see* **Note 8**) or Fibronectin/laminin/collagen.

5. 3% BSA in PBS solution.

6. 0.1% crystal violet (*see* Subheading 2.5, **item 9**).

7. Dimethyl Sulfoxide (DMSO).

8. Plate reader. Measure absorbance at 570 nm.

9. Fibronectin, laminin, collagen.

3   Methods

To rule out that SNCG in the culture medium could be due to cell death, the cell viability was measured by trypan blue dye exclusion or measuring the release of lactate dehydrogenase (LDH) using a CytoTox 96 Non-Radioactive Cytotoxicity Assay kit. We carried out a kinetics experiment to analyze the correlation of SNCG levels with the cell viability in the supernatant of HT-29 colon cancer cell line at different time points. SNCG was characterized as an unconventional secretion protein using inhibitors of distinct secretion pathways. Finally, we investigate the effect of extracellular SNCG on tumor cells.

3.1   Cell Culture for Analysis of Intracellular and Extracellular SNCG from Cancer Cells

Here we show that the extracellular and intracellular SNCG levels were positively correlated.

1. Put HT-29, HCT-116, and Lovo cells in exponential phase in 6-well culture plate (5×10^5 cells/well) in triplicate and incubate overnight.

2. Aspirate the cell media, wash the cells three times with 1 mL of serum-free media/well/time, and add 1 mL of serum-free media/well and incubate for 1 h.

3. Collect and mix the supernatant.

4. Filter the culture supernatant through the 0.2 μm nitrocellulose filters and keep on ice.

5. Use the filtered medium directly for ELISA or concentrate by centrifugation in Amicon Ultra-3 centrifugal filters at 4 °C for western blot analysis.

6. Wash the corresponding cells with PBS for two times and homogenize in 100 μL of SDS-PAGE loading buffer.

7. Load appropriate volume of the cell lysates onto the gel for western blot analysis.

An example of the results from this method can be found in Fig. 1a.

3.2 Kinetics of Extracellular SNCG from HT-29 Cells

1. Put HT-29 cells in exponential phase in 10-mm culture plate (4×10^6 cells/plate) and incubate overnight.

2. Aspirate the cell media, and wash the cells three times with 4 mL of serum-free media.

3. Add 7 mL of serum-free media per plate and incubate.

4. Collect 300 μL of the supernatant per plate and replenish with equal volume of warm serum-free media at different time points.

5. Filter and analyze the culture supernatants (see Subheading 3.1, steps 4 and 5).

An example of the results from this method can be found in Fig. 1b.

3.3 Screening of the Secretion Pathway Using Chemical Inhibitors

To understand the mechanism of SNCG secretion, we investigate whether chemical inhibitors targeting distinct pathways could affect SNCG secretion. HT-29 cells were treated with indicated agents for 1 h and SNCG levels in the medium were evaluated by ELISA and western blot. We ensured that the concentrations of agents did not induce cytotoxicity to avoid leakage of SNCG from apoptotic or dying cells (see Note 9).

1. Put HT-29 cells in exponential phase in 12-well plates (1×10^5 cells/well) and allow to grow overnight.

2. Aspirate the cell media, and wash the cells three times with serum-free media.

3. Maintain the cells in the presence of 500 μL of serum-free media with the following agents, respectively, for 1 h. Final concentration: 2 μg/mL of BFA, 100 μM of Glyburide, 5 nM of DMA, 20 mM of NH_4Cl, 10 μM of LY294002, 10 μM of U0126, 50 μM of SP600125, 10 μM of Akt inhibitor.

4. Collect the media and filter through the 0.2 μm nitrocellulose filters.

5. Concentrate the supernatants by the Amicon Ultra-3 centrifugal filters at $13,000 \times g$ at 4 °C.

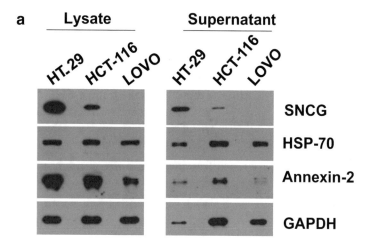

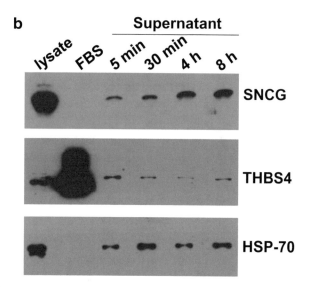

Fig. 1 Extracellular SNCG positively correlates with intercellular SNCG level in tumor cells and SNCG secreted naturally from tumor cells. Intracellular and extracellular SNCG levels were evaluated by western blot (**a**) and SNCG levels in HT-29 cell lysate (10 μg total protein, *lane 1*), fetal calf serum (0.5 μL FCS, *lane 2*), and serum-free cell culture supernatant collected at indicated time points (10 μg total protein/lane, **lane 3–6**) were shown (**b**). In (**b**), HSP-70 and thrombospondin 4 (THBS4) were used as controls as HSP-70 was secreted by unconventional secretion pathway and fetal calf serum contained plenty of THBS4 protein. The results suggest that the release of SNCG into the culture medium was resulted from cell secretion, but not from cell death or fetal calf serum contamination

6. Calculate protein concentration with protein assay kit.

7. Detect and quantify extracellular proteins by ELISA or/and western blot.

8. Cell viability was determined by measuring the release of LDH.

3.4 The Sandwich ELISA

1. Coat High-binding polystyrene microplates overnight at room temperature with 1 μg/mL of anti-SNCG mAb 42#.

2. Wash the plate three times with PBST. Block the plate with 1% BSA for 2 h at room temperature.

3. Twofold serially diluted SNCG standards and blank (diluent), and cell culture supernatants were added, 50 μL/well, respectively, and 50 μL of HRP-labeled anti-SNCG mAb 1# (0.3 μg/mL) was applied to all wells.

4. Incubate the plates for 1 h at room temperature with gentle shaking.

5. Wash plates with PBST for four times.

6. Dispense 100 μL/well of TMB substrate solution to all wells and incubate plates at room temperature for 15 min.

7. Stop the enzyme reaction with 100 μL/well of stop solution.

8. Determine the optical density of each well using a microplate reader at 450 nm.

3.5 Western Blot

1. Once the samples have been sufficiently separated (*see* **Note 10**), transfer to a nitrocellulose membrane using standard western blot techniques.

2. Rinse the membrane with deionized water.

3. Block the membrane in blocking solution for 2 h at room temperature.

4. Incubate the membrane in primary antibody overnight (*see* **Note 4**) at 4 °C.

5. Wash the membrane three times for 6 min in PBST with shaking.

6. Incubate with secondary antibody for 1 h at room temperature.

7. Wash the membrane four times for 6 min in PBST with shaking.

8. Detect proteins using enhanced chemiluminescence (ECL) and visualize by exposure to X-ray film.

3.6 Migration and Invasion Assays

1. Place the cell culture inserts with 8 μm pores into empty wells of the 24-well plate using sterile forceps. For the invasion assay, add delicately 60 μL of fourfold diluted Matrigel solution to the insert (*see* **Note 8**). Then place the plates in a 37 °C incubator for about 4 h to ensure polymerization of the Matrigel.

2. Fill the lower chamber with 800 μL of medium containing 10% FCS.

3. Cells were trypsinized, counted, and cell suspension was prepared (*see* **Note 11**).

4. Transfer HCT-116 or Lovo cells of $5–10 \times 10^4$ cells for migration and $5–10 \times 10^5$ cells for invasion into 1.5 mL Eppendorf tube.

5. Centrifuge at $80 \times g$ for 3 min.

6. Aspirate the supernatant.

7. Add 200 µL of serum-free media containing 1.2 µM of GST or GST-SNCG (*see* **Note 10**) to resuspend the pellet and transfer to each upper chamber (*see* **Note 12**).

8. Transfer the upper chamber (**step 7**) to the 24-well plate (**step 2**) without trapping air bubbles beneath the membranes.

9. Incubate for an appropriate period at 37 °C (*see* **Note 13**).

10. Aspirate the media inside the insert, and put the insert into another 24-well plate with precooled methanol to fix cells at 4 °C for 20 min.

11. Wash the insert three times with PBS.

12. Stain cells with 0.1% crystal violet at room temperature for 30 min.

13. Wash the insert three times with PBS.

14. Remove the nonmigrating/invading cells from the upper surface of the membrane by "scrubbing" using cotton swabs.

15. Rinse the insert three times with PBS.

16. Peel the membrane and mount it on a slide with mounting medium.

17. Place a cover slip on top of the membrane and apply gentle pressure to expel any air bubbles.

18. Collect images of the migrated cells under the microscope at 20× magnification (depending on cell density).

19. Count migrated or invaded cells through the membrane (*see* **Note 14**). Six fields per filter were counted.

20. Data was presented as migrated/invaded cells per field.

The representative data is shown in Fig. 2.

3.7 Adhesion Assay

1. Wells of 96-well plates were coated in triplicate with 50 µL of the diluted matrigel solution (1:40 diluted in PBS) or fibronectin, laminin, or collagen overnight at 4 °C.

2. Add 200 µL of 3% BSA-PBS solution to each well, and incubate for 2 h at RT.

3. During incubation, prepare the cell suspension (2×10^4/well) in serum-free media containing 1.2 µM of GST or GST-SNCG, using the same methods as for the cell migration assay (*see* Subheading 3.5, **steps 6–8**).

4. Discard the 3% BSA-PBS from each well.

5. Add 50 µL of cell suspension to each well in triplicate and incubate at 37 °C for 1–2 h.

6. Wash the cells three times with PBS (*see* **Note 15**).

7. Cell confluence (%) was measured with the CloneSelect Imager system (Molecular Devices).

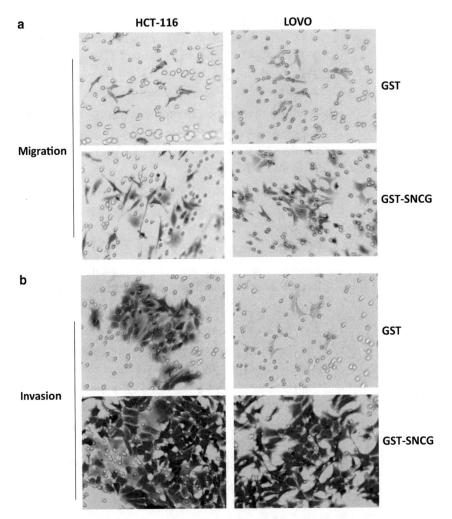

Fig. 2 Representative results of the effect of extracellular SNCG on tumor cells migration and invasion. Transwell cell migration assay (**a**) and Transwell cell invasion assay (**b**) were performed with exogenously added GST (**a** and **b**, *upper panel*) or GST-SNCG (**a** and **b**, *lower panel*) into HCT-116 (*left panel*) or Lovo cell (*right panel*) culture medium, respectively. Twenty-four (for migration assay) or 48 (for invasion assay) hours after incubation, cells were fixed in precooled methanol and stained with crystal violet. Magnification of objective lens, 20×

The representative results can be found in Fig 5b in ref. 9 (*see* **Note 16**).

4   Notes

1. HT-29 cells were used as a model system, as the intracellular and extracellular SNCG levels were readily detected by SNCG sandwich ELISA. Because of the morphologic feature, HT-29 cells were not suitable for migration and invasion assays. HCT-116 and Lovo cells were used in migration and invasion assays.

2. Some researchers used 5,5-(N-N-Dimethyl)-amiloride hydrochloride (DMA) as an exosome inhibitor [13, 14]. As an inhibitor of Na+/H+ exchange, DMA is known to affect endosomal maturation and trafficking, and also known to have pleiotropic effects on the cells. We used siRNA to knock down Rab27 because this protein is specifically involved in exosomes secretion [9].

3. For some phosphorylated proteins, 5% BSA in 0.05 M TBS containing 0.1% Tween-20 should be used for blocking the membranes and diluting primary antibodies, as skim milk in PBS could interfere with detection.

4. To improve sensitivity in western blot, antibodies should always be incubated overnight.

5. Leaderless proteins Annexin II and HSP-70 are ubiquitously expressed in cancer cells and secreted by unconventional secretion pathway [15, 16]; therefore they were used as controls for characterization of SNCG's unconventional secretion.

6. The anti-SNCG mAb 1# has high specificity and sensitivity, interacts with endogenous SNCG from cancer cells, and recognizes the epitope located in the C-terminal of SNCG protein. In the sandwich ELISA, the coated anti-SNCG mAb 42# recognizes the N-terminal epitope of SNCG. The characterization and application of anti-SNCG monoclonal antibodies can be found in ref. 7.

7. Compared with SNCG-immunodepleted medium, conditioned medium from HT-29 cells decreased adhesion of tumor cells to matrix, correlating with an increased migration and invasion into matrigel. Similarly, the addition of recombinant SNCG to fresh medium increased migration and invasion (*see* Fig. 5 in ref. 9).

8. Matrigel is liquid at 4 °C, but rapidly polymerizes into gel when warmed at room temperature. Make aliquots of the desired volume using chilled pipette tips and tubes, and store the aliquots at –20 °C. Avoid repeated freeze/thaw cycles. To invade, cells must across a two-dimensional surface and a three-dimensional extracellular matrix. It is generally accepted that this process requires cells not only to migrate but also to proteolyze or alter their local extracellular matrix (ECM) microenvironment.

9. To ensure the concentrations of the chemical agents did not induce cytotoxicity to cells, cell viability was determined by trypan blue dye exclusion or measuring the release of LDH.

10. Monitor the migration of the molecular weight markers to ensure that the size corresponding to the protein of interest is at least halfway into the gel, to increase resolution in this region.

11. The number of cells depends on the cell types; in the case of HCT-116 or Lovo cells, use $5-10 \times 10^4$ cells/well for migration and $1-10 \times 10^5$ cells/well for invasion.

12. The pipette tip should not touch the membrane of the chamber in order to avoid breaking the membrane. Between each pipetting step, resuspend the cells in order to avoid settling and to keep the cell suspension homogenous.

13. The incubation time depends on cell type and cell numbers that should be determined experimentally. In most cases, 24 h for migration assay or 48 h for invasion assay.

14. Migrated or invaded cells can be counted directly under a microscope. However, it is easier to count cells after taking digital images using a camera.

15. Since there are several washing steps involved in this assay and the cells are not fixed, washes can result in removal of cells from the microplates. Gentle washing is therefore essential.

16. Results can vary depending upon the cell types and the conditions (such as matrix coated, incubation time, cell-seeding density, and the medium).

Acknowledgement

This work was supported by the National Natural Science Foundation of China (No. 81272410).

References

1. Shi YE, Chen Y, Dackour R et al (2010) Synuclein gamma stimulates membrane-initiated estrogen signaling by chaperoning estrogen receptor (ER)-alpha36, a variant of ER-alpha. Am J Pathol 177:964–973

2. Liang W, Miao S, Zhang B et al (2015) Synuclein γ protects Akt and mTOR and renders tumor resistance to Hsp90 disruption. Oncogene 34:2398–2405

3. Li M, Yin Y, Hua H et al (2010) The reciprocal regulation of gamma-synuclein and IGF-I receptor expression creates a circuit that modulates IGF-I signaling. J Biol Chem 285:30480–30488

4. Pan ZZ, Bruening W, Giasson BI et al (2002) Gamma-synuclein promotes cancer cell survival and inhibits stress- and chemotherapy drug-induced apoptosis by modulating MAPK pathways. J Biol Chem 277:35050–35060

5. Zhang H, Kouadio A, Cartledge D et al (2011) Role of gamma-synuclein in microtubule regulation. Exp Cell Res 317:1330–1339

6. Li Z, Sclabas GM, Peng B et al (2004) Overexpression of synuclein-gamma in pancreatic adenocarcinoma. Cancer 101:58–65

7. Liu C, Guo J, Qu L et al (2008) Applications of novel monoclonal antibodies specific for synuclein-gamma in evaluating its levels in sera and cancer tissues from colorectal cancer patients. Cancer Lett 269:148–158

8. Liu C, Ma H, Qu L et al (2012) Elevated serum synuclein-gamma in patients with gastrointestinal and esophageal carcinomas. Hepatogastroenterology 59:2222–2227

9. Liu C, Qu L, Lian S et al (2014) Unconventional secretion of synuclein-γ promotes tumor cell invasion. FEBS J 281:5159–5171

10. Nickel W, Rabouille C (2009) Mechanisms of regulated unconventional protein secretion. Nat Rev Mol Cell Biol 10:148–155

11. Villarreal L, Méndez O, Salvans C et al (2013) Unconventional secretion is a major contributor of cancer cell line secretomes. Mol Cell Proteomics 12:1046–1060

12. Zhang M, Schekman R (2013) Cell biology. Unconventional secretion, unconventional solutions. Science 340:559–561

13. Merendino AM, Bucchieri F, Campanella C et al (2010) Hsp60 is actively secreted by human tumor cells. PLoS One 5, e9247

14. Savina A, Furla'n M, Vidal M et al (2003) Exosome release is regulated by a calcium-dependent mechanism in K562 cells. J Biol Chem 278:20083–20090

15. Lancaster GI, Febbraio MA (2005) Exosome-dependent trafficking of HSP-70: a novel secretory pathway for cellular stress proteins. J Biol Chem 280:23349–23355

16. Danielsen EM, van Deurs B, Hansen GH (2003) "Nonclassical" secretion of annexin A2 to the lumenal side of the enterocyte brush border membrane. Biochemistry 42:14670–14676

Part IV

Secretome Isolation from Plant and Animal Samples to Identify Leaderless Secretory Proteins (LSP)

Chapter 16

Characterization of the Tumor Secretome from Tumor Interstitial Fluid (TIF)

Pavel Gromov and Irina Gromova

Abstract

Tumor interstitial fluid (TIF) surrounds and perfuses bodily tumorigenic tissues and cells, and can accumulate by-products of tumors and stromal cells in a relatively local space. Interstitial fluid offers several important advantages for biomarker and therapeutic target discovery, especially for cancer. Here, we describe the most currently accepted method for recovering TIF from tumor and nonmalignant tissues that was initially performed using breast cancer tissue. TIF recovery is achieved by passive extraction of fluid from small, surgically dissected tissue specimens in phosphate-buffered saline. We also present protocols for hematoxylin and eosin (H&E) staining of snap-frozen and formalin-fixed, paraffin-embedded (FFPE) tumor sections and for proteomic profiling of TIF and matched tumor samples by high-resolution two-dimensional gel electrophoresis (2D-PAGE) to enable comparative analysis of tumor secretome and paired tumor tissue.

Key words Tumor microenvironment, Tumor interstitial fluid, Tissue secretome, Tissue section, Proteomic profiling, 2D-PAGE

1 Introduction

Tumor interstitial fluid (TIF) accumulates aberrantly externalized components, mainly proteins, that are released by tumor and stromal cells through various mechanisms, including classical secretion, non-classical secretion, exosome-mediated secretion, and membrane protein shedding [1, 2]. Accordingly, the term "tumor secretome" currently refers to all proteins/peptides secreted, shed, or leaking from the tumor tissue under a given condition in a relatively local tissue space. These externalized proteins/peptides act as mediators of tumor-host communication within the local tumor microenvironment and play a decisive role in processes promoting tumorigenesis, such as tumor growth, invasion, and angiogenesis [3, 4].

Interstitial fluid forms the interface between circulating bodily fluid and intracellular fluid, and provides an environment that enables the exchange of ions, proteins, and nutrients between various components, such as cells, within the interstitial space.

Andrea Pompa and Francesca De Marchis (eds.), *Unconventional Protein Secretion: Methods and Protocols*, Methods in Molecular Biology, vol. 1459, DOI 10.1007/978-1-4939-3804-9_16, © Springer Science+Business Media New York 2016

This fluid is not static, but is continually being refreshed and recollected by lymphatic channels. The TIF bathes the tumor and its stromal components, and therefore provides several important advantages towards the discovery of potential biomarkers for cancer. Given that tumor secretome proteins accumulate at higher concentrations close to their source, proximal lesion sampling and omic profiling of tumor-associated fluid are promising approaches for identifying novel diagnostic and therapeutic targets. However, analysis of the protein composition of TIF is rather problematic, and may be attributed to technical difficulties such as (a) low concentrations of proteins upon fluid recovery due to dilution steps and (b) masking and contamination of protein samples by structural or other normally non-secreted proteins released into the extracellular space as a result of cell damage, lysis, or death. These circumstances highlight the importance of establishing appropriate procedures for tissue preparation and fluid collection to avoid, or at least to control for, possible contamination and provide reliable and representative native interstitial fluid for downstream proteomic analysis by the technique of choice.

Several methods have been developed for the recovery of interstitial fluids in vivo from various tissues (reviewed by [2, 5]). These include insertion of glass capillaries [6], implanted chambers [7], implanted wicks [8, 9], microdialysis [10], and capillary ultrafiltration through a semipermeable membrane [11, 12]. The in vivo TIF harvesting techniques are based on invasive procedures and, therefore, may affect the structure of surrounding tissue resulting in bleeding, inflammation, and wound repair—the processes which may influence the proteome composition of the recovered TIF. In addition to these in vivo protocols, two ex vivo TIF sampling methods have been utilized for the preparation of interstitial fluids from excised tissue: (a) tissue centrifugation at low G-forces [13–17] and (b) passive elution from excised tissue [18].

In this chapter, we describe the method for recovering TIF from tumor and nonmalignant tissues based on passive extraction of fluid from surgically dissected tissue specimens that was first conducted by Celis and colleagues using human mammary tumors [18]. This procedure is now regarded as the most appropriate method to collect interstitial fluids of suitable quality and amount for the analysis of the tumor secretome and identification of candidate cancer biomarkers [2]. The same procedure was successfully used to harvest interstitial fluid from normal breast epithelial tissue [18] and from mammary adipose tissue [19] for the purpose of defining the molecular phenotypes underlying epithelial normalcy and determining the role of adipocytes in the breast tumor microenvironment. Recently, this method was also successfully applied to the isolation of TIF from other types of human malignancies including bladder transitional carcinoma [20], renal cell carcinoma [21], liver cancer [22], and ovarian cancer [23].

In addition, we include two complementary protocols for the preparation, sectioning, and H&E staining of snap-frozen and FFPE tumor tissue blocks and for proteomic profiling of TIF and matched tumor samples by 2D-PAGE, which we established for the characterization and reciprocal comparison between tumor-associated fluids and paired tumor specimen [18]. Together, these procedures will enable the researcher not only to efficiently isolate high yields of TIF, but also to evaluate and prepare the TIF sample for downstream proteomic analysis.

2 Materials

Prepare all solutions using ultrapure water (prepared by purifying deionized water to attain a sensitivity of 18 MΩ cm at 20 °C) and analytical grade reagents. Prepare all solutions and reagents and store at 4 °C (unless indicated otherwise). Diligently follow all waste disposal regulations when discarding waste materials. No chemical preservatives, such as sodium azide, were used.

2.1 Materials for Tissue Dissection and TIF Isolation

1. Dissection instruments: Scissors, forceps, disposable scalpels.
2. Phosphate-buffered saline, 1× (PBS).
3. 10 ml Plastic conical tubes.
4. 8 cm Petri dish.
5. Centrifuge.
6. Pasteur pipettes.
7. Humidified CO_2 incubator set to 37 °C.
8. 6 cm Petri dish.

2.2 Materials for H&E Staining of Snap-Frozen Breast Tissue Sections and FFPE Breast Tissue Sections

1. Cryostat set to −20 °C.
2. Cryosectioning chuck.
3. OCT compound (Tissue-Tek).
4. Microscope slides.
5. Slide rack.
6. 70 % Ethanol.
7. 96 % Ethanol.
8. 99 % Ethanol.
9. 100 % Ethanol.
10. Mayer's hematoxylin solution.
11. Eosin solution.
12. Glass cover slips.
13. Mounting medium (Pertex).

14. 10% Formalin.

15. Tissue cassette.

16. Xylene.

17. Paraffin.

18. Microtome.

19. Water bath set to 40 °C.

20. Microscope slides.

21. Air oven set to 60 °C.

2.3 Materials for TIF Sample Preparation for 2D-PAGE

1. Lyophilizer.

2. O'Farrell's lysis solution: 9.8 M Urea, 2% (w/v) NP40, 2% ampholines pH 7–9, 100 mM dithiothreitol (DTT). To make 50 ml, add 29.42 g, 10 ml of a 10% stock solution of NP-40, 1 ml of ampholytes pH 7–9, and 0.771 g of DTT to glass cylinder. Dissolve carefully at room temperature and complete to 50 ml with deionized water. The solution should not be heated. Aliquot in 2 ml portions and keep at –20 °C.

2.4 Materials for 2D-PAGE: First-Dimension Separation by Isoelectrofocusing

1. Glass tubes (140 mm in length, 2 mm inside diameter).

2. Parafilm.

3. Tube gel casting stand.

4. Vacuum pump.

5. 10 ml syringe and long needle.

6. Whatman 3MM paper.

7. Electrophoresis tube gel module.

8. Power supply.

9. Ultrapure urea.

10. 30% Acrylamide stock solution (acrylamide/N,N'-methylene-bis-acrylamide = 29/1).

11. 10% NP-40.

12. Carrier ampholytes, pH range 5–7.

13. Carrier ampholytes, pH range 3.5–10.

14. 10% Ammonium persulfate.

15. TEMED.

16. Overlay solution: 8 M Urea, 1% ampholytes pH 7–9, 5% (w/v) NP-40, and 100 mM DTT. To make 25 ml, add 12.012 g of urea, 0.25 ml of ampholytes pH 7–9, 12.5 ml of a 10% stock solution of NP-40, and 0.386 g of DTT. After dissolving, complete to 25 ml with distilled water. The solution should not be heated. Aliquot in 2 ml portions and keep at –20 °C.

17. 20 mM NaOH.

18. 20 mM H_3PO_4.

19. 35 mm Tissue culture dishes.

20. Equilibration solution: 0.06 M Tris–HCl, pH 6.8, 2% SDS, 100 mM DTT, and 10% glycerol. To make 250 ml, add 15 ml of a 1 M stock solution of Tris–HCl, pH 6.8, 50 ml of a 10% stock solution of SDS, 3.857 g of DTT, and 28.73 ml of glycerol (87% concentration). After dissolving, complete to 250 ml with distilled water. Store at room temperature.

2.5 Materials for 2D-PAGE: Second-Dimension Gel Electrophoresis (SDS-PAGE)

1. Glass plates and matching notched plates, 16.5 cm×20 cm; notch 2 cm deep and 13 cm wide.

2. Polystyrene spacers (1 cm×20 cm, 1 mm thick).

3. Polystyrene spacers (3 cm×13 cm, 1 mm thick).

4. Fold-back clamps.

5. 100 ml filter flask.

6. Magnetic stir bar.

7. Vacuum pump.

8. Whatman 3MM paper.

9. Scalpel.

10. Electrophoresis chamber for slab gel.

11. Power supply.

12. 40% Acrylamide solution.

13. 2% N,N'-methylene-bis-acrylamide (BIS) solution.

14. 10% SDS.

15. 1.5 M Tris–HCl, pH 8.8.

16. 1.0 M Tris–HCl, pH 6.8.

17. 10% Ammonium persulfate.

18. N,N,N',N-tetramethylethylenediamine (TEMED).

19. Agarose solution: 0.06 M Tris–HCl, pH 6.8, 2% SDS, 100 mM DTT, 10% glycerol, 1% agarose, and 0.002% bromophenol blue. To make 250 ml, add 15 ml of a 1 M stock solution of Tris–HCl, pH 6.8, 50 ml of a 10% stock solution of SDS, 3.857 g of DTT, 28.73 ml of glycerol (87% concentration), 2.5 g of agarose, and 2.5 ml of a 0.2% stock solution of bromophenol blue. Add distilled water and heat in a microwave oven. Complete to 250 ml with distilled water and aliquot in 20 ml portions while the solution is still warm. Keep at 4 °C.

20. Electrode buffer: To make 1 l of a 5× solution, add 30.3 g of Tris base, 144 g of glycine, and add 50 ml of 10% SDS solution. Complete to 1 l with distilled water.

21. 10 ml syringe with bent needle.

22. Fixative solution: Ethanol:acidic acid:water (40:5:55).

3   Methods

This section describes the procedure for recovering TIF from breast tumor tissue together with several accompanying protocols regarding processing and basic H&E staining techniques for frozen and FFPE tissue sections (*see* **Note 1**). We also provide a protocol for high-resolution two-dimensional gel electrophoresis (2D-PAGE), which we established in our laboratory to characterize the proteomes of cells, tissues, and fluids and is suitable for downstream analysis of the TIF proteome (*see* **Note 2**).

3.1 Breast Tissue Dissection

1. Surgically remove breast tissue by mastectomy (*see* **Note 3**).

2. Under the supervision of an experienced pathologist, carefully examine all or part of the dissected breast tissue, and select the tumor lesion(s) (usually surrounded by healthy tissues and fat) to be used for TIF recovery and other accompanying experiments (Fig. 1A).

3. Dissect a homogenous piece of tumor (about 0.5 g) from the tissue mass and place it into a 10 ml conical tube containing 3 ml of PBS. Place the tube in an ice bath and transport immediately to the laboratory.

4. Upon arrival in the laboratory, transfer the tumor sample from the tube to an 8 cm Petri dish. Thoroughly cleanse the tumor specimen of nonmalignant tissue (mainly stroma and fat) using a disposable scalpel and fine forceps (*see* **Note 4**) (Fig. 1B).

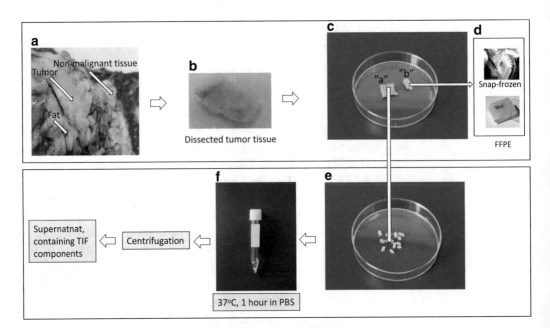

Fig. 1 Workflow for recovering of TIF from fresh breast tumor tissue

5. With the aid of the disposable scalpel, divide the tissue into two halves and label them specimen "a" (intended for TIF preparation) and specimen "b" (intended for snap-frozen and/or FFPE tissue blocks) (Fig. 1C, D).

3.2 Isolation of TIF from Breast Tumor Tissue

1. Place a clean, fresh piece of tumor sample weighing about 0.25 g (i.e., specimen "a," *see* Subheading 3.1, **step 5**) (*see* **Note 5**) into a Petri dish of an appropriate size (Fig. 1C, E).

2. Neatly cut the tissue sample into small pieces (about 1 mm³ or less) using a disposable scalpel and thin forceps. Try to make an incision using only one motion of the scalpel. Avoid multiple repetitive incisions in one area (Fig. 1E) (*see* **Note 6**).

3. Place the small fragments of tissue into a 10 ml conical tube containing 1 ml of PBS (Fig. 1F) (*see* **Note 7**).

4. Shake the tube manually and then centrifuge the sample at $100 \times g$ for 1 min. With the aid of an elongated Pasteur pipette, carefully aspirate and discard the supernatant containing tissue and cellular debris generated by sample manipulation in **step 2** (*see* Subheading 3.2, **step 2**; *see* **Note 8**).

5. Add 1 ml of PBS to the pellet, shake the tube manually, and incubate at 37 °C in a humidified CO_2 incubator for 1 h (*see* **Note 9**).

6. Centrifuge the tube at $100 \times g$ for 2 min (*see* **Note 10**).

7. Transfer the supernatant to another 10 ml conical tube using an elongated Pasteur pipette.

8. Centrifuge the tube at $2000 \times g$ for 20 min at 4 °C.

9. The final supernatant (TIF sample) will contain a protein concentration typically ranging from 1 to 4 mg/ml. Isolate a 0.1 ml aliquot for 2D gel electrophoresis. Divide the remainder of the TIF sample into smaller aliquots suitable for further analysis to minimize future repetitive freeze/thaw cycles. Store the samples at –80 °C.

10. Lyophilize the 0.1 ml aliquot of the TIF sample (*see* Subheading 3.5, **step 1**) and then store at –20 °C before proceeding with 2D-PAGE (*see* **Note 2**).

3.3 H&E Staining of Snap-Frozen Breast Tissue Sections

1. Place the frozen tissue sample (i.e., "b," *see* Fig. 1C) on a suitable cryosectioning chuck. Coat the tissue with fresh OCT compound and orientate the sectioning surface of the tissue block. Place the chuck with the fixed tissue block into the cryostat chamber set to –20 °C.

2. Cut sections to 8 µm thickness using the cryostat set to –20 °C. For optimal cutting of soft tissues, adjust the temperature in the cryostat chamber (*see* **Note 11**).

3. Transfer the sections onto glass slides suitable for H&E staining by touching the tissue section gently to the slide surface (*see* **Note 12**).

4. Allow the slides to air-dry for about 20 min at room temperature.

5. Place the slides to be stained in a slide rack and fix in 96 % ethanol for 1 min.

6. Rinse the slides in deionized water.

7. Immerse the slide rack into Mayer's hematoxylin solution for 3 min.

8. Rinse the slides in running tap water for 2 min.

9. Immerse the slide rack into eosin solution for 45 s.

10. Rinse the slides in deionized water.

11. Immerse the slide rack briefly and sequentially in 70 % ethanol, 96 % ethanol, and 99 % ethanol. Allow the slides to air-dry for about 20 min at room temperature.

12. Place a small drop of mounting media onto the section. Carefully lower a glass cover slip on top of the mounting media. Ensure that the cover slip comes as close to the section as possible.

13. Examine and photograph the stained slides under a light microscope.

14. Cover the remaining frozen tissue block with additional OCT to avoid freeze-drying, and store the sample at –80 °C for future re-processing as necessary.

3.4 H&E Staining of FFPE Breast Tissue Sections

1. Place a piece of freshly dissected tissue (i.e., "b," *see* Fig. 1C, <3 mm thick) into an appropriate small plastic container filled with 5 ml of 10 % formalin, and incubate for 24–36 h at room temperature (*see* **Note 13**).

2. Following fixation, transfer the sample to a tissue cassette and dehydrate the tissue by sequential washes in 70, 96, and 99 % ethanol for 1 h each. Complete the dehydration by incubating the sample in two changes of 100 % ethanol for 1 h each (*see* **Note 14**).

3. Immerse the dehydrated tissue into the clearing agent, xylene, for 1 h. Repeat with fresh xylene.

4. Immerse the tissue into three changes of melted paraffin at 60 °C for 1 h each.

5. After the final paraffin immersion, transfer the tissue cassette to an embedding instrument. Embed the tissue in a paraffin block. The paraffin tissue block can be stored at room temperature for years.

6. Cut the paraffin-embedded tissue block to sections of 5–8 μm thickness on a microtome. Transfer the sections to a bath containing deionized water at 40 °C. Ensure that the sections are fully unfolded on the surface of the water.

7. With the aid of a small brush, transfer the sections onto glass slides.

8. Place the glass slides into an air oven set to 60 °C overnight (*see* **Note 15**).

9. Allow the slides to air-dry at room temperature for 1 h. Store the slides at 4 °C until needed for H&E staining (*see* **Note 16**).

10. Place the slides to be stained in a slide rack and wash in xylene twice for 5 min each.

11. Immerse the slide rack sequentially into the following ethanol washes: 99 % ethanol twice for 3 min each, 96 % ethanol twice for 3 min each, and 70 % ethanol once for 1 min. Finally, wash the slides in deionized water for 1 min.

12. Immerse the slide rack into Mayer's hematoxylin solution for 10 min.

13. Rinse the slides in running tap water for 5 min.

14. Immerse the slide rack into eosin solution for 1 min.

15. Rinse the slides briefly in running tap water and then in deionized water.

16. Dip the slide rack about 20 times each in 70, 96, and then 99 % ethanol.

17. Place a small drop of mounting media (Pertex) onto the section. Carefully lower a glass cover slip on top of the mounting media. Ensure that the cover slip comes as close to the section as possible.

18. Examine and photograph the stained slides under a light microscope (*see* **Note 17**; Fig. 2).

3.5 TIF Sample Preparation for 2D-PAGE

1. Solubilize the freeze-dried TIF (*see* Subheading 3.2, **step 10**) in 0.1 ml of O'Farrell's lysis solution. Pipette up and down (avoid foaming). Store at −20 °C until ready for further use.

2. To prepare the sample from a tissue, cut 20–30 serial sections from a frozen block (*see* Subheading 3.3, **step 2**), collect them into Eppendorf tubes, and solubilize in 0.1 ml of O'Farrell's lysis solution. Store at −20 °C until ready for further use.

3.6 2D-PAGE: First-Dimension Separation by IEF

1. Mark the glass tubes selected for isoelectrofocusing (IEF) with a line 12.5 cm from the bottom. Enclose the bottom end of the tube with Parafilm and stand the tube in a gel casting rack.

2. For 12 first-dimension IEF gels, mix 5.49 g urea; 1.3 ml of 30 % acrylamide solution (*see* Subheading 2.6, **item 10**); 2 ml

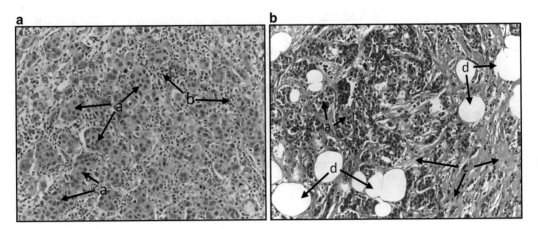

Fig. 2 H&E staining of FFPE sections of two breast tumors (A and B) with different intratumor morphological features. Tumor A is characterized by a high presence of malignant epithelial cells (*A*) and infiltrating lymphocytes (*B*). Tumor B is characterized by a lower level of malignant cells (*A*) and by pronounced presence of stroma (*C*) and adipocyte cells (*D*)

10% NP-40; 2 ml H$_2$O; 0.40 ml carrier ampholytes, pH 5–7; and 0.133 ml carrier ampholytes, pH 3.5–10. Swirl the solution gently until the urea is dissolved. The solution should not be heated. Add 15 μl of 10% ammonium persulfate and 10 μl of TEMED, mix gently, and degas using a vacuum pump.

3. Draw up the gel solution with a 10 ml syringe attached to a long needle. Insert the tip of the needle to the bottom of the tube and slowly fill to the mark. Avoid air bubbles.

4. Overlay the gel solution with 10 μl of glass-distilled water and allow to polymerize for 45 min.

5. Take the tubes from the gel casting rack, remove the Parafilm using a scalpel, remove excess liquid from the upper half of the tube by shaking, and dry using a thin strip of Whatman 3MM paper. Insert the tubes into the upper chamber of the electrophoresis tube gel module.

6. Add 10 μl of O'Farrell's lysis solution and then 10 μl of overlay solution into the tubes. Fill the tubes and the upper chamber (−) with 20 mM NaOH. Pre-run the gels at room temperature for 15 min at 200 V, 30 min at 300 V, and then 60 min at 400 V. Remove the tubes and wash the top of the gels with deionized water. Remove excess water from the surface of the gels with a thin strip of Whatman 3MM paper.

7. Apply 40 μl of the TIF sample (*see* Subheadings 3.2, **step 10**, and Subheading 3.5, **step 1**) or tumor sample (*see* Subheading 3.5, **step 2**) and then 10 μl of overlay solution into the tubes. Fill the tubes and the upper chamber (−) with 20 mM NaOH and fill the lower chamber with 20 mM H$_3$PO$_4$. Run for 19 h at 400 V at room temperature.

8. Once the run is complete, remove the gels from the tubes with a syringe filled with glass-distilled water and place them into 35 mm tissue culture dishes containing 3.5 ml of equilibration solution. Store at –20 °C until ready for further use.

3.7 2D-PAGE: Second-Dimension Gel Electrophoresis (SDS-PAGE)

1. Assemble the glass plate (16.5 cm × 20 cm) with the notched plate and spacers greased with Vaseline. Hold the assembled plates together using fold-back clamps. Mark a line 2.5 cm from the top of the notched plate.

2. Prepare the 15 % resolving solution for one gel by mixing the following solutions in a 100 ml filter flask containing a magnetic stirrer: 8.6 ml 40 % acrylamide solution, 0.9 ml BIS solution, 0.2 ml 10 % SDS, 5.7 ml 1.5 M Tris–HCl, pH 8.8, and 7.4 ml deionized water.

3. Add 113 μl of 10 % ammonium persulfate and 5 μl of TEMED to the solution just before degassing using a vacuum pump. Pour the solution between the assembled plates to the marked line and overlay carefully with deionized water. Allow the gel to polymerize for approximately 1 h.

4. Prepare the 5 % stacking solution for one gel by mixing the following solutions: 0.6 ml 40 % acrylamide solution, 0.6 ml BIS solution, 0.05 ml 10 % SDS, 0.6 ml 1.0 M Tris–HCl, pH 6.8, and 3.1 ml deionized water. Remove excess liquid and dry the top of the gel with a strip of Whatman 3MM paper.

5. Add 40 μl of 10 % ammonium persulfate and 2 μl of TEMED to the stacking gel solution just before degassing using a vacuum pump. Carefully pour the solution on top of the resolving gel and insert a polystyrene spacer (3 cm × 13 cm) a few millimeters between the assembled plates. Allow the gel to polymerize for approximately 1 h.

6. Following polymerization, remove the top and bottom spacers and clean the space between the two glass plates with a strip of Whatman 3MM paper.

7. Prepare the agarose solution in a microwave oven and immediately cover the top of the stacking gel with a small amount of agarose to fill in the space left by the spacer.

8. Remove the 35 mm tissue culture dishes containing the first-dimension gel in 3.5 ml of equilibration solution from the freezer and allow to thaw at room temperature.

9. Remove the first-dimension gel from the equilibration buffer and carefully place it across the top of the second-dimension gel with the aid of plastic tweezers and a piece of Parafilm. Cover the first-dimension gel with 2–3 ml of melted agarose.

10. Clamp the gel plates to the electrophoresis chamber for slab gel filled with electrode buffer. There should be enough

electrode buffer in the upper chamber to completely cover the agarose. Remove air bubbles at the bottom of the gel using a 10 ml syringe attached to a bent needle.

11. Connect the electrodes (upper, –; lower, +) to the power supply. Run the gels at 10 mA for 4 h and at 3 mA overnight at room temperature (until the tracking dye has reached 1 cm from the bottom). At the end of the run, turn off the power supply, disassemble the plates using a spatula, and discard the stacking gel by cutting it away from the resolving gel with a scalpel.

12. Place the gel in the fixative solution (*see* **Note 18**).

13. Silver stain the gel using a protocol that is compatible with mass spectrometry as described elsewhere (*see* ref. 24) (Fig. 3; *see* **Note 18**).

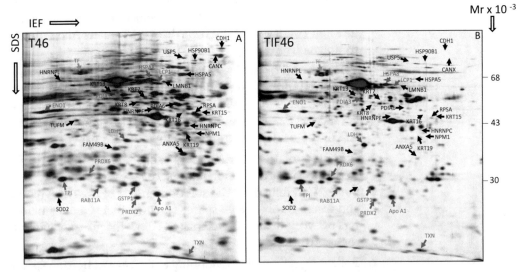

Fig. 3 2D gel protein profiling of interstitial fluid recovered from breast carcinoma, and tumor tissue obtained from the same patient (patient 46) after mastectomy. (**a**) IEF 2D gel of tumor 46. (**b**) IEF 2D gel of TIF 46. *Black arrows* show the proteins which are not detected or are present at trace level in TIF as compared in tumor: KRT7—cytokeratin 7; KRT8—cytokeratin 8; KRT13—cytokeratin 13; KRT13—cytokeratin 15; KRT16—cytokeratin 16; KRT17—cytokeratin 17; KRT19—cytokeratin 19; LMNB1—lamin B1; HNRNPL—heterogeneous nuclear ribonucleoprotein L; HNRNPC—heterogeneous nuclear ribonucleoproteins C1/C2; HNRNPF—heterogeneous nuclear ribonucleoprotein F; NPM1—nucleophosmin; CDH1—cadherin-1, CANX—calnexin; HSP90B1—endoplasmin; USP5—ubiquitin carboxyl-terminal hydrolase 5; PDIA6—protein disulfide-isomerase A6; HSPA5—78 kDa glucose-regulated protein; ANXA5—annexin A5; RPSA—40S ribosomal protein SA; FAM49B—protein FAM49B; TF—serotransferrin; SOD2—superoxide dismutase [Mn], mitochondrial; and TUFM—elongation factor Tu, mitochondrial. *Blue arrows* indicate the proteins which are highly presented or at comparable level in both tumor and TIF: LCP1—plastin; PDIA3—protein disulfide-isomerase A3; HSPA8—heat-shock cognate 71 kDa protein; RAB11A—Ras-related protein Rab-11A; PRDX2—peroxiredoxin-2; PRDX6—peroxiredoxin-6; LDH—L-lactate dehydrogenase B chain; ENO1—alpha-enolase; APOA1—apolipoprotein A-I; TXN—thioredoxin; TPI1—triose-phosphate isomerase; and GSTP1—glutathione S-transferase P. Proteins were identified by MALDI TOF MS–MS (reproduced with permission from Elsevier)

4   Notes

1. Human tumors exhibit highly heterogeneous features including various tumor subtypes, histologies, proportions of malignant and stromal cells, gene expression patterns, genotypes, and metastatic and proliferative potentials. The histological heterogeneity of tumor samples introduces significant challenges to the interpretation of tumor secretome data. Therefore, even the initial information regarding the morphological features of the tissue specimen from which TIF is recovered is absolutely essential for the proper interpretation of tumor secretome data. Histopathological examination of the tumor sample by microscopic imaging of H&E-stained frozen or FFPE sections should be performed by an expert pathologist. Therefore, the tumor specimen is divided into two separate samples, one of which should be used to prepare the snap-frozen and/or FFPE blocks for H&E staining and/or immunohistochemistry (*see* Subheading 3.1, **step 5**). H&E staining of frozen sections offers a fast and simple method to obtain initial histopathological information and rapidly evaluate the quality of the tumor sample (proportion of malignant epithelia, presence of stroma and adipose tissue, infiltrated lymphocytes, and other morphological parameters) before processing the tissue for TIF isolation.

2. 2D-PAGE was selected as the method of choice to separate complex protein mixtures present in TIF because it remains widely regarded as the "gold standard" for comprehensive proteomic analysis (*see* refs. 25, 26). Moreover, 2D-PAGE enables the comparison of the levels of secreted and structural protein components in a single run, and therefore can provide a quality check of the TIF content for the extent of proteolysis (*see* ref. 18). Western blotting analysis can also be performed to evaluate the relative levels of structural and secreted components of the TIF proteome as an alternative method. In Western blotting, proteins are probed for using specific antibodies and expression is estimated in relation to consistent loading controls, so-called housekeeping genes. The samples loaded on the gel are normalized according to protein concentration. However, this method of TIF analysis has limitations because it can only provide information about selected individual proteins one experiment at a time rather than an unbiased view of the proteome within a single experiment.

3. This step is always carried out by the surgeon in the operating room of the hospital. Fresh tissue specimens can be obtained from different regions of the breast: tumor or nonmalignant/adipose tissue located distally from or adjacent to the tumor. There are two main types of surgical biopsies: (a) incisional

biopsy removes just enough of the suspected area to make a diagnosis and (b) excisional biopsy removes the entire tumor (mastectomy) or abnormal area, with or without removing an edge of normal breast tissue (this will depend on the reason underlying excisional biopsy). Fresh tissue samples are often very fragile and can easily become damaged during the removal and handling process. Therefore, it is important to handle tissues carefully as soon as possible after resection in the operating room and immediately transport them to the laboratory for further processing.

4. Try to remove any obvious nonmalignant tissue from the tumor sample as thoroughly as possible so that no visible portion of the nonmalignant area extends past the margin of the tumor sample. This step is critical because the presence of nearby non-tumor tissue will affect TIF composition and therefore will greatly complicate the interpretation of the results. The goal is to isolate a morphologically or anatomically homogeneous sample.

5. TIF can be recovered from a relatively small tissue specimen, such as from a core needle biopsy of breast tissue obtained in a hospital by a radiologist or surgeon for diagnostic purposes. In this case, reduce the volume of PBS in which the pieces of tissue are contained accordingly (*see* Subheading 3.2, **step 3**). The TIF yield will be proportional to the size or amount of starting material.

6. Multiple repetitive incisions in one area may cause significant damage to the tissue and lead to the accumulation of excessive debris.

7. Other types of physiological saline buffer can be used instead of PBS for TIF harvesting through passive incubation. Teng and colleagues (*see* ref. 27) investigated five different buffer systems (PBS, Dulbecco's modified Eagle's medium [DMEM], and three organ transplantation preservative solutions: Celsior solution S [CS], histidine-tryptophan-ketoglutarate [HTK], and a proprietary solution of the University of Wisconsin [UW]) and concluded that there were no significant differences in the protein composition of TIF harvested in these five solutions. However, additional components endogenously present in DMEM, UW, CS, and HTK as compared to PBS may affect some downstream analysis and should be eliminated by additional sample preparation steps as described elsewhere. In general, PBS is the most suitable and convenient solution for recovering TIF proteins for proteomic analysis.

8. Adipose tissue is a major component of the breast (especially in older women), and breast tumor cells are often detected interdigitating with and spreading through the peripheral fat tissue [28]. Consequently, it is often impossible to thoroughly clean

a tumor sample of fat, and pieces of tumor sample adhering to a significant proportion of adipose tissue will be found floating in suspension after centrifugation due to the low buoyant density of fat. In such cases, it is recommended that the liquid phase is collected/aspirated by placing an elongated Pasteur pipette between the pelleted and surfaced layers of tissue.

9. In our experience, the time of incubation can be increased up to 12 h without leading to substantially increased proteolysis (*see* ref. 18), but may result in a higher TIF yield.

10. This additional centrifugation step is recommended to remove debris more thoroughly (*see* Subheading 3.2, **step 8**).

11. Soft tumor tissue containing fat may require a lower temperature for proper cutting.

12. Tissue sections should be transferred to slides within 1 min after cutting to keep them from drying out.

13. Formalin is a hazardous chemical capable of causing skin, eye, and respiratory tract irritation. All procedures with formalin should be handled under a working fume hood.

14. **Steps 2–5** ("tissue processing") can be performed manually, but many current laboratories are equipped with automated tissue processing instruments to enable more rapid processing and provide higher quality outcomes.

15. This step allows even penetration of paraffin into the tissue section.

16. FFPE tissue processing is also the most common method for immunohistochemistry (IHC). Because each antigen may require its own optimized set of IHC steps, it is beyond the scope of this chapter to include a comprehensive protocol encompassing all IHC circumstances. Consequently, the researcher is responsible for optimizing the conditions for each antigen of interest. Many protocols for IHC staining as well as useful tips and troubleshooting guides can be found at the IHC World website (http://www.ihcworld.com/).

17. Upon H&E staining, DNA and RNA will be stained blue or violet/blue, whereas the cytoplasm, collagen, keratin, and red blood cells will be stained pink. Hematoxylin is positively charged and binds to anions, such as nucleic acids. Eosin is negatively charged and binds to all positively charged groups in proteins (i.e., amino acids), and is therefore used as a background or contrast stain.

18. Gels can also be used directly for Western blotting procedures. In this case, the gel does not undergo fixation and proteins are transferred electrophoretically to adsorbent membranes such as nitrocellulose or polyvinylidene difluoride, as described elsewhere (*see* ref. 29).

Acknowledgements

Preparation of this manuscript was supported by the Danish Cancer Society through the Danish Cancer Society Research Center budget and by grants from the John and Birthe Meyer Foundation and EurocanPlatform.

References

1. Baronzio G, Parmar G, Baronzio M, Kiselevsky M (2014) Tumor interstitial fluid: proteomic determination as a possible source of biomarkers. Cancer Genomics Proteomics 11:225–237

2. Gromov P, Gromova I, Olsen CJ, Timmermans-Wielenga V, Talman ML, Serizawa RR, Moreira JM (2013) Tumor interstitial fluid: a treasure trove of cancer biomarkers. Biochim Biophys Acta 1834:2259–2270

3. Schaaij-Visser TB, de Wit M, Lam SW, Jiménez CR (2013) The cancer secretome, current status and opportunities in the lung, breast and colorectal cancer context. Biochim Biophys Acta 1834:2242–2258

4. Paltridge JL, Belle L, Khew-Goodall Y (2013) The secretome in cancer progression. Biochim Biophys Acta 1834(11):2233–2241

5. Wiig H, Tenstad O, Iversen PO, Kalluri R, Bjerkvig R (2010) Interstitial fluid: the overlooked component of the tumor microenvironment? Fibrogenesis Tissue Repair 3:12

6. Sylven B, Bois I (1960) Protein content and enzymatic assays of interstitial fluid from some normal tissues and transplanted mouse tumors. Cancer Res 20:831–836

7. Gullino PM, Clark SH, Grantham FH (1964) The interstitial fluid of solid tumors. Cancer Res 24:780–794

8. Aukland K, Fadness HO (1973) Protein concentration in interstitial fluid collected from rat skin by a wick method. Acta Physiol Scand 88:350–358

9. Stohrer M, Boucher Y, Stangassinger M, Jain RK (2000) Oncotic pressure in solid tumors is elevated. Cancer Res 60:4251–4255

10. Dabrosin C (2005) Microdialysis – an in vivo technique for studies of growth factors in breast cancer. Front Biosci 10:1329–1335

11. Huang CM, Ananthaswamy HN, Barnes S, Ma Y, Kawai M, Elmets CA (2006) Mass spectrometric proteomics profiles of in vivo tumor secretomes: capillary ultrafiltration sampling of regressive tumor masses. Proteomics 6: 6107–6116

12. Stone MD, Odland RM, McGowan T, Onsongo G, Tang C, Rhodus NL, Jagtap P, Bandhakavi S, Griffin TJ (2010) Novel in situ collection of tumor interstitial fluid from a head and neck squamous carcinoma reveals a unique proteome with diagnostic potential. Clin Proteomics 6:75–82

13. Wiig H, Aukland K, Tenstad O (2003) Isolation of interstitial fluid from rat mammary tumors by a centrifugation method. Am J Physiol Heart Circ Physiol 284:H416–H424

14. Wiig H, Berggreen E, Borge BA, Iversen PO (2004) Demonstration of altered signaling responses in bone marrow extracellular fluid during increased hematopoiesis in rats using a centrifugation method. Am J Physiol Heart Circ Physiol 286:H2028–H2034

15. Semaeva E, Tenstad O, Bletsa A, Gjerde EA, Wiig H (2008) Isolation of rat trachea interstitial fluid and demonstration of local cytokine production in lipopolysaccharide-induced systemic inflammation. J Appl Physiol 104:809–820

16. Brekke HK, Oveland E, Kolmannskog O, Hammersborg SM, Wiig H, Husby P, Tenstad O, Nedrebø T (2010) Isolation of interstitial fluid in skin during volume expansion: evaluation of a method in pigs. Am J Physiol Heart Circ Physiol 299:H1546–H1553

17. Haslene-Hox H, Oveland E, Berg KC, Kolmannskog O, Woie K, Salvesen HB, Tenstad O, Wiig H (2011) A new method for isolation of interstitial fluid from human solid tumors applied to proteomic analysis of ovarian carcinoma tissue. PLoS One 6, e19217

18. Celis JE, Gromov P, Cabezón T, Moreira JM, Ambartsumian N, Sandelin K, Rank F, Gromova I (2004) Proteomic characterization of the interstitial fluid perfusing the breast tumor microenvironment: a novel resource for biomarker and therapeutic target discovery. Mol Cell Proteomics 3:327–344

19. Celis JE, Moreira JM, Cabezón T, Gromov P, Friis E, Rank F, Gromova I (2005) Identification of extracellular and intracellular signaling components of the mammary adipose tissue and its interstitial fluid in high risk breast cancer patients: toward dissecting the molecular circuitry of epithelial-adipocyte stromal cell interactions. Mol Cell Proteomics 4:492–522

20. Gromov P, Moreira JM, Gromova I, Celis JE (2008) Proteomic strategies in bladder cancer: from tissue to fluid and back. Proteomics Clin Appl 2:974–988

21. Teng PN, Hood BL, Sun M, Flint MS, Bateman NW, Dhir R, Bhargava R, Richard SD, Edwards RP, Conrads TP (2010) Differential proteomic analysis of renal cell carcinoma tissue interstitial fluid. J Proteome Res 10:1333–1342

22. Sun W, Ma J, Wu S (2010) Characterization of the liver tissue interstitial fluid (TIF) proteome indicates potential for application in liver disease biomarker discovery. J Proteome Res 9:1020–1031

23. Wang TH, Chao A, Tsai CL, Chang CL, Chen SH, Lee YS, Chen JK, Lin YJ, Chang PY, Wang CJ, Chao AS, Chang SD, Chang TC, Lai CH, Wang HS (2010) Stress-induced phosphoprotein 1 as a secreted biomarker for human ovarian cancer promotes cancer cell proliferation. Mol Cell Proteomics 9:1873–1884

24. Gromova II, Celis JE (2006) Protein detection in gels by silver staining: a procedure compatible with mass spectrometry. In: Celis JE, Carter N, Hunter T, Simons K, Small JV, Shotton D (eds) A cell biology. Laboratory handbook, 3rd edn. Elsevier, New York, NY, pp 219–223

25. Rogowska-Wrzesinska A, Le Bihan MC, Thaysen-Andersen M, Roepstorff P (2013) 2D gels still have a niche in proteomics. J Proteomics 88:4–13

26. Oliveira BM, Coorssen JR, Martins-de-Souza D (2014) 2DE: the phoenix of proteomics. J Proteomics 104:140–150

27. Teng PN, Rungruang BJ, Hood BL (2010) Assessment of buffer systems for harvesting proteins from tissue interstitial fluid for proteomic analysis. J Proteome Res 9:4161–4169

28. Gromov P, Gromova I, Bunkenborg J, Cabezon T, Moreira JM, Timmermans-Wielenga V, Roepstorff P, Rank F, Celis JE (2010 Feb) Up-regulated proteins in the fluid bathing the tumour cell microenvironment as potential serological markers for early detection of cancer of the breast. Mol Oncol 4(1):65–89. doi:10.1016/j.molonc.2009.11.003, Epub 2009 Nov 23

29. Celis JE, Moreira JMA, Gromov P (2006) Determination of antibody specificity by western blotting. In: Celis JE, Carter N, Hunter T, Simons K, Small JV, Shotton D (eds) A cell biology. Laboratory handbook, 3rd edn. Elsevier, New York, NY, pp 527–532

Chapter 17

Vacuum Infiltration-Centrifugation Method for Apoplastic Protein Extraction in Grapevine

Bertrand Delaunois, Fabienne Baillieul, Christophe Clément, Philippe Jeandet, and Sylvain Cordelier

Abstract

The apoplastic fluid moving in the extracellular space external to the plasma membrane provides a means of delivering molecules and facilitates intercellular communications. However, the apoplastic fluid extraction from *in planta* systems remains challenging and this is particularly true for grapevine (*Vitis vinifera* L.), a worldwide-cultivated fruit plant. Here, we describe an optimized vacuum-infiltration-centrifugation method to extract soluble proteins from apoplastic fluid of grapevine leaves. This optimized method allows recovering of the grapevine apoplastic soluble proteins suitable for mono- and bi-dimensional gel electrophoresis for further proteomic analysis in order to elucidate their physiological functions.

Key words Apoplastic fluid extraction, Vacuum-infiltration-centrifugation method, VIC method, Secretome, Apoplastic protein, Grapevine leaves, *Vitis vinifera*

1 Introduction

Important physiological processes such as development, intercellular communications, or defense mechanisms take place in the apoplast [1]. The apoplast is defined as the extracellular matrix or plant cell wall and the intercellular spaces where the apoplastic fluid (AF) circulates [2]. The AF plays a key role in intercellular and intracellular communications and is composed of many substances including nutrients, polysaccharides, secondary metabolites, and secreted proteins. The apoplastic proteins (or secretome) are involved in different physiological and biological processes related to growth regulation, cell wall maintenance, and responses to biotic and abiotic stresses [3, 4, 5].

Despite their biological significance, investigations on apoplastic proteins are hampered due to their low abundance compared to intracellular protein concentrations. Moreover, the AF extraction from in planta systems is far from easy and remains challenging. This is particularly true for a recalcitrant plant like grapevine

Andrea Pompa and Francesca De Marchis (eds.), *Unconventional Protein Secretion: Methods and Protocols*, Methods in Molecular Biology, vol. 1459, DOI 10.1007/978-1-4939-3804-9_17, © Springer Science+Business Media New York 2016

regarding its polyphenols and polysaccharide contents. The most commonly used technique for AF extraction is the vacuum-infiltration-centrifugation (VIC) method involving two critical steps: vacuum-infiltration with appropriate extraction buffer and centrifugation [2, 6, 7]. Here, we present an optimized VIC method to extract soluble proteins from AF of grapevine leaves suitable for gel electrophoresis analysis. The VIC method to recover the AF of grapevine leaf has been optimized from results obtained in barley [8] and tomato [9] resulting in a protein sample enriched in apoplastic proteins [10].

During protein extraction with TCA/acetone buffer, the precipitation of sugars present in the AF sample leads to a viscous extract not suitable for gel electrophoresis. To clean up proteins from leaf AF before further processing, a procedure with TCA/acetone combined to a phenol extraction step was performed [11, 12]. Denatured proteins and other hydrophobic proteins are soluble in phenol or agglomerate at the phenol-water interface, unlike the small molecules and nucleic acids, which are soluble in the aqueous phase. The combined use of TCA/acetone precipitation and phenol-based extraction improves sample quality, resulting in a better protein separation and in reduced background and streaking on electrophoresis gels.

Since high centrifugation speed may cause damage to cell walls and membranes leading to contamination of the AF by inner cell components, Western blot analysis has been performed on AF sample to check the contamination level by other compartment proteins and to confirm the sample enrichment in AF proteins. Specific antibodies of RuBisCo, the most abundant protein found in plant leaves, exclusively localized in cytoplasmic and chloroplastic compartments [13] have been used as marker of the AF quality to check the contamination level of intracellular proteins. The RuBisCo large subunit (cytoplasmic marker) could not be detected by Western blot analysis in the AF sample, which confirms that this VIC method seems to be suitable to recover a highly enriched AF protein sample from grapevine leaf [10].

The VIC method has been optimized to allow protein recovery from grapevine AF suitable for 2D-PAGE analyses and to establish a well-defined proteomic map of grapevine leaf apoplastic soluble proteins. This proteomic map has been released in the public World-2DPAGE database to be used as interactive reference map and provides a comprehensive overview of the most abundant proteins present in the grapevine apoplast [10].

2   Materials

All solutions were prepared with deionized water unless indicated otherwise and analytical grade reagents. All solvents and hazardous products must be handled in a fume hood with adequate protective equipment. All material and solution waste must be discarded following waste disposal regulations.

2.1 Plant

Vegetative cuttings of *V. vinifera* L. (cv. Chardonnay clone 7535) were obtained from healthy pruned canes of grapevine (Vranken Pommery, Reims vineyard, France) [14]. Cuttings were planted in 0.5 L pots containing loam, transferred in growth chamber at 20/26 °C (night/day) with a 16-h light period (500 μmol/m²/s¹), and relative humidity of 70%, and irrigated with tap water. Ten-week-old leaves were used to perform apoplastic fluid extraction.

2.2 Apoplastic Fluid Extraction

1. Ice-cold deionized water: Store several 1 L glass bottles at 4 °C.
2. Ice-cold infiltration buffer: 150 mM Tris–HCl, pH 8.5, containing 6 mM 3-[(3-cholamidopropyl) dimethylammonio]-1-propanesulfonate (CHAPS) (*see* **Note 1**). Add about 900 mL deionized water to a 1 L glass beaker (*see* **Note 2**). Weigh 18.2 g Tris and 3.69 g of CHAPS. Mix and adjust pH with HCl or NaOH (*see* **Note 3**). Transfer to a graduated cylinder. Fill up to 1 L with water. Store at 4 °C.
3. Soft towel paper.
4. Large glass beaker. The size should fit in the vacuum chamber.
5. Vacuum chamber linked to a vacuum pump to apply vacuum at 80 kPa (600 mmHg).
6. Nylon mesh filter with 11 μm² pore size.
7. 20 mL plastic syringe without plunger.
8. 50 mL Falcon tube or equivalent.
9. Refrigerated high-speed centrifuge with swing rotor for 50 mL Falcon tubes and fixed-angle rotor for 2 mL tubes.

2.3 Apoplastic Protein Extraction

1. Ice-cold 10% (w/v) TCA/acetone solution: Add about 800 mL acetone to a 1 L glass beaker (*see* **Note 2**). Weigh 100 g trichloroacetic acid and mix. Transfer to a 1 L graduated cylinder. Fill up to 1 L with acetone. Store at 4 °C.
2. Ice-cold 80% (v/v) acetone: Add 800 mL acetone to a 1 L graduated cylinder and fill up to 1 L with deionized water. Mix well and store at 4 °C.
3. Dense SDS solution: 100 mM Tris–HCl, pH 8, containing 30% (w/v) sucrose, 2% (w/v) SDS, 5% (v/v) 2-mercaptoethanol. Add about 700 mL deionized water to a 1 L glass beaker (*see* **Note 2**). Weigh 12.1 g Tris, 300 g of sucrose, 20 g of sodium dodecyl sulfate, and 50 mL of 2-mercaptoethanol. Mix and adjust pH with HCl or NaOH (*see* **Note 3**). Transfer to a graduated cylinder. Fill up to 1 L with water.
4. Phenol solution, equilibrated with 10 mM Tris–HCl, pH 8.0 and 1 mM EDTA.

5. Ice-cold MeOH/100 mM ammonium acetate: Add about 800 mL methanol (MeOH) to a 1 L glass beaker (*see* **Note 2**). Weigh 7.7 g ammonium acetate and mix. Transfer to a 1 L graduated cylinder. Fill up to 1 L with methanol. Store at 4 °C.

2.4 Western Blot Analysis Components

1. Solubilization buffer: 50 mM Tris–HCl, pH 6.8, containing 10 % (w/v) sucrose, 1 % (w/v) SDS, 5 % (v/v) 2-mercaptoethanol. Add about 700 mL deionized water to a 1 L glass beaker (*see* **Note 2**). Weigh 6.05 g Tris, 100 g of sucrose, 10 g of SDS, and 50 mL of 2-mercaptoethanol. Mix and adjust pH with HCl or NaOH (*see* **Note 3**). Transfer to a graduated cylinder. Fill up to 1 L with water.

2. Bradford reagent.

3. Spectrophotometer.

4. Protein standard: Bovine serum albumin solution (BSA, 2 mg protein/mL).

5. Laemmli loading buffer 3×: 150 mM Tris–HCl, pH 6.8, containing 30 % glycerol, 6 % (w/v) SDS, 0.5 % (w/v) bromophenol blue, and 0.15 % (w/v) dithiothreitol (DTT). Add about 10 mL deionized water to a 50 mL glass beaker (*see* **Note 2**). Weigh 473 mg Tris, 6 mL of glycerol, 1.2 g of SDS, and 0.1 g bromophenol blue. Mix and adjust pH with HCl or NaOH (*see* **Note 3**). Transfer to a graduated cylinder. Fill up to 20 mL with water. Add and mix 30 mg of DTT just before use.

6. 4–20 % precast SDS-polyacrylamide gel (Amersham ECL Gel 4–20 % 10 wells; GE Healthcare, Sweden).

7. Gel electrophoresis system with power supplier.

8. TGS buffer: 25 mM Tris–HCl, pH 8.3, containing 192 mM glycine and 0.1 % (w/v) SDS. Add about 700 mL deionized water to a 1 L glass beaker (*see* **Note 2**). Weigh 3.025 g Tris, 14.4 g of glycine, and 1 g of SDS. Mix and adjust pH with HCl or NaOH (*see* **Note 3**). Transfer to a graduated cylinder. Fill up to 1 L with water.

9. Polyvinylidene difluoride (PVDF) membrane.

10. Gel-blot transfer system.

11. TBST buffer: 20 mM Tris–HCl, 500 mM NaCl at pH 7.5, 0.05 % (v/v) Tween-20 containing 3 % (w/v) of powdered milk. Add about 700 mL deionized water to a 1 L glass beaker (*see* **Note 2**). Weigh 2.42 g Tris, 29.2 g of NaCl, 0.5 mL Tween-20, and 30 g of powdered milk. Mix and adjust pH with HCl or NaOH (*see* **Note 3**). Transfer to a graduated cylinder. Fill up to 1 L with water.

12. Anti-RbcL primary antibody: Rabbit polyclonal antibody targeting RbcL/Rubisco large subunit, form I and form II (Ref# AS03 037; Agrisera Antibodies, Sweden).

13. Goat anti-rabbit IgG horseradish peroxidase-conjugated secondary antibody.

14. Chemiluminescent substrate (ECL).

15. Chemiluminescence detection analysis system.

---

3 Methods (*see* Note 4)

3.1 Apoplastic Fluid Extraction

Apoplastic fluids were collected by an adapted VIC method [8, 9].

1. Harvest enough fully expanded leaves from the middle of the green grapevine shoots to obtain approximately 75 g fresh weight material after removing the middle vein using a scalpel (*see* **Note 5**).

2. Cut leaves into pieces of 1 cm² and transfer them into a large glass beaker filed with ice-cold water (*see* **Note 6**).

3. Rinse grapevine leaf pieces twice in ice-cold deionized water (*see* **Note 7**).

4. Quickly dry leaf pieces between two sheets of soft paper towel and transfer them into a large glass beaker filed with ice-cold infiltration buffer (*see* **Note 8**).

5. Transfer the glass beaker into the vacuum chamber linked to the vacuum pump and apply vacuum at 80 kPa (600 mmHg) during 10 min (*see* **Note 9**).

6. Remove the infiltrated leaf pieces from the infiltration solution and wash them twice in ice-cold deionized water (*see* **Note 7**).

7. Quickly dry leaf pieces between two sheets of soft paper towel. Transfer them onto a square of nylon mesh filter and arrange them in bundles (*see* **Note 10**).

8. Place each bundle into 20 mL syringe (*see* **Note 11**) and then in a 50 mL Falcon tube.

9. Centrifuge the Falcon tubes at 4 °C at $7500 \times g$ for 30 min.

10. After centrifugation, remove the syringe and withdraw the collected apoplastic fluid into a new fresh 2 mL tube (*see* **Note 12**).

3.2 Apoplastic Protein Extraction

1. Add on apoplastic fluid sample 2 volume of ice-cold 10 % (w/v) TCA/acetone solution and store overnight at −20 °C for protein precipitation (*see* **Note 13**).

2. Centrifuge the tubes at 4 °C at $10,000 \times g$ for 5 min.

3. Discard the supernatant and wash the pellet with ice-cold 10 % (w/v) TCA/acetone solution.

4. Centrifuge the tubes at 4 °C at $10,000 \times g$ for 5 min.

5. Repeat **steps 3** and **4** twice with ice-cold 80 % (v/v) acetone (*see* **Note 14**).

6. Resuspend wet pellets in 0.8 mL of dense SDS solution at room temperature (RT).

7. Add 0.8 mL phenol at RT and vortex samples for 1 min (*see* **Note 15**).

8. Centrifuge the tubes at RT at $10,000 \times g$ for 5 min.

9. Collect the phenolic fraction and transfer it in a 15 mL Falcon tube.

10. Add 5 volumes of ice-cold MeOH/0.1 M ammonium acetate solution and store overnight at –20 °C for protein precipitation.

11. Centrifuge the tubes at 4 °C at $10,000 \times g$ for 5 min.

12. Discard the supernatant and wash the pellet with ice-cold MeOH/0.1 M ammonium acetate solution.

13. Centrifuge the tubes at 4 °C at $10,000 \times g$ for 5 min.

14. Repeat **steps 12** and **13** once with ice-cold MeOH/0.1 M ammonium acetate solution.

15. Repeat **steps 12** and **13** twice with ice-cold 80% (v/v) acetone.

16. Let the protein pellet air-dry and store at –80 °C for further analyses.

3.3 Assessment of Cytoplasmic Contamination by Western Blot Analysis

1. Resuspend protein pellet in solubilization buffer (*see* **Note 16**). Quantify the protein concentration using the BSA standard and the Bradford assay commonly used in the laboratory [15].

2. Calculate the volume required to get 20 μg of protein and add half of the volume of Laemmli loading buffer. Load the 20 μg of protein extracts with Laemmli loading buffer on a 4–20% precast SDS-polyacrylamide gel (*see* **Note 17**).

3. Run electrophoresis at 180 V during 1 h with TGS buffer until the dry front reaches the bottom of the gel.

4. After electrophoresis, open the precast gel with a flat spatula, remove the staking part of the gel, and wash the resolving part with deionized water (*see* **Note 18**).

5. Place the electrophoresis gel with the PVDF membrane in the transfer system for the time indicated by the manufacturer's recommendations.

6. Following protein transfer, place the PVDF membrane in a 9×12 cm plastic box using flat forceps. Incubate the membrane in TBST + 3% milk buffer during 1 h at RT to saturate the membrane.

7. Remove the saturation buffer and incubate the PVDF membrane in TBST + 3% milk buffer with the anti-RbcL primary antibody (Agrisera Antibodies, Sweden) at dilution 1:5000 for large subunit RuBisCo detection during 1 h at RT.

8. Wash the membrane three times for 10 min in TBST + 3 % milk buffer to remove the aspecifically binded primary antibody.

9. Incubate the PVDF membrane in TBST + 3 % milk solution with the goat anti-rabbit IgG horseradish peroxidase-conjugated secondary antibody at 1:10,000 dilution during 1 h at RT.

10. Wash the membrane three times for 10 min in TBST + 3 % milk buffer to remove the aspecifically binded secondary antibody.

11. Incubate the membrane with the chemiluminescent substrate for 5 min and transfer it in a plastic membrane protector. Remove all air bubbles between the plastic sheet and the surface of the membrane.

12. Reveal signal by fluorography in optimizing the analysis conditions according to the manufacturer's recommendations.

4 Notes

1. CHAPS is used in the process of protein extraction and purification as a non-denaturing solvent to facilitate the solubilization of some apoplastic proteins which can be sparingly soluble or insoluble in aqueous solution due to their native hydrophobicity.

2. Having large volume of water in the glass beaker helps to dissolve powder reagent relatively easily, allowing the magnetic stir bar to go to work immediately.

3. Concentrated HCl (6 N) or NaOH (5 N) can be used at first to narrow the gap from the starting pH to the required pH. From then on it is better to use HCl (1 N) or NaOH (1 N) with lower ionic strengths to avoid a sudden drop in pH below or above the required pH.

4. All procedures were carried out at room temperature. Ice-cold solutions were used quickly to avoid warming. They were then quickly stored back in cold conditions. Ice-cold solutions can be kept for a couple of weeks. However, the extraction yields were higher when solutions were prepared the day before and stored in cold conditions.

5. To reach 75 g of fresh weight, around 20 randomized 10-week-old grapevine cutting leaves were needed.

6. Grapevine leaf infiltration is difficult since leaves are waxy and not very pulpy. The middle vein was removed and leaves were cut into pieces to increase the accessibility of the infiltration buffer.

7. It was useful to use a strainer to facilitate leaf pieces rinsing. The rinsing of grapevine leaf pieces prevents contamination deriving from other cell compartments.

8. The composition of the infiltration buffer was designed to facilitate the protein solubilization and to preserve as much as possible the plasmalemma integrity [6, 7].

9. Infiltration was carried out 3–4 times within 10 min until leaves became glassy in appearance.

10. Use 20 cm square of nylon mesh filter. To arrange them in bundle, place leaf pieces in a half of the nylon mesh. Fold the second half onto leaf pieces and roll the nylon mesh to form the bundle. Be careful with the bundle diameter so that it can fit into the syringe. Usually the 75 g of leaf pieces were divided into six parts to form six bundles.

11. Remove the plunger of the syringe and insert first the folded side of the bundle into the syringe.

12. Each sample was individually collected in 2 mL tube but it is possible to pool all samples in 15 mL Falcon tubes. If the protein extraction is not performed the following day, store the samples at −20 °C until use. At this step, sample can be used to analyze diverse apoplastic components like ions, sugars, or secondary metabolites. To perform the protein extraction, let slowly thaw tubes on ice the day before and carry out the first step of protein extraction overnight.

13. All the following steps are performed in cold conditions unless otherwise specified. Usually, samples were kept on ice bucket.

14. This purification step with phenol is essential to remove polyphenols and sugars that could interfere with protein separation during electrophoresis.

15. After the last centrifugation step, the residual acetone was carefully removed but be careful to keep the pellet wet.

16. The protein pellet is quite difficult to resuspend in adequate buffer for protein quantification. Do not hesitate to strongly shake the tube with vortex. The sample sonication during 10 min also facilitates the protein resuspension. An aliquot was used for protein quantification according to the manufacturer's instructions with BSA as a standard. In our hand, the soluble protein extraction yield reached approximately 5–6 μg per gram of leaf fresh weight for AF sample. However, if the protein quantification is not required for further analysis, it is easier to resuspend the protein pellet in buffer containing detergent such as loading buffer (Laemmli buffer which contains DTT and SDS) or urea/thiourea buffer (7 M urea/2 M thiourea).

17. Load a protein molecular weight marker to estimate the approximate molecular weight of proteins and a total protein extract as a positive control. The marker (usually colored) will also help identifying the loading order of samples.

18. Do not forget to wear vinyl gloves.

Acknowledgment

The authors are grateful to Vranken Pommery (Reims, France) for access to their vineyard. This study was supported by the "Region Champagne Ardennes" and by "Comité Champagne" (Epernay, France) through the project VINEAL 2 for the Ph.D. grant of B.D. and by AROCU (Association Recherche Oenologique Champagne et Université).

References

1. Sakurai N (1998) Dynamic function and regulation of apoplast in the plant body. J Plant Res 111:133–148
2. Agrawal GK, Jwa N-S, Lebrun M-H, Job D, Rakwal R (2010) Plant secretome: unlocking secrets of the secreted proteins. Proteomics 10:799–827
3. Alexandersson E, Ali A, Resjö S, Andreasson E (2013) Plant secretome proteomics. Front Plant Sci 4:9. doi:10.3389/fpls.2013.00009
4. Doehlemann G, Hemetsberger C (2013) Apoplastic immunity and its suppression by filamentous plant pathogens. New Phytol 198:1001–1016
5. Delaunois B, Jeandet P, Clément C, Baillieul F, Dorey S, Cordelier S (2014) Uncovering plant-pathogen crosstalk through apoplastic proteomic studies. Front Plant Sci 5:249. doi:10.3389/fpls.2014.00249
6. Lohaus G, Pennewiss K, Sattelmacher B, Hussmann M, Muehling KH (2001) Is the infiltration-centrifugation technique appropriate for the isolation of apoplastic fluid? A critical evaluation with different plant species. Physiol Plant 111:457–465
7. Witzel K, Shahzad M, Matros A, Mock HP, Muhling KH (2011) Comparative evaluation of extraction methods for apoplastic proteins from maize leaves. Plant Methods 7:48. doi:10.1186/1746-4811-7-48
8. Rohringer R, Ebrahim-Nesbat F, Wolf G (1983) Proteins in intercellular washing fluids from leaves of barley (*Hordeum vulgare* L.). J Exp Bot 34:1589–1605
9. Shabab M, Shindo T, Gu C, Kaschani F, Pansuriya T, Chintha R, Harzen A, Colby T, Kamoun S, van der Hoorn RAL (2008) Fungal effector protein AVR2 targets diversifying defense-related Cys proteases of tomato. Plant Cell 20:1169–1183
10. Delaunois B, Colby T, Belloy N, Conreux A, Harzen A, Baillieul F, Clément C, Schmidt J, Jeandet P, Cordelier S (2013) Large-scale proteomic analysis of the grapevine leaf apoplastic fluid reveals mainly stress-related proteins and cell wall modifying enzymes. BMC Plant Biol 13:24. doi:10.1186/1471-2229-13-24
11. Röhrig H, Schmidt J, Colby T, Bräutigam A, Hufnagel P, Bartels D (2006) Desiccation of the resurrection plant *Craterostigma plantagineum* induces dynamic changes in protein phosphorylation. Plant Cell Environ 29:1606–1617
12. Wang W, Scali M, Vignani R, Spadafora A, Sensi E, Mazzuca S, Cresti M (2003) Protein extraction for two-dimensional electrophoresis from olive leaf, a plant tissue containing high levels of interfering compounds. Electrophoresis 24:2369–2375
13. Johnson X (2011) Manipulating RuBisCO accumulation in the green alga, Chlamydomonas reinhardtii. Plant Mol Biol 776:397–405
14. Lebon G, Duchene E, Brun O, Clement C (2005) Phenology of flowering and starch accumulation in grape (*Vitis vinifera* L.) cuttings and vines. Ann Bot 95:943–948
15. Bradford MM (1976) A rapid and sensitive method for the quantitation of microgram quantities of protein utilizing the principle of protein-dye binding. Anal Biochem 72:248–254

Isolation of Exosome-Like Vesicles from Plants by Ultracentrifugation on Sucrose/Deuterium Oxide (D₂O) Density Cushions

Christopher Stanly, Immacolata Fiume, Giovambattista Capasso, and Gabriella Pocsfalvi

Abstract

Exosomes are nanovesicles of endocytic origin that are about 30–100 nm in diameter, surrounded by a lipid bilayer membrane, and contain proteins, nucleic acids, and other molecules. Mammalian cells- and biological fluids-derived exosomes have become the subject for a wide range of investigations in biological and biomedical sciences. More recently, a new interest is on the verge of rising: the presence of nanovesicles in plants. Lipoprotein vesicles from apoplastic fluid and exosome-like vesicles (ELVs) from fruit juice have been isolated and shown that they could be loaded with drugs and uptaken by recipient cells. In order to explore and analyze the contents and functions of ELVs, they must be isolated and purified with intense care. Isolation of ELVs can be a tedious process and often characterized by the co-purification of undesired contaminants. Here we describe a method which isolates ELVs based on their buoyant density. The method utilizes differential centrifugation in step 1 and 1 and 2 M sucrose/deuterium oxide double-cushion ultracentrifugation in step 2, to purify two diverse ELV subpopulations. In this method fruit juice is used as an example of starting material, although this protocol can be used for the isolation of vesicles from apoplastic fluid too. The quality and the quantity of ELV preparations have been found appropriate for downstream biological and structural studies, like proteomics, transcriptomics, and lipidomics.

Key words Extracellular vesicles, Exosome-like vesicles, Apoplastic vesicles, Plant tissue, Juice, Purification, Isolation, Ultracentrifugation, Differential centrifugation, Sucrose gradient centrifugation

1 Introduction

Mammalian extracellular vesicles (EVs), including exosomes, microvesicles, and apoptotic bodies, are spherical, phospholipid bilayer protected particles composed of bioactive molecules, including RNAs, DNAs, proteins, and lipids. Despite that they were discovered decades ago, it has only recently been revealed that EVs mediate a novel form of cell-to-cell communication, and thus represent an important and rapidly growing research field in

Andrea Pompa and Francesca De Marchis (eds.), *Unconventional Protein Secretion: Methods and Protocols*, Methods in Molecular Biology, vol. 1459, DOI 10.1007/978-1-4939-3804-9_18, © Springer Science+Business Media New York 2016

biology and translational medicine [1, 2]. Secretion and uptake of EVs appear to be evolutionary conserved processes. Besides mammalian cells, accumulated evidence suggests that exosome-like structures are secreted by plant cells too [3]. Two classes of EVs have been isolated from plants so far: apoplastic vesicles from rice shoot [4], sunflower seed [3, 5], and tomato leaves [6]; and exosome-like vesicles (ELVs) from fruit juices such as grape [7] and grapefruit [8]. Several studies have demonstrated that the complexity of the dynamic events occurs along the endosomal trafficking pathways [9], and emerging data supports the multivesicular body-mediated secretion mechanism of apoplastic vesicles in plant cells [3, 10]. Much investigation remains to be done to understand whether and how extramural plant vesicles participate in intercellular communication and plant defence. In contrast to the mammalian EVs, there is only limited information available about the bio-cargo of plant-derived nanoparticles. Purification of EVs is an essential first step not only for the structural analysis of EV molecular constituents but also for downstream biological and clinical applications.

Isolation of EVs is based on their specific physical and chemical properties such as size, morphology, buoyant density, charge, solubility, surface protein, and membrane lipid compositions. Centrifugation, filtration, precipitation, and chromatography-based separation techniques are used to isolate mammalian EVs. Most of the reported studies on plant-derived vesicles [5, 6] have used the differential centrifugation (DC) procedure. DC-based purification of EVs involves a series of low-velocity centrifugation steps followed by a single high-velocity ultracentrifugation step. The pellet at the end of the high-velocity centrifugation is the crude EV sample used for further analysis. DC purified samples can contain a heterogeneous population of vesicles and often with co-purified impurities. In order to purify vesicles with narrower particle size distribution and at an increased level of purity, after DC, the crude EV sample can be subjected to a density-gradient centrifugation (DGC) step. DGC separates mixture of particles based on their buoyant density. In DGC, the sample is generally layered on the top of a gradient material having a progressively increasing density. Vesicles with different buoyant densities distribute themselves into bands which occupy different positions in the gradient. Conventional gradient ultracentrifugation uses high salt concentration and subsequent purification step(s) to eliminate them. The high salt concentration can affect negatively the integrity of the vesicles and also their biological activities. Therefore, for EV separation sucrose or iodixanol (OptiPrep™) is the most frequently used gradient materials. There are two basic ways to perform DGC according to the means of preparation of the gradient: continuous and stepwise (discontinuous) gradient centrifugations. Linear sucrose gradient has been used, for example, in the purification of clathrin-coated vesicles and mammalian EVs at high purity. One of the most critical points of continuous DGC is the reproducible preparation of the linear sucrose gradient. Stepwise gradient ultra-centrifugation, on the other hand, has been used for the isolation of EVs from cell culture supernatant, and more recently, to isolate grapes and grapefruit juice derived ELVs using 8, 30, 45, and 60% sucrose solutions [7, 8].

Here, we describe a protocol which applies a sucrose double-cushion centrifugation to concentrate vesicles with discrete buoyant densities into two fractions (Fig. 1). The protocol uses sucrose/

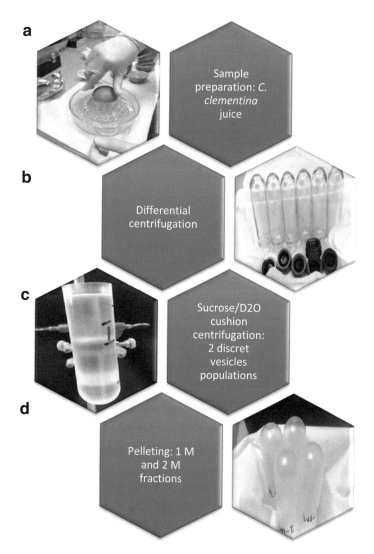

Fig. 1 The main steps of the isolation of ELVs from fruit juice: (**a**) preparation of the fruit juice, (**b**) differential centrifugation including a series of low velocity and one ultracentrifugation steps, (**c**) gradient centrifugation on sucrose/D_2O double-cushion, and (**d**) washing steps

deuterium oxide (D_2O) gradient material which has been shown to improve particle separation. Swinging bucket rotor is used for the centrifugation and the gradient is formed by itself avoiding the gradient preparation. It has been used to separate nucleus and different lipoprotein fractions, and also in estrogen receptor purification. A similar method has been described for the successful isolation of urinary EVs for subsequent quantitative proteomics studies [11, 12]. The method described here has been implemented for the successful extraction and purification of ELVs from different fruit juices, such as clementine, grapes, coconut water, and feijoa. The yields in terms of ELV-related protein concentration measured after the procedure in the two fractions were as follow: 20 µg proteins in 1 mL of starting clementine juice in the 1 M fraction (lower density vesicle population) and 4 µg proteins in 1 mL of starting juice in the 2 M fraction (higher density vesicle population). Depending on the starting material the final protein yield however can considerably differ. The protocol can be used to isolate vesicles from apoplastic fluids too. Apoplastic fluid can be prepared from different plant tissues, like leaf [4, 6], roots, and seeds [5] by methods based on a vacuum infiltration and centrifugation procedure [13].

2   Materials

Prepare all solutions using ultrapure water (18 MΩ cm conductivity and 2 ppb Total Organic Carbon at 25 °C) and analytical or better grade reagents.

2.1   Reagents and Solutions

1. Proteinase inhibitor cocktail: 1 M sodium azide, 100 mM phenylmethylsulfonyl fluoride (PMSF), 1 mM leupeptin. Store stock solutions at 4 °C up to 2 months.
   To prepare the inhibitor cocktail add 1.67 mL 1 M Sodium Azide, 2.5 mL of 100 mM PMSF, and 0.5 mL of 1 mM leupeptin stock solutions and mix well (this volume should be sufficient for 500 mL of sample). Prepare the cocktail just before use (*see* **Note 1**).

2. PBS: Phosphate Buffered Saline: 137 mM NaCl, 2.7 mM KCl, and 10 mM phosphate buffer solution (pH 7.4 at 25 °C).

3. 1 M Tris–HCl buffer pH 8.6.
   Filter by 0.22 µm syringe filter. Store at 4 °C up to 2 months. To prepare 20 mM Tris–HCl dilute the 1 M Tris–HCl stock 50 times.

4. 1 M Tris–HCl/D_2O stock solution: 1 M Tris base, D_2O.
   Dissolve Tris in 40 mL of D_2O and mix well, pH is usually at 9.0. Adjust pH to 8.6 with 1 N HCl and bring final volume to 50 mL with D_2O and filter by 0.22 µm syringe filter. Store at 4 °C up to 2 months (*see* **Note 2**).

5. 1 M sucrose-20 mM Tris–HCl/D$_2$O: add 1 M Tris–HCl/D$_2$O stock solution to 2.05 g sucrose in a tube and bring the final volume to 6 mL with D$_2$O. Prepare fresh just before use.

6. 2 M sucrose-20 mM Tris–HCl/D$_2$O: add 1 M Tris–HCl/D$_2$O stock solution to 2.73 g sucrose in a tube and bring the final volume to 4 mL with D$_2$O. Prepare fresh just before use.

2.2 Equipment and Disposable Required

1. Plastic laboratory container.
2. Ceramic knife.
3. Clear glass bottles.
4. Manual juice squeezer.
5. Whatman filter papers.
6. Filter funnel.
7. Disposable, sterile, vacuum operated bottle-top filtration system with 0.45 µm membrane filter.
8. pH Meter.
9. High precision analytical weighting balance.
10. Vortex.
11. Micropipettes.
12. Magnetic stirrer.
13. Benchtop centrifuge or floor-standing centrifuge with fixed angle rotor capable of reaching $15,000 \times g$ centrifugal force.
14. 50 mL conical centrifuge tubes.
15. Ultracentrifuge. Beckman Optima L-XP series ultracentrifuge was used with 70 Ti fixed angle rotor and SW-28 Ti swinging bucket rotor with appropriately sized polycarbonate and polyallomer tubes.
16. Forceps.
17. 0.22 µm syringe filter.
18. 10 mL syringes fitted with extension tubes.
19. Retort stand.
20. Syringe pump.
21. 18-G needle.
22. Water proof marker.
23. Refrigerator.

3 Methods

3.1 Sample Preparation

The method describes the preparation of fruit juice sample for subsequent ELV isolation protocol. Fruit juice can easily be obtained from all citrus fruits, like *C. maxima* (pomelo), *C. limon* (lemon),

C. sinensis (sweet orange), *C. reticulata* (mandarin), and C. *clementina* (clementine). In this protocol we started about 1 kg of citrus fresh fruit to obtain 250 mL of fruit juice. Other than citrus fruits, fruits like coconut or grapes can also be used, though the preparation of the juice and the quantity of the juice obtained from different fruits can be different. Apoplastic fluids can also be used as a starting material for the isolation of a special class of EVs from plants (i.e., apoplastic vesicles). Recently, Joosten et al. have described a protocol using vacuum infiltration-centrifugation for the purification of apoplastic fluid [13]. In the case of purification of apoplastic vesicles from apoplastic fluid, the procedure starts directly at **step 5** of Subheading 3.1. The volume of sample required depends on the nature of the plant material used for the preparation of the apoplastic fluid and the concentration of apoplastic vesicles in the sample.

1. Place fresh fruit in a clean plastic laboratory container and gently rinse in cold running tap water for 5 min to remove dirt. Repeat this step twice and allow the fruit to air dry.

2. Peel off the outer protective layer (exocarp) of the fruit.

3. Cut the fruit in half vertically using a clean knife and slowly squeeze the juice using a hand juice squeezer. Discard the left-over albedo and pulp and transfer the juice into a clean glass bottle. Repeat this step until the required volume of juice (in our protocol 250 mL) is obtained. The squeezing should be done as slowly as possible (*see* **Note 3**).

4. Dilute the sample using PBS to 500 mL final volume and add immediately the protease inhibitor cocktail.

5. Filter the sample by gravity filtration through a Whatman filter paper placed on a filter funnel and collect the filtrate in a clean glass laboratory bottle. This process can be slow and the filters have to be changed regularly (*see* **Note 4**) Discard the pellet.

6. Filter the sample obtained in **step 5** using a disposable, sterile, vacuum operated bottle-top filtration system with 0.45 μm pore sizes membrane microfilter. Discard the pellet.

7. Measure the pH of the filtrate using a pH meter. Adjust the pH of the sample using 1 M Tris–HCl (pH 8.6) buffer (*see* **Note 5**). Use the filtrate in the subsequent purification Subheading 3.2.1.

3.2 Differential Centrifugation

The first step in the isolation of distinct subpopulations of plant ELVs is DC. DC involves a series of low-velocity centrifugations which is performed at room temperature followed by a high-velocity ultracentrifugation at 4 °C. The setup of the DC procedure is time-consuming and labor-intensive, and changes in the various experimental parameters (rotor type, centrifugal force, time, temperature, etc.) can alter the yield and the quality of the sample. After the procedure, the resulting high-velocity centrifugation pellet can be stored at 4 °C or solubilized for the second step of the ELVs purification as described under Subheading 3.3.

3.2.1 Low-Velocity Centrifugation

1. Transfer the filtrate into 50 mL conical centrifuge tubes (*see* **Note 6**). For 500 mL of sample ten tubes are required. Centrifuge at $400 \times g$ for 20 min to remove cells and other large debris. Any benchtop low-velocity centrifuge that can accommodate conical centrifuge tubes of the required volume can be used.

2. Carefully transfer the supernatant to a clean 50 mL conical centrifuge tube (*see* **Note 7**) and discard the pellet.

3. Centrifuge at $800 \times g$ for 20 min to remove cellular debris. Any benchtop low-velocity centrifuge that can accommodate conical centrifuge tubes of the required volume can be used.

4. Carefully transfer the supernatant into a 36 mL polyallomer tube with screw tap. Using a waterproof marker, mark one side of each centrifuge tube and carefully place the tubes into the rotor in such a way that the mark is facing out as a reference to where the pellet will be following centrifugation.

5. Centrifuge at $15,000 \times g$ for 20 min to remove cellular debris. A centrifuge with fixed angle rotor is used. Collect the supernatant (*see* **Notes 8** and **9**).

6. Place the resulting supernatant from **step 5** into clean polycarbonate ultracentrifuge tube. Sign the tube as mentioned in **step 4** and proceed to the high-velocity centrifugation in Subheading 3.2.2.

3.2.2 High-Velocity Centrifugation

1. Progressing from **step 6** of Subheading 3.2.1, centrifuge the supernatant in the labeled polycarbonate ultracentrifuge tube at $200,000 \times g$ for 60 min at 4 °C to pellet the crude ELV fraction. Beckman Optima L-XP ultracentrifuge with 70 Ti fixed angle rotor is used (*see* **Notes 10** and **11**).

2. Carefully discard the supernatant without disturbing the pellet. Continue this step and repellet a fresh sample into the same tubes. Considering a starting volume of 500 mL, the 26.3 mL volume of polycarbonate tube and that the rotor can accommodate eight tubes for one run (corresponding to a total volume of 210.4 mL), repelleting will occur 2–3 times into the same tube. The resulting pellet contains the crude ELVs enriched fraction (*see* **Note 12**) which can be used for biological and structural analysis but note that the quality of this fraction is usually not suitable for "omics" studies.

3. Resuspend the pellet in a small volume of 20 mM Tris (pH 8.6) (*see* **Note 13**), vortex rigorously for at least 10 min in order break all aggregation, and obtain a suspension of primary particles. At this step, the sample can be left overnight at 4 °C.

4. Transfer the sample into a polyallomer tube with screw cap and bring the sample to a volume of 28 mL with 20 mM Tris (pH 8.6). Vortex rigorously and leave the sample on ice for 1 h.

5. Centrifuge at 15,000 × g for 20 min at 4 °C. A centrifuge with fixed angle rotor is used. This low-velocity centrifugation step resulted in improving the quality of the preparation by removing aggregates and other insoluble impurities. Discard the pellet.

6. Carefully transfer the supernatant to a polyallomer ultracentrifuge tube (*see* **Note 10**) and proceed to the second purification and separation step of ELVs.

3.3 Double-Cushion Ultracentrifugation for the Separation of Different Subpopulations of ELVs

This protocol uses discontinuous sucrose/deuterium oxide (D$_2$O) density cushions. D$_2$O has a distinct density but chemically similar to water.

1. Sucrose double-cushion is prepared by the under-layering method (*see* **Note 14**). The protocol uses the sucrose/D$_2$O gradient material in two different concentrations (densities), 1 and 2 M (*see* **Note 15**). First, progressing from **step 6** of Subheading 3.2.2, the polyallomer tube containing 28 mL of sample is held upright in a tube stand. A 10 mL syringe fitted with an extension tube is filled with 1 M sucrose/D$_2$O solution and placed on a syringe pump. Care is taken to avoid air bubbles and the tube is inserted into the sample down to the bottom of the centrifuge tube. 5.2 mL of 1 M Sucrose/D$_2$O is layered by syringe pump using a velocity about 1 mL/min. After loading, gently take the tube out, fill the syringe with 2 M sucrose/D$_2$O and layer 3.5 mL (*see* **Note 16**). Once the sucrose/D$_2$O cushions are layered discrete interfaces should be visible. Mark the layers and balance the tubes (*see* **Notes 10, 11,** and **17**). The ultracentrifuge tube filled up with samples should be handled and loaded into the rotor very carefully.

2. The sample is centrifuged using Beckman Optima L-XP ultracentrifuge with SW 28 Ti swinging bucket rotor at 110,000 × g for 3 h at 4 °C (*see* **Note 18**).

3. Immediately after the run the tubes containing the samples should be carefully removed from the rotor taking great care not to disturb the layers of sucrose. Cloudy band(s) will be visible. For fraction collection the tube should be held steady and upright by a clamp stand. A tiny hole should be introduced into the very bottom of the tube using a fine needle. The hole should be just big enough to allow the sucrose solution to drip out at approximately one drop per second. Fractions of 1 M sucrose/D$_2$O and 2 M sucrose/D$_2$O are collected in labeled polyallomer ultracentrifuge tubes and the soluble fraction is collected in a 50 mL conical centrifuge tube (*see* **Notes 19** and **20**).

4. Add 35 mL of 20 mM Tris (pH 8.6) to the 1 M sucrose/D$_2$O and 2 M sucrose/D$_2$O fractions and centrifuge at 110,000 × g

for 1 h at 4 °C to pellet the vesicles. The Beckman Optima L-XP ultracentrifuge with SW 28 Ti swinging bucket rotor is used (*see* **Notes 10** and **11**). This step is necessary to remove sucrose and D_2O from the sample (*see* **Note 21**).

5. Repeat **step 4** with the 1 and 2 M pellets obtained in **step 4**. Carefully remove the supernatant. The two pellets contain the two distinct vesicle fractions.

6. The two pellets containing the two distinct vesicle fractions can be used for downstream analysis or alternatively conserved as they are at –20 °C.

7. For downstream analysis like physiochemical characterization (transmission electron microscopy, particle size determination) determination of the protein concentration, lipid analysis, vesicle lysis, RNA and protein profiling, etc. resuspend the pellet in a small volume (typically 50–100 µL) of Tris–HCl pH 8.6 or other solution/buffer preferred for the analysis.

4   Notes

1. The lipid bilayer surrounding the exosome-like vesicles protects their cargo from extracellular and extravesicular enzymes. But that is not the case with the outer surface proteins of the vesicles. To protect them from the enzymatic hydrolysis of proteases, it is advisable the use of proteases inhibitors during the preparation. Fruits contain different proteases in different amounts. Ready to use protease inhibitor cocktail for plant cell and tissue extract can be obtained from a number of typical biochemical suppliers.

2. D_2O can be purchased from typical chemical vendors at 99.8 atom%D purity, which is suitable for these protocols.

3. Once the juice is extracted from the fruits, it is important to add the protease inhibitors and to proceed to exosome-like vesicles isolation immediately.

4. To speed up the filtration process gravity filtration can be replaced by vacuum filtration.

5. Universal pH paper can also be used to measure the pH of the filtrate. Other buffer solutions used in molecular biology or general biochemistry applications can also be used to keep the pH of the sample above 7. After buffering the sample should be checked for possible precipitation.

6. Depending on sample volume smaller or bigger conical centrifuge tube can also be used.

7. For some samples where heavy sedimentation is observed, it is recommended to repeat the initial centrifugation steps.

8. Attention must be paid while removing supernatant after centrifugation as sometimes the pellet is not tight and may contaminate the sample during decantation.

9. The pellet after the $15,000 \times g$ centrifugation step can be collected and used as a crude microvesicle enriched sample for subsequent biological assays and/or analytical and structural studies.

10. Always balance the ultracentrifuge tubes within a mass difference of ±1 mg.

11. For the selection of the ultracentrifuge tube and rotor consult https://www.beckmancoulter.com/wsrportal/wsr/research-and-discovery/products-and-services/centrifugation/tubes-and-adapters/index.htm.

12. The appearance of the pellet obtained after the high-velocity centrifugation depends on the starting material used. Pellet can be very small and translucent.

13. The volume of buffer used in this stem depends on the volume of the pellet. Add enough buffer to cover the pellet.

14. The layering is performed by under-layer, i.e., placing the sample solution first followed by the 1 M sucrose/D_2O and then the 2 M sucrose/D_2O cushions loading always from the bottom of the tube until filling the tube entirely. Under-layer the samples with double-cushion should be done precisely and quickly. Pay attention not to mix the layers. Over-layering method can also be used in this step. In the over-layering method, a fixed volume of the 2 M cushion is taken at the bottom followed by a gentle pipetting of 1 M cushion and the sample.

15. To our experience nano-sized vesicles usually float in the upper part of the 1 M sucrose cushion which has densities of 1.1270 g/mL. The 2 M sucrose cushion has a density of 1.2575 g/mL and usually bands bigger vesicles. The concentration of sucrose used in the cushion can be modified according to the density and particle size characterizing the vesicle population(s) aimed to be isolated.

16. The high viscosity of 2 M sucrose solution could cause loading difficulties into the syringe. Load the syringe from the top in order to avoid loading difficulties.

17. Use a solution containing table sugar for the balance tubes.

18. Higher velocity and longer time centrifugation can also be applied. However, these changes may influence the purification yield.

19. Prepare and label the different tubes for fraction collection before the piercing of the tube.

20. Collection of the fractions, especially if visible bands are observed, can be done also by piercing the ultracentrifuge tube from the side just at the lower layer of the band(s) using a syringe.

21. Exosome-like vesicle pellets are often so small that they are not visible to the naked eye. However from the marking you have placed on the ultracentrifuge tube, you will know where the small pellet will have formed and so progress accordingly with care.

References

1. Yáñez-Mó M, Siljander PR-M, Andreu Z, Zavec AB, Borràs FE, Buzas EI, Buzas K, Casal E, Cappello F, Carvalho J, Colás E, Cordeiro-da Silva A, Fais S, Falcon-Perez JM, Ghobrial IM, Giebel B, Gimona M, Graner M, Gursel I, Gursel M, Heegaard NHH, Hendrix A, Kierulf P, Kokubun K, Kosanovic M, Kralj-Iglic V, Krämer-Albers E-M, Laitinen S, Lässer C, Lener T, Ligeti E, Linē A, Lipps G, Llorente A, Lötvall J, Manček-Keber M, Marcilla A, Mittelbrunn M, Nazarenko I, Nolte-'t Hoen ENM, Nyman TA, Driscoll L, Olivan M, Oliveira C, Pállinger É, del Portillo HA, Reventós J, Rigau M, Rohde E, Sammar M, Sánchez-Madrid F, Santarém N, Schallmoser K, Stampe Ostenfeld M, Stoorvogel W, Stukelj R, Van der Grein SG, Vasconcelos MH, Wauben MHM, De Wever O (2015) Biological properties of extracellular vesicles and their physiological functions. Journal of Extracellular Vesicles 4. doi:10.3402/jev.v4.27066

2. Yoon YJ, Kim OY, Gho YS (2014) Extracellular vesicles as emerging intercellular communicasomes. BMB Rep 47(10):531–539. doi:10.5483/BMBRep.2014.47.10.164

3. Regente M, Pinedo M, Elizalde M, de la Canal L (2012) Apoplastic exosome-like vesicles: a new way of protein secretion in plants? Plant Signal Behav 7(5):544–546

4. Song Y, Zhang C, Ge W, Zhang Y, Burlingame AL, Guo Y (2011) Identification of NaCl stress-responsive apoplastic proteins in rice shoot stems by 2D-DIGE. J Proteomics 74(7):1045–1067

5. Regente M, Corti-Monzón G, Maldonado AM, Pinedo M, Jorrín J, de la Canal L (2009) Vesicular fractions of sunflower apoplastic fluids are associated with potential exosome marker proteins. FEBS Lett 583(20):3363–3366. doi:10.1016/j.febslet.2009.09.041

6. Gonorazky G, Laxalt AM, Dekker HL, Rep M, Munnik T, Testerink C, de la Canal L (2012) Phosphatidylinositol 4-phosphate is associated to extracellular lipoproteic fractions and is detected in tomato apoplastic fluids. Plant Biol 14(1):41–49. doi:10.1111/j.1438-8677.2011.00488.x

7. Ju S, Mu J, Dokland T, Zhuang X, Wang Q, Jiang H, Xiang X, Deng ZB, Wang B, Zhang L, Roth M, Welti R, Mobley J, Jun Y, Miller D, Zhang HG (2013) Grape exosome-like nanoparticles induce intestinal stem cells and protect mice from DSS-induced colitis. Mol Ther 21(7):1345–1357

8. Wang Q, Zhuang X, Mu J, Deng Z-B, Jiang H, Zhang L, Xiang X, Wang B, Yan J, Miller D, Zhang H-G (2013) Delivery of therapeutic agents by nanoparticles made of grapefruit-derived lipids. Nat Commun 4:1867. doi:10.1038/ncomms2886

9. Reyes FC, Buono R, Otegui MS (2011) Plant endosomal trafficking pathways. Curr Opin Plant Biol 14(6):666–673. doi:10.1016/j.pbi.2011.07.009

10. An Q, van Bel AJ, Huckelhoven R (2007) Do plant cells secrete exosomes derived from multivesicular bodies? Plant Signal Behav 2(1):4–7

11. Pocsfalvi G, Raj DAA, Fiume I, Vilasi A, Trepiccione F, Capasso G (2015) Urinary extracellular vesicles as reservoirs of altered proteins during the pathogenesis of polycystic kidney disease. Proteomics Clin Appl 9(5-6):552–567. doi:10.1002/prca.201400199

12. Raj DA, Fiume I, Capasso G, Pocsfalvi G (2012) A multiplex quantitative proteomics strategy for protein biomarker studies in urinary exosomes. Kidney Int 81(12):1263–1272. doi:10.1038/ki.2012.25

13. Joosten MAJ (2012) Isolation of apoplastic fluid from leaf tissue by the vacuum infiltration-centrifugation technique. In: Bolton MD, Thomma BPHJ (eds) Plant fungal pathogens, vol 835, Methods in molecular biology. Humana, New York, NY, pp 603–610. doi:10.1007/978-1-61779-501-5_38

INDEX

A

Acylation 192, 196
Adhesion assay 219, 223–224
Animal cells 4, 12, 18, 31–41, 47, 55
Apoplastic fluid extraction 251, 253
Apoplastic protein extraction 249–256
Apoplastic vesicles 56, 260, 264
Arabidopsis lyrata 92, 93, 96
Arabidopsis thaliana 15, 82, 92, 93, 95
ATP-binding cassette (ABC) transporter 37, 105, 150, 217
Autophagosome 39, 55, 56, 192, 193, 197
Autophagy 32–35, 38–40, 92

B

β-glucuronidase (GUS) 150–159
Biochemistry 4
Biomarker 40, 232
Brassica napus 92, 99
Brefeldin A (BFA) 32, 50, 81, 83, 206, 217

C

Cancer cell migration 215
Candida albicans 176
Carcinoma 232, 242
Caspase-1 .... 34, 41, 135–146
Chemical genomics 58
Chemiluminescence 87, 209, 253
Coat protein complex I (COPI) 5, 13–16, 35, 72
Coat protein complex II (COPII) 5, 11–16, 31, 35, 106, 205
Confocal microscopy 194–195, 199, 200
Conventional secretion 40, 47, 49, 53, 208, 211
Cryptococcus neoformans 176
Cystic fibrosis transmembrane conductance regulator
(CFTR) 32, 33, 35, 36, 38, 40, 41, 105–113, 115–118, 120, 121, 123, 124
Cytokines 32, 34, 136

D

Dictyostelium discoideum 150
Differential centrifugation (DC) 260, 261, 264–266

E

Ebola virus (EBOV) 205
Ectosomes 32, 37–38, 40
Electron microscopy 4, 55, 178
Endoglycosidase H (EndoH) 76–77, 108, 117, 118
Endomembrane trafficking 58, 59
Endoplasmic reticulum (ER) 4, 6, 9–11, 31, 47, 67, 72, 81, 105, 117
Enzyme-linked immunosorbent assay
(ELISA) 129–132, 139, 142–144, 146, 164, 165, 167, 169–170, 186, 217, 220, 221, 224, 225
Exocyst-positive organelle (EXPO) 51, 52, 54–56, 59
Exosome-like vesicles 41, 56, 259–268
Exosomes 32–34, 37, 39, 48, 49, 51, 54, 55, 92, 99, 162, 215, 225, 259
Extracellular vesicles (EVs) 41, 161–173, 175–182, 185–188, 259

F

Fibroblast growth factor-1 (FGF-1) 127, 130–132
Fluorescence recovery after photobleaching
(FRAP) 192, 194, 199, 202
Fucosidase 109, 117, 118
Fungi 150

G

Gel chromatography 165, 168
Glycan processing 107
Glycosylation 16, 32, 49, 50, 55, 76, 108–109, 117–118
Golgi 3–18, 31, 32, 34, 35, 38, 47–52, 54–59, 67, 72, 76, 105–113, 115–118, 120, 121, 123, 124, 192, 197, 206
Golgi-mediated protein traffic 72–77
Golgi reassembly stacking protein (GRASP) 34, 38–40, 49, 106
Grapevine leaves 250
Green fluorescent protein (GFP) 82, 192

H

Heparinase II 129–131, 133
High-throughput microscopy 109
Hydrophilic acylated surface protein B (HASPB) 41, 191–202
Hygromycin phosphotransferase (HYG^R) 50, 81–90

Andrea Pompa and Francesca De Marchis (eds.), *Unconventional Protein Secretion: Methods and Protocols*, Methods in Molecular Biology, vol. 1459, DOI 10.1007/978-1-4939-3804-9, © Springer Science+Business Media New York 2016

I

Inflammasome.. 34, 136, 144, 146
Interleukin-1β (IL-1β) 33, 34, 149
Invasion assay 218–219, 222–224, 226
Isolation...................................... 41, 59, 68–69, 82, 136, 137,
139–141, 144, 145, 161, 172, 233, 260–262, 264, 267

K

Keratinocytes.. 135–141, 143–145

L

Leaderless proteins 48, 59, 135, 136, 225
Leishmania 41, 191–202
Live cell imaging 191–202
Low molecular weight fibroblast growth factor-2 (LMW
FGF-2)... 127, 128, 130, 133

M

4-Methylumbelliferyl-β-d-glucuronide
(MUG) 150, 151, 153
Migration assay 218–219, 223, 224, 226
Mitomycin C... 138–140, 145
Multivesicular bodies (MVBs)......................... 37, 51, 53, 92

N

NaClO$_3$ 129–131
Nanoparticle tracking analysis (NTA).................... 162, 163,
165–167
Neuraminidase..109, 118

P

Paraformaldehyde (PFA)....................................70, 76, 78, 93,
94, 109, 185, 198, 200
Pichia pastoris 15, 39, 150
Plant juice... 260–263, 267
Plant tissue 14, 71, 150, 262
Plasma membrane (PM)............................ 4, 5, 16, 18, 31–35,
37–40, 47–57, 81, 92, 99, 105–124, 127, 128, 150, 183,
188, 192, 194, 197, 202, 205, 215
PNGase treatment.. 108, 117
Pollen–pistil interactions 92
Protein secretion................................. 3–18, 31–33,
47, 50, 149
Proteomic analysis 55, 59, 232,
233, 243, 244
Protoplast isolation....................................... 68–69, 71, 73,
78, 82–85
Pulse-chase analysis..77
Purification......................................77, 113, 167, 255, 256,
260, 264, 266, 268

R

Regulated secretion.............................. 4, 18, 37, 135
Ribosome...4, 7–9, 11, 31, 49

S

Saccharomyces cerevisiae..................................9–11, 15, 39, 40,
150, 176, 183, 184
Secretome 40, 41, 215, 216, 231–235,
237–239, 241–245, 249
Secretory granules.. 4, 17, 18
Secretory traffic 105, 117
Secretory vesicles (SVs) 4, 5, 15–17, 49,
52, 56–58, 67, 92, 149
Self-incompatibility..91, 92
Signal peptide...........................7, 8, 34, 48–50, 54, 127,
135, 136, 149–151, 206, 215
Signal recognition particle (SRP)................5, 8, 10
siRNA transfection.. 138, 141–144
SNARE 11, 14, 15, 17, 35, 38–40, 54, 59, 106
Stress inducible protein1 (STI1) 41, 161–173
Sucrose gradient centrifugation 260
Synuclein-γ (SNCG)..215–225

T

Tissue secretome 41
Tissue section 233–234, 236, 238–239, 245
Tobacco (Nicotiana tabacum)...71
Traffic chemical inhibition 58, 59
Trans Golgi network (TGN)........................4, 5, 15–18, 31,
35, 47, 56, 57, 206, 211
Translocon 3, 5, 9–11, 37
Transmission electron microscopy (TEM) 56, 58,
92, 178, 180–182, 184, 267
Trichloroacetic acid (TCA) precipitation 82, 83,
163, 169, 172, 250, 253
Trypsin 7, 129, 139–141,
145, 209, 211
Tumor interstitial fluid (TIF)........................... 41, 231–235,
237–239, 241–245
Tumor microenvironment...231, 232
Tunicamycin 68, 70, 77
Two-dimensional gel electrophoresis
(2D-PAGE)...........233–235, 237, 239, 241–243, 250

U

Ultracentrifugation............................163, 165–167, 170–172,
179, 206–209, 211, 259–268
Ultramicrotome 98, 181
Ultraviolet B (UVB) irradiation............... 136, 139, 144, 146
Ustilago maydis..41, 149–158

V

Vacuum-infiltration-centrifugation (VIC)
 method 250, 253
Vesicle proteomics 59
Vesicles 31, 32, 37–39, 259–261, 264, 267, 268
Virus...206, 207
Vitis vinifera...251
VP40 ....41, 205, 206, 208–212

W

Western blot 82–87, 107–108, 110–111, 114–116, 124,
 127, 142–144, 153, 158, 161, 165, 167, 169, 171, 172,
 206, 209–211, 217–218, 220–222, 225, 250, 252–255
Wortmannin 55, 58

Y

Yeast Extracellular Vesicles........................ 175–182, 185–188

Printed in the United States
By Bookmasters